Introduction to
PLANT & SOIL
SCIENCE AND TECHNOLOGY
Second Edition

Top row, from left—cotton boll, corn ears, apples, soybeans
Bottom row, from left—mature wheat, kenaf flower, asparagus, apricots

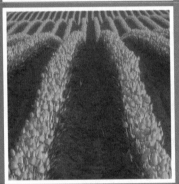

RONALD J. BIONDO

Field Advisor, Agricultural Education
Countryside, Illinois

JASPER S. LEE

Agricultural Educator
Lee and Associates
Demorest, Georgia

Introduction to
PLANT & SOIL
SCIENCE AND TECHNOLOGY

Second Edition

AgriScience & Technology Series

Jasper S. Lee — Series Editor

Interstate Publishers, Inc.
Danville, Illinois

Introduction to
PLANT & SOIL SCIENCE AND TECHNOLOGY
Second Edition

Library of Congress Control Number: 2001096384

ISBN 0-8134-3216-2

1 2 3 4 5 6 7 8 9 10   06 05 04 03 02 01

Order from

Interstate Publishers, Inc.

510 North Vermilion Street
P.O. Box 50
Danville, IL 61834-0050

Phone: (800) 843-4774
Fax: (217) 446-9706

Email: info-ipp@IPPINC.com
World Wide Web: www.interstatepublishers.com

Preface

Plants are important. They provide food, clothing, and other products for people, food for other animals, and promote a quality environment. Producing more plants and their products is essential as the Earth's population continues to rise. Meeting the needs of people poses a huge challenge for future agriculturalists.

Planting improved crop varieties on fertile soil and providing good cultural practices promotes production. The crops planted include highly useful genetically-enhanced organisms (GEOs). Increasingly, the GEOs will be planted for their specific benefits in meeting human needs. It is likely that human well being cannot be achieved without them.

Using the Earth's resources to meet human needs on a long-term basis requires greater efficiency. Sustainable resource use will be more widely practiced in the future. Education in plant and soil science and technology will be essential. This is where the 2nd Edition of this book comes in.

Introduction to Plant and Soil Science and Technology, 2nd Edition, builds on the highly successful 1st Edition. The 1st Edition was a book that brought plant and soil science together when it was released. The approach was well received by students and teachers throughout the United States. The latest technology and science principles were integrated throughout the book. Practical production experiences were included. Curriculum input was used from throughout the United States to organize the content. Student- and teacher-friendly approaches were used to make the book a welcomed, exciting addition. All of these have been enhanced in the 2nd Edition.

The format of this book is that of Interstate's AgriScience and Technology (IST) Series. Student-friendly examples are used. Teacher-friendly sequencing has been used. Many color photographs and line art illustrations have been included. When possible, students have been used in the photographs to enhance meaning. "Career Profiles" and "Connections" are used in each chapter to provide a tie-in with careers and industry.

The authors are excited about the 2nd Edition of the book. They believe that you will be also.

Acknowledgments

The authors of *Introduction to Plant and Soil Science and Technology* are indebted to many individuals for their help. Some of them are acknowledged here. Others are acknowledged throughout the book.

Professor Luther Talbert of the Department of Plant Sciences at Montana State University is acknowledged for his assistance with crops and insect pests for the 2nd Edition. Professor Van Shelhamer, also of Montana State University and an agricultural educator, is acknowledged for his assistance. William I. Segars, Extension Agronomist at the University of Georgia is acknowledged for reviewing both the 1st and 2nd Editions.

Special acknowledgment goes to the dedicated individuals who reviewed the manuscript for the first edition: Bob Keenan, Director of the Maryland Center for Agriculture Science and Technology; Gail Komoto, Agriculture Instructor at Sumner High School, Washington; Carl Reed, Agriculture Instructor at Barrington High School, Illinois; Diana Loschen, Agriculture Instructor at Tri-Point High School, Illinois; and Steve Millett, Plant Pathologist at the University of Wisconsin, Madison.

Individuals acknowledged for assistance with photographs include: Jeremy Bradley of Chase High School, North Carolina; Amanda Patrick and Tracy Westrom of Piedmont College; Megan Cox of Peoria High School, Arizona; Joey Tomlinson and Steve Chumbley of Sandra Day O'Conner High School, Texas; Doak Stewart of James Madison High School, Texas; Anne Stewart Clark of Fleming County High School, Kentucky; Wilbur H. Palmer of Alvirne High School, New Hampshire; John Waller of North Kitsap High School, Washington; Keturah Harley and Anita Y. Daniels of the Agricultural Research Service, USDA, Beltsville, Maryland; David Frazier, Snyder High School, Texas; Kevin Fochs, Livingston High School, Montana; Bill Johnson, Joliet Junior College, Illinois; Ron Olson, Soil Testing Service, Inc., Illinois; T.C. Nelson, National Cotton Council of America, Memphis; Bill Agerton, Potash & Phosphate Institute, Georgia; Floyd Giles, University of Illinois; Erendira Sanchez, Ball Seed, Inc., Illinois; Ryan Solberg, Abbott Labs, Illinois; Ken Kashian,

Illinois Farm Bureau; Gene Redlin, Poplar Farms, Illinois; Miriam Volle, Cantigny Golf, Illinois; John Lancey, Spectra Physics Laserplane, Inc., Ohio; Mark D. Branstetter, New Holland North America, Inc., Pennsylvania; Kelly Limbaugh, AGCO Incorporated, Georgia; John A. Webster, Bush Hog Division of Allied Products Corporation, Alabama; and James Lytle, Shawn Askew, and Marco Nicovich, Mississippi State University.

Other individuals acknowledged include Christy Smith, Michigan State University (formerly of Arizona); John Morgan, Yavapai College, Arizona; Brad Dodson, California Department of Education; Doug Anderson and Mike White, PBL High School, Illinois; Michael Barros, Hawaii Department of Education; and Larry Bottorff, South Putnam High School, Indiana.

Acknowledgment also goes to Wendy Biondo of Naperville, Illinois, for her assistance with photography and in proofreading sections of the manuscript of the 1st Edition.

The authors are most grateful to the staff of Interstate Publishers, Inc., for their commitment to quality learning tools in agricultural education. Interstate President Vernie Thomas is specially acknowledged. Dan Pentony and Bob Hofmann of Interstate Publishers, Inc., are also acknowledged. Interstate personnel closely involved with the production of this book who are acknowledged here include Kim Romine, Rita Lange, and Jane Weller.

Contents

PART THREE: FUNDAMENTALS OF SOIL SCIENCE

PART FOUR: PLANT PESTS

PART FIVE: MEETING HUMAN NEEDS WITH PLANTS

Part One

Role of Plant and Soil Science

JUST how important are plants and soil? Sometimes we fail to think about their importance. All living organisms—plants, animals, and others—depend on processes carried out by plants for food and as a source of energy. Soil assures plant growth. We must find and use ways to maintain long-term productivity. Soil that loses fertility is no longer productive. It won't produce abundant crops.

We increasingly turn to research to answer questions about plants and soil. We use education to prepare people for successful careers in plant and soil science.

CHAPTER 1

Plant and Soil Science in a Productive Agriculture

OBJECTIVES

This chapter introduces the needs of people and the important role of agriculture in meetings these needs. It has the following objectives:

1 Describe how agriculture meets basic human needs.

2 Explain the role of plant and soil science.

3 Name important crop groups and give examples of each.

4 Describe the role of research in plant and soil science.

5 Identify career and entrepreneurship opportunities and expectations.

TERMS

agriculture	forestry	plant science
agronomy	grain crop	pomology
cereal grain	herb	research
clothing	horticultural crop	shelter
direct plant source	hunger	soil
entrepreneurship	indirect plant source	soil conservation
experiment	landscape horticulture	soil science
fiber crop	medicinal plant	spice
field crop	oil crop	sugar crop
floriculture	olericulture	tree farm
food	ornamental crop	turf
food crop horticulture	ornamental horticulture	
forage	plant domestication	

1-1. Plants are interesting to study and provide many useful products.

PEOPLE need food, clothing, and shelter to live. Plants help satisfy these essential human needs. Larger amounts of plant products will be needed as the earth's population continues to increase. The increased demand makes greater productivity from agriculture essential.

Today, the earth has slightly over 6 billion people. The United Nations predicts that this could increase by 50 percent and reach 9 billion by the year 2050. More immediately, North America now has about 435 million people who must be fed, clothed, and housed. This number is increasing daily!

How can food, clothing, and shelter production keep up with demand? Can we have a more productive agriculture on a long-term basis? We must have a level of production that will meet the needs of people.

MEETING HUMAN NEEDS

Humans use many products that are directly or indirectly from plants. These products meet the basic needs of humans for survival. The availability of these products is the outcome of a large agricultural industry.

AGRICULTURE

Agriculture is the science of growing crops and raising animals. The crops are used directly by humans or are used to produce animals, which are used by humans. Agriculture has changed considerably over the years.

Agriculture in North America has developed the capacity to produce large amounts of crops. The crops are used to meet needs on this continent; in addition, they are exported worldwide to other nations that do not have a productive agricultural system.

1-2. Salsa and chips are popular snacks. The ingredients are made from plant products.

Agricultural Abundance

Highly productive agriculture has resulted from the efforts of people to improve their quality of life. Why does North America have high agricultural productivity? Several factors are important:

- **Natural resources**—Fertile soil and good supplies of water are important natural resources needed to produce crops.

- **Agricultural research**—Research has found new and better ways of producing crops. Losses to pests have been greatly reduced.

- **Agricultural education**—Education has been used to inform crop producers about the findings of research and the benefits of new technology. The educational system has been carried to the practical level so that new technology is put to use.

- **Government role**—People in North America have been productive and value the freedom to pursue their own goals. Favorable government policies have encouraged agricultural productivity. Capitalism and the work ethic of the people have been major factors. People know that a key to success is hard work.

World Hunger

People in some nations are not as fortunate as those in North America. Hunger is a problem. **Hunger** is the discomfort caused by a need for food. The food supply is so low that people do not have proper nourishment. They suffer malnutrition and are never able to reach their potential. Estimates are that 40 million people die each year from starvation or diseases linked to malnutrition around the world.

Hunger is a much bigger problem in nations that do not have a productive agriculture. Reasons for a lack of agricultural production vary. The reasons are often the lack of one or more of the four factors that contribute to productivity in North America. They do not have natural resources, research, education, and government policies that support production.

Areas of the world where hunger is a major problem include parts of Africa, Asia, and Central and South America. Production of food in these countries has increased in recent years. The amount produced has not been adequate to keep up with population growth.

Some people in North America suffer from hunger. Their situation is due to a lack of income to buy what they need. Various government programs help meet these needs.

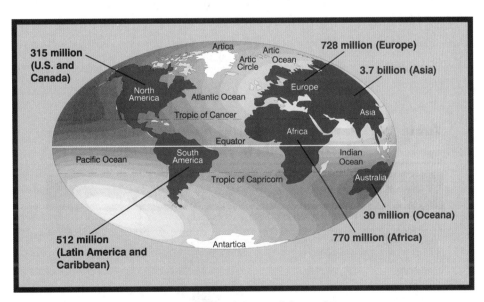

1-3. Population of the earth by regions.

MEETING BASIC HUMAN NEEDS

Humans have three basic needs: food, clothing, and shelter. Plants help meet these needs by producing products that are used as food, to make clothing, and to construct housing.

Plants are direct and indirect sources of these needs. A **direct plant source** means that the plants or plant products are used by humans. An **indirect plant source** means that the plants are used as animal feed and humans use the animal or its product. Some people contend that more efficient use is made of natural resources by using products that come directly from plants. Other people contend that this is not so because many animals live off plants that are not used directly by humans, such as low-quality pasture grasses. What do you think?

Food

Food is the solid and liquid material humans eat. Food provides the nutrients that humans need to grow and have healthy lives. Not only should food be available, but it should be wholesome and promote health.

Americans also expect their food to be readily available in easy-to-use forms. They want to go to a supermarket and get food products that require little preparation before eating. Convenience is important to people today. With careers, family needs, and other commitments, many people do not have much time to spend preparing food.

People have become more concerned about nutrition. They want foods that promote good health. Those obtained directly from plants are often cited as health-promoting foods. Broccoli and carrots are two examples (they have large amounts of vitamins and minerals). Genetically modified plants are being used to get unique health benefits. An example is canola. It is used in foods to reduce cholesterol in the human body.

Table 1-1. Per Person Annual Average Consumption of Selected Foods in the United States

Foods	Amount Consumed pounds (kilograms)
Direct Plant Origin	
Vegetables (fresh, frozen, and canned)	219.6 (99.6)
Fresh fruit and melon	111.9 (50.8)
Potatoes	127.2 (57.7)
Rice, flour, and cereal products	184.3 (83.6)
Indirect Plant Origin	
Dairy products and milk	564.5 (256.1)
Eggs	29.3 (13.3)
Fish and shellfish (wild harvest)	14.8 (6.7)
Chicken and turkey	56.8 (25.8)
Beef, pork, veal, lamb, and mutton	112.4 (51.0)

*Conversion from pounds to kilograms involves multiplying the weight in pounds by 0.45359 (the weight of a pound in kilograms). If kilograms are known, multiply the number of kilograms by 2.2046 (the weight of a kilogram in pounds).

Clothing

Clothing includes the garments, accessories, and ornaments worn on the human body. Garments are the coverings that people wear to protect the body and give it a certain appearance, such as skirts and jeans. Accessories are worn to supplement the basic clothing, such as ties and scarfs. Ornaments are worn for a certain appearance or status, such as rings and pins.

Plants are important sources of clothing materials. Most clothing made from natural materials comes from a direct plant source, such as cotton and linen. Wool, which is from an indirect plant source, is also used in clothing. Of course, nylon, rayon, and other artificial materials are widely used in clothing.

Shelter

Shelter includes our homes and the items in our homes that make them comfortable. Shelter protects humans from the weather and harm.

Wood and wood products are widely used in building construction. Lumber is used to construct frames and to give buildings strength to withstand the weather and support weight. Plywood and particle board are used to construct

1-4. Clothing projects the image that we want other people to have of us. (Courtesy, National Cotton Council)

CAREER PROFILE

Agronomist

An agronomist studies crops and soils to learn better ways of managing them. Most agronomists specialize in a particular crop or soils area. This photograph shows an agronomist checking an ear of corn for development and pest damage.

Agronomists have college degrees in agronomy, plant pathology, or a closely related area. Many agronomists complete masters degrees and some complete doctorates. Practical experience with crops and soils is beneficial.

Agronomists are employed with private agribusinesses, government agencies, universities, and research stations. Some are involved in international development. A few are entrepreneurs and operate consulting or service businesses for land owners. (Courtesy, U.S. Department of Agriculture)

1-5. Building construction often involves using many wood products.

walls connecting the frame materials. Fine woods are used to make furniture and other products in the home.

ROLE OF PLANT AND SOIL SCIENCE

Plants and soils are often associated in the minds of people. Here is why: plants have roots in the soil and get nutrients for growth from the soil. Plants will not grow without the essential nutrients. Plant growth may be poor in soil that does not have sufficient nutrients.

By understanding plant and soil science, people can improve the capacity of soil to provide the nutrients that plants need. This results in greater crop yields.

PLANT SCIENCE

Plant science is the study of the structure, functions, growth, and protection of plants. Usually, plant science includes three broad areas: field crops, horticultural crops, and forestry.

Field Crops

Field crops include plants grown in large fields and used for oil, fiber, grain, and similar products. Field crops, such as corn and wheat, are often grown for their seed, but other parts may also be used, such as the fiber from cotton.

Agronomy is the specialized area of plant science that deals with field crops. Agronomy includes the relationships between plants and the soil.

Major tasks in growing field crops include preparing the seedbed; planting the seed; pest management (controlling weeds, insects, diseases, and other pests); and harvesting the crop. Sometimes, irrigation may be used to supplement the moisture available to the crop. Marketing also plays a big role in getting products to consumers.

Horticultural Crops

Horticultural crops are grown for food, comfort, and beauty. In some cases, horticulture appears to overlap with field crops. Horticultural crops are not grown on vast areas of land. However, some locations in North America devote land to horticultural crops. The two major areas of horticulture are ornamental and food crop production.

Ornamental Horticulture. *Ornamental horticulture* is growing and using plants for their beauty. It includes floriculture, landscape horticulture, and interiorscaping.

1-6. Corn is a crop grown throughout much of North America.

- **Floriculture** is the production and use of plants for their flowers and foliage (stems and leaves). The roses used on Valentine's Day and chrysanthemums in homecoming corsages are examples of floriculture.

- **Landscape horticulture** is growing and using plants to make the outdoor environment more appealing. It includes shrubs, flowering plants, and lawn areas.

- **Interiorscaping** is using plants inside buildings to create an attractive interior environment. Offices, foyers, and other places inside buildings often have plants for beauty and appeal.

-7. The end product of floriculture is attractive flowers, such as these roses.

Food Crop Horticulture. *Food crop horticulture* is growing plants for food. Food crop horticulture is often divided into two areas: olericulture and pomology.

- **Olericulture** is the science of producing vegetable crops, such as tomatoes, snap beans, asparagus, and artichoke. Olericulture is a large, productive enterprise in some locations, such as in South Florida, the Rio Grande Valley of Texas, and several valleys in California.

- **Pomology** is the science of producing fruits and nuts. It includes growing, harvesting, and marketing the crops. Examples of fruit include the citrus crops–oranges and limes; apples; cherries; strawberries, and blueberries. Nuts include pecans, walnuts, and pistachios.

1-8. The tomato is an important vegetable crop.

1-9. Pistachio nuts are produced in the San Joaquin Valley of California.

Forestry

Forestry is the science of growing trees and producing wood products. Lumber, paper, plywood, furniture, and similar products are made from trees. In addition, some trees provide specialty products, such as syrup from the maple, rosin from the pine, and oil from the tung tree. Forestry involves caring for native forests and tree farming.

Tree farms are cultured forests. They have been carefully planned and established. Considerable work goes into assuring good high-quality timber. Large tracts of land are used for tree farming in some areas of the North America, such as the Pacific Northwest and the South. Many different kinds of trees are grown on tree farms, including pine, walnut, and fir.

SOIL SCIENCE

Soil science is the study of the structure, composition, fertility, use, and protection of soil. Knowing about soil helps produce crops more abundantly and efficiently. Understanding soil also helps prevent damaging it by using poor practices.

Soil is the top layer of the earth's crust. These few inches contain essential nutrients that support plant life. Plant roots grow in the soil to get the nutrients and anchor the plant. Tests can be made of soil to determine its contents and how to make it more productive.

1-10. Fertile soil is essential for crop production. Note the excellent condition of this recently planted field.

Soil conservation is using cropping practices to assure long-term productivity and sustainability. Damage and loss are prevented. Soil can be improved with some practices, such as planting cover crops. These crops are grown on the land and plowed back into the soil to increase its productivity.

IMPORTANT PLANTS AND THEIR USES

Many different plants are grown. Hundreds are produced in North America. A few plants are harvested wild, such as poke salet and sassafras roots. Long ago people found that they were better able to have a dependable food supply if they cultured plants. This resulted in plant domestication.

Plant domestication is removing plants from their native wild environment and growing them under controlled conditions. Thousands of years ago, all plants were wild. Today, nearly all plants used for food, clothing, and shelter are domesticated.

The important field and horticultural crops in North America can be divided into seven categories: grain crops; sugar and oil crops; fiber crops; vegetable, fruit, and nut crops; forage; turf and ornamentals; and other crops. In addition, several crops do not fit into one of these categories and are listed as "other."

GRAIN CROPS

Grain crops include plants grown for their edible seeds not including the horticultural crops. Grain crops provide many important foods, including the cereal grains. Foods from grain crops are a part of nearly every meal we eat!

Cereal grain is the seed of grass-type plants grown for food and animal feed. Often, grain and cereal are viewed as the same. Important grain crops include rice, corn, wheat, oats, barley, rye, and sorghum. These crops are used to make many food products. Rice is one of the world's oldest crops and a major food in many countries. Some grain crops undergo considerable processing, such as making flour from wheat and breakfast cereal from oats.

1-11. Wheat is a major crop in many areas of North America. This shows mature wheat being harvested with a combine. (Courtesy, Case Corporation)

1-12. Sugar beets are grown for sugar in areas of the western United States. This shows a maturing beet on a California farm.

SUGAR AND OIL CROPS

Sugar and oil crops are produced for two important food commodities: sweeteners and vegetable oil. Some crops are grown exclusively for these products.

Sugar crops are used as a source of sucrose. We commonly use sucrose as table sugar to sweeten foods or beverages. Sucrose is a carbohydrate that provides energy for physical activity by the body. The major sugar crops are sugar beets and sugar cane. Other sources of sugar are honey, maple syrup, and some kinds of sorghum.

Oil crops are plants grown for the vegetable oil contained in their seed and fruit. The seeds of about 40 crops are used to make oil, with soybeans, cotton, canola, and corn among the most important. Other common oil seeds include sunflower, safflower, peanut, coconut, linseed, and palm. Only one tree fruit is widely used in making oil—the olive.

FIBER CROPS

Fiber crops are grown for the fiber produced in their fruit, leaves, or stems. Fibers are tiny threadlike structures used in manufacturing cloth, paper, and other materials.

Cotton is the major crop grown for fiber. Flax, kenaf, hemp, jute, and ramie are also grown for fiber. Kenaf is the focus of considerable research to study its potential in making paper and cloth.

Cotton and flax are also used in making oil. The cotton material in the boll around the seeds is the fiber; the seeds are used to make vegetable oil. Flax cloth is known as linen and its oil is linseed.

1-13. Cotton is produced in bolls on plants.

VEGETABLE, FRUIT, AND NUT CROPS

Vegetable, fruit, and nut crops are primarily grown for food. Many different plants are grown in this group.

CONNECTION

Quality Flowers

Flowers are important to people. Blossoms, foliage, and other plant parts are used to decorate homes and offices, express emotion, and brighten our day. Flower growers carefully tend plants to assure an attractive and long-lasting flowering plant.

A bouquet or potted flowering plant requires high-quality flowers. No one wants flowers with broken stems, damaged buds, or wilted petals. The flowers must be kept in the proper environment. Lack of water, heat, and careless handling can damage flowers so that they are worthless.

This shows healthy Easter lilies with fully formed buds that will soon bloom. (Courtesy, Doug Anderson, Paxton, Illinois)

1-14. Apples are grown in some locations.

1-15. Baled hay covered with plastic to prevent damage by the weather.

The vegetable crops may be grown for different parts of the plant, such as leaves (lettuce), stems (celery), seed (bean), and roots (beet). Examples of common vegetables include potatoes, broccoli, carrots, and cabbage.

The fruits include cranberries, apricots, peaches, apples, and grapes. Some of these grow on large trees; others grow on small plants and vines.

The nut crops include macadamia, almond, walnut, pecan, and filbert. These are often known as tree nuts. Some trees grow very large and live for many years, such as the pecan.

Forage

Forage crops are primarily grasses and legumes. Grasses are nonwoody plants that have parallel veins in their leaves, with fescue and Bermudagrass being two examples. Legumes are broadleaf plants that have the potential of fixing nitrogen from the air in the soil, with clover and vetch being examples. *Forage* is the leaves and stems of plants used for animal feed. It is most nutritious while the plants are still young and before seed mature. Forage may be grown in a pasture or field and eaten (grazed) by cattle or horses. Some forage crops are cut and dried, stored, and fed later, such as hay. Others are cut green and stored, such as silage.

Ornamental and Turf Crops

Ornamental crops include flowers, shrubs, vines, and other species grown for their beauty and personal appeal. Many different plant species are included. The species are often adapted to climates in local growing areas.

A large horticultural industry has emerged to produce and market ornamental crops. Some forms of the industry are found in nearly all the towns and cities of North America. Increasingly, cut flowers are being imported from Central and South America and other areas of the earth. Automated greenhouse systems promote efficient year-round production.

Turf comprises plants used to present a pleasing appearance and protect the soil. The plants are low-growing, fine-leaved grasses used in lawns and to cover the ground to prevent soil erosion. Turf is used on athletic fields, golf courses, and other locations where aesthetics are important.

1-16. Turf is carefully maintained in most public areas.

OTHER CROPS

Many other crops are grown for various uses. These include the beverage crops, herbs and spices, and medicinal plants.

The beverage crops include coffee, tea, and cocoa. These crops typically grow in tropical areas, with Hawaii and Puerto Rico leading in coffee production in the United States.

Herbs and spices are grown in small quantities throughout North America. Most herbs and spices have little food value, but do improve the flavor of food and make it more attractive. Distinguishing between herbs and spices is difficult. Even with the best explanations, some overlap exists.

An **herb** is a non-woody plant with leaves, seed, or other parts used as medicine or to enhance food. Fennel seeds are used in seasoning some foods.

A **spice** is an aromatic plant part (seed, leaves, etc.) that is used to season food. Spices give food a desired taste or smell. Most spices grow in tropical climates. Common spices are dill, paprika, and pepper.

1-17. Ornamental plants and turf are combined to provide a pleasing environment at this resort in Hawaii.

1-18. Bananas grow in tropical climates.

A few places have large production areas for herbs and spices. Many people plant herb gardens in flower pots at their homes.

Plants are important in many human medicines. Those used in making medicine are known as **medicinal plants**. Steroids, pain killers, ointments, and medicines are derived from plants. One example of a medicinal plant is the pacific yew. The bark of the pacific yew is stripped off and used to make taxol—a drug that holds promise in treating cancer.

1-19. Tea leaves are being examined at the Charleston Tea Plantation in South Carolina. (Courtesy, U.S. Department of Agriculture)

THE ROLE OF RESEARCH

Research is seeking answers to questions using systematic methods of investigation. In plant and soil science, the questions often deal with how to gain more efficient production. Methods are used that allow data to be collected, analyzed, and reported. The findings are used to improve the practices that are followed in plant production.

The findings of research are used to develop new methods or products. This is often referred to as development, or the process of creating something that is new.

EXPERIMENTATION

Experimentation is an important part of research. Carefully designed experiments are a part of plant and soil research. An **experiment** is action to demonstrate or discover a truth. It typically answers "what happens if I do something?"

An example is using an experiment to determine the effects of fertilizers on plants. The scientific method would be followed. Once the problem has been identified, a hypothesis would be formed. An experiment would be designed to test the hypothesis. With plants, several test plots might be used. Each might be given different rates or kinds of fertilizer. Observations are made as the plants grow. Careful measurements are made at harvest to assess the influence of fertilizer on yield. Conclusions are formed and the research is reported.

Once an action has been proven experimentally, demonstration plots may be used to show the public the effects of the new technology that has been studied.

AGRICULTURAL RESEARCH

Agricultural research is used to answer questions about the practices to use in agriculture

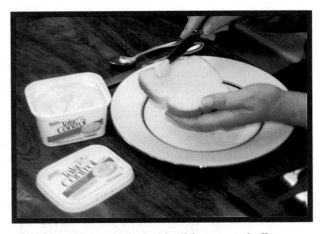

1-20. Margarine produced with oil from genetically enhanced canola promotes human health by lowering cholesterol levels.

1-21. A researcher is studying grain sorghum heads. (Courtesy, Agricultural Research Service, USDA)

The scientific method is a process for solving scientific questions. The steps are:

1. **Identify a problem**
2. **Get information about the problem**
3. **Form a hypothesis**
4. **Design an experiment**
5. **Conduct an experiment and collect data**
6. **Analyze the data and form conclusions**
7. **Report the results**

1-22. Seven steps in the scientific method.

and related areas. It includes plant research as well as research with animals, natural resources, marketing, and many other areas in providing food, fiber, and shelter.

Research is used to test new crops, new cultural practices, and other conditions. The end result is always to gain more efficient and effective production.

Agricultural research is carried out by private companies, government agencies, and others. A system of agricultural experiment stations exists in the states. These are in conjunction with the land-grant colleges and universities. The U. S. Department of Agriculture also maintains the Agricultural Research Service.

CAREER AND ENTREPRENEURSHIP OPPORTUNITIES

There are many career opportunities in plant and soil science. In some cases, the opportunities are as employees of agribusinesses or agencies. In other cases, individuals can be entrepreneurs and form and operate their own businesses. Choosing the direction to take requires good information. School counselors, agriculture teachers, and others have information. Each chapter of this book has at least one career profile that highlights an occupation in soil and plant science.

Employers want people who are:

- **at work on time**
- **willing to learn new ways of doing work**
- **on-task—they are productive**
- **loyal**
- **honest and dependable**
- **accurate**
- **good team members**
- **happy and easy to get along with**
- **aware of and follow procedures, including those in safety**
- **efficient and do not waste resources or abuse equipment**
- **good citizens and have personal lives that do not conflict with work**

1-23. Examples of employer expectations.

SUPERVISED EXPERIENCE

Students gain valuable insight and develop jobs skills through supervised experience. Supervised experience is carried out as part of enrollment in agriculture classes. It is a planned experience involving work in an agribusiness, on a farm, or elsewhere. The experiences are related to class content and the interests of students. In some cases, students may own a production enterprise, such as a grain crop or ornamental plants in a greenhouse.

EMPLOYER EXPECTATIONS

People who work for others are expected to meet the expectations of their employers.

These are known as work habits and involve job productivity as well as community citizenship skills. The expectations are to help workers in being more productive in their jobs.

ENTREPRENEURSHIP

Entrepreneurship is owning and operating a business. The person who owns it is an entrepreneur. The goal is to produce a good or service resulting in a profit for the owner. In plant and soil science, the businesses may include a wide range of enterprises, such as farms, flower shops, soil testing services, custom harvesting, seed and fertilizer businesses, and processing and otherwise marketing crops and other produce.

WORK SAFETY

Practicing safety at work is essential. Safety means that the injuries and losses are prevented. Employers expect workers to follow appropriate safety rules. Employers are expected to inform workers of safety practices and provide some safety equipment. Employees are expected to follow appropriate safety practices, including the use of personal protective equipment (PPE). Examples of PPE include eye safety, hearing protection, hand safety, and protective clothing.

Safety Rules:

- **Know the work you are doing and its possible hazards**
- **Always follow safe practices**
- **Anticipate dangers and take steps to avoid them**
- **Use personal protective equipment (PPE)**
- **Keep equipment and tools in good condition**
- **Remove hazards from the work environment**
- **Alert other people to hazards**
- **Read and follow instructions**
- **Never take unnecessary risk**

1-24. General safety rules.

REVIEWING

MAIN IDEAS

The basic needs of humans are met by the direct or indirect products of plants. Modern agricultural practices have resulted in an abundance of food, clothing, and shelter materials. Natural resources, agricultural research, education, capitalism, and a strong work ethic among the people are also important in a productive agricultural system.

Plant science is the study of the structure, functions, growth, and protection of plants. Plant science includes field crops, horticultural crops, and forestry.

Soil science is the study of the structure, composition, fertility, and use of soil. It includes soil conservation and soil fertility to assure that future productivity is maintained.

Plants can be grouped based on the products that they produce. Common groups include grain crops; sugar and oil crops; fiber crops; vegetable, fruit, and nut crops; forage crops; ornamental and turf crops; and other crops.

Research is used to improve plants and the products they produce. Experiments are carried out by private industry, government agencies, and others to find new and improved methods.

A wide range of career opportunities are available in areas of plant and soil science. Supervised experience can be used to explore the careers and begin developing job skills. Employers expect workers to be productive and work safely.

QUESTIONS

Answer the following questions using complete sentences and correct spelling.

1. What has contributed to high agricultural productivity in North America?

2. What are the basic needs of humans? How does plant and soil science help meet these needs?

3. What is plant science?

4. What are the three major crop areas of plant science? Briefly describe each.

5. What is soil science?

6. What is plant domestication?

7. What are the six categories of field and horticultural crops? Briefly explain each.

EVALUATING

Match the term with the correct definition. Write the letter by the term in the blank that is provided.

a. oil crop
b. cereal grain
c. forage
d. soil
e. agriculture
f. field crop
g. ornamental horticulture
h. agronomy
i. medicinal plants
j. olericulture

1. _____ science of growing crops and raising animals

2. _____ area of plant science that deals with field crops

3. _____ plants grown in large fields and used for oil, fiber, grain, and similar products

4. _____ growing and using plants for their beauty

5. _____ top layer of the earth's crust

6. _____ plants grown for making vegetable oil

7. _____ seed of grass-type plants grown for food and animal feed

8. _____ plants used for animal feed

9. _____ science of producing vegetables crops

10. _____ plants used for making medicine

EXPLORING

1. Draw a map of your state on poster board. Show the locations where the most important plant crops are grown. Use a symbol that represents each crop. Use reference materials for assistance in finding where crops are grown. Contact the Cooperative Extension Service, a college with an agriculture program, or the state department of agriculture for information.

2. Make a list of the foods you eat in one day. Determine which are from a direct plant source and which are from an indirect plant source. (All animal products are indirect plant sources.)

3. Make a tour of a field or horticulture crop farm. Have the manager/owner describe the kinds of crops produced and the cultural practices that are used. Prepare a written report on your findings. Give an oral report in class. Another option is to take photographs of the activities that are underway and prepare a bulletin board or poster for your school.

CHAPTER 2

Sustaining Plant and Soil Productivity

OBJECTIVES

This chapter is about the importance of sustainable agricultural practices. It has the following objectives:

1 Explain sustainable resource use.

2 Describe the role of plant and soil conservation.

3 Explain how quality of life is influenced by sustainable resource use.

4 Describe the role of science and technology in plant production.

5 Explain the use of precision technology in agriculture.

TERMS

conservation
erosion
eutrophication
geographic information
   system (GIS)
global positioning system
   (GPS)
ground truthing
irrigation

machinery controller
nonrenewable natural
   resource
precision farming (PF)
prescription farming
quality of life
remote sensing
renewable natural resource
science

sensor
spatial variability
sustainable agriculture
sustainable agriculture
   system
sustainable resource use
technology
water conservation

PLANT and soil productivity is fragile. It can be easily damaged or destroyed. When this happens, it can no longer be used to produce the crops people need to have a good life.

People do not want to run out of food, clothing, and shelter. It is frightening to think of not having what we need. If we act wisely, we can provide for our needs and for those of future generations.

A major goal in agriculture today is to assure that the basic needs of people will be met in the future. Practices are used to protect the soil, water, and air. Fuel-efficient machinery is used to protect the supply of nonrenewable fossil fuels.

2-1. Maintaining productivity often requires the use of fertilizer and other materials.

SUSTAINABLE RESOURCE USE

Sustainable resource use is using resources so they last a long time. It involves reusing, renewing, and recycling resources to the fullest extent possible. Soil and plant resources will last a long time if proper conservation practices are used.

2-2. Rain water has washed away much of the fertile top soil on this land.

KINDS OF RESOURCES

The earth has two kinds of natural resources: renewable resources and nonrenewable resources. People need to use both kinds carefully to sustain use for a long time.

Renewable natural resources are resources that can be replaced when used up. Renewing often takes a long time, but it can be done. Care is needed to assure that a resource is not used up before it is renewed. Soil, air, water, and wildlife are renewable natural resources. People in agriculture carry out many practices to help renew our resources.

Nonrenewable natural resources cannot be replaced when they are used. Once used, they are gone! Examples of nonrenewable natural resources include minerals and fossil fuels. Some minerals are in very short supply, such as copper and gold. Fossil fuels are used each time petroleum, coal, or natural gas are burned.

SUSTAINABLE AGRICULTURAL PRACTICES

Agriculture involves using many resources. Some can be renewed; others cannot. More attention is now being given to having adequate resources for future generations. Sustainable agriculture is important in this effort.

Sustainable Agriculture

Sustainable agriculture is using practices that assure the ability to produce crops and livestock in the future. Most of the emphasis is on crop production, especially field

2-3. Planting corn in strips on this hillside has helped prevent erosion.

crops. Sustainable agriculture involves using resources so they are conserved and renewed as best possible.

Conservation practices are followed in the use of natural resources. Soil is managed to reduce loss and maintain fertility. Wastes are properly disposed of so the land is not degraded by dumping them improperly. Pesticides are used only as needed to prevent damage and pesticide buildup in the soil.

Sustainable Agriculture Systems

Sustainable agriculture usually requires more than one practice in plant and soil science. A single practice often does not stand alone, nor is it satisfactory. Practices are grouped together. Appropriate technology must be used. Using multiple practices in combination results in a *sustainable agriculture system*. Production practices are selected to have good yields of crops. For example, using no-till cropping involves leaving crop stubble on the land, removing excess stalks, not plowing the land, and using specially designed planting equipment.

Sustainable agriculture systems typically involve four components:

2-4. No-till soybeans are beginning to grow through corn crop residue on this Illinois farm. (Courtesy, U.S. Department of Agriculture)

- **Rotating crops**—Crop rotation is planting land to different crops on alternating

2-5. The lady bug is helpful in biological pest control.

2-6. Using genetically improved crops can reduce the need for pesticides. (Potatoes have been developed that produce their own insect killer to control the Colorado potato beetle.)

years. The same crops are not grown on the land year after year. An example is to rotate soybeans and corn.

- **Using biological pest control**—Biological pest control is using organisms or biological processes to manage pests, such as weeds and insects. Chemical pesticides are not used, or are used on a very limited basis. Predatory insects, bacteria, and fungi can be used in biological control.

- **Preventing disease**—Preventing a disease outbreak in a crop is far easier than curing it after an outbreak. Plant diseases can be prevented in several ways. Producers can begin by selecting disease-free seed. Sanitizing equipment that moves from one field to another prevents the spread of disease. Providing proper nutrients for plant growth also prevents disease.

- **Using improved crops**—New crops are being developed that resist disease, insects, and drought. These new crops are the products of genetic engineering or other genetic improvements. Using a potato that produces its own insect killer eliminates the need for using a chemical pesticide.

CONSERVATION

Conservation is using resources to assure that some will be available in the future. With plant and soil conservation, the emphasis is on future productivity. The resources are used so the capacity to produce in the future is not damaged.

Conservation should not be confused with preservation. In agriculture, resources are to be wisely used. Preservation is impossible because it implies that the resource will not be used.

Most of the emphasis in agriculture is on soil and water conservation.

Soil CONSERVATION

Soil conservation is using soil so the damage or loss is very little or none. Often, soil may be improved while it is being used, if it is used properly. By improving the soil, the potential for future production is increased.

Soil productivity is lost to erosion, pollution, nutrient deficiencies, and a pH that makes nutrients unavailable. Soil conservation involves taking steps to prevent these from degrading the soil.

The major cause of soil loss is erosion. **Erosion** is the washing or wearing away of the soil. It occurs when soil particles are lost by wind, water, or other natural actions. The particles lost first are from the topsoil that contains the nutrients that plants need for growing.

The next time you see a stream with muddy water you are seeing evidence of soil erosion. Tiny soil particles are floating away in the water. Land where they originated has lost some of its fertility. (Later chapters in this book will address soil conservation in more detail.)

2-7. Severe gully erosion results from poor practices and abuses the land.

Water CONSERVATION

Water conservation is using crop production practices that make efficient use of available moisture and prevent the loss of moisture. Some practices can be used that increase moisture that goes into the soil. Covering the soil with mulch forms a barrier that keeps moisture from evaporating from the soil.

In some climates, water in the soil is inadequate to get an acceptable crop yield. **Irrigation** is the artificial application of water to encourage plant growth and productivity. It should be used only when water is needed and only in the amount needed by the crop. If you see water flowing out of a field that is being irrigated, too much water is being applied.

Quality irrigation water is needed or the soil can be degraded. Irrigation water can deposit salt and other minerals and materials on the land. These pollute the land and destroy the ability to produce crops. Removing salt from land is an expensive process.

2-8. Irrigation is being used to promote the growth of a forage crop in Montana.

QUALITY OF LIFE

Quality of life is having a good environment in which to live. This means that the needs of people are met. They have adequate wholesome food. The water is safe to drink and the air is free of excess pollution. The people are healthy and enjoy living.

CAREER PROFILE

SOIL CONSERVATIONIST

A soil conservationist studies soil and develops methods of preventing soil damage. Most soil conservationists plan and implement land use practices to prevent erosion and protect the soil. This photograph shows two soil conservationists reviewing aerial photographs of farms in the Yuba City, California, area.

Soil conservationists have college degrees in areas of agriculture, such as agronomy, horticulture, and agricultural education. Many get masters and doctors degrees in related areas. Practical experience using land surveying, geographical information systems, remote sensing, global positioning systems, and crop production are beneficial.

Most soil conservationists work for government agencies, especially the Natural Resource Conservation Service of the U.S. Department of Agriculture. Others work with universities, research stations, and agribusinesses. Some are entrepreneurs and provide assistance to land owners. (Courtesy, U.S. Department of Agriculture)

Living involves using resources. People require resources to carry out life processes. These processes also release wastes into the environment. Ways of using some of these wastes help in promoting sustainable agriculture. One example is the use of sludge (a product of sewage) on land to improve fertility and other characteristics. Exercise caution in using sludge. It sometimes contains high amounts of heavy metals, such as cadmium and mercury. Sludge is usually not considered good fertilizer for vegetable crops. (For additional details, refer to *Environmental Science and Technology*, which is available from Interstate Publishers, Inc.)

2-9. Studying water quality helps people in using plant and soil resources. (A kit is being used to test the water in this stream for pollution.)

PROTECTING THE ENVIRONMENT

Sustainable resource use protects the environment from damage. Excess fertilizer, pesticides, engine emissions, and other substances get into the environment from agricultural activities.

Fertilizer use should be adjusted to the level needed as determined by careful analysis of the soil. Excess fertilizer enters water runoff and may get into streams and lakes. The fertilizer nutrients, especially nitrogen and phosphorus, can destroy the ecosystem in the stream or lake. **Eutrophication** is a condition that develops when the water in lakes is changed by increased nutrients. The water no longer provides a good environment for the fish and other organisms that once inhabited the area. A dense growth of aquatic plants, algae, and other organisms suggests eutrophication. The desired species can no longer live in the lake.

With fewer pesticides used in sustainable agricultural systems, the environment has lower levels of pesticides. Some pesticides move about in the air and may reach residential areas. Other pesticides get into drinking water.

Sustainable agriculture involves using fewer inputs to work the land. This saves limited fossil fuel and reduces emissions from internal combustion engines.

PROVIDING FOR THE FUTURE

With sustainable agriculture, the ability to provide adequate food, clothing, and shelter in the future is more likely to occur. Soil and water are protected from loss. The environment is protected from pollution. Scarce fossil fuels are used more sparingly, assuring future availability.

ROLE OF SCIENCE AND TECHNOLOGY

Science and technology help sustain agriculture and the environment. Understanding the basics of plant and soil science and keeping informed of developments in technology help assure profitable and sustained crop production.

SCIENCE

Science is knowledge of the environment around us gained by systematic study. Scientists are continually studying many areas of plants and soils. They are searching for better ways of gaining profitable high crop yields and sustaining the environment.

2-10. Science is used in studying the root system of soybean plants. (Courtesy, Agricultural Research Service, USDA)

TECHNOLOGY

Using the findings of science improves plant production and soil efficiency. **Technology** is the practical use of science. Scientific procedures are used to provide information; technology is the use of the information in plant and soil. Many new areas of technology will shape the future of crop production.

Important areas in agriculture involve applying science. Agronomy, horticulture, plant pathology, entomology, and forestry all apply vast amounts of science in developing new agricultural technologies. In recent years, computers, satellite sensing, genetic engineering, and other areas have emerged as important technologies in crop production.

PRECISION TECHNOLOGY

Sustainable agriculture is enhanced by using practices that are specific to a particular site. This has resulted in new site-specific crop management technologies.

Precision farming and prescription farming are tools for crop producers, turf managers, and others. The goal is to use environmentally sound production agriculture practices. The National Environmentally Sound Production Agriculture Laboratory at Tifton, Georgia, is focusing on precision farming. Montana State University has made major efforts in developing technologies and educating people about precision methods.

PRECISION FARMING

Precision farming (PF) is an information- and technology-based crop management system. It is intended to assure profitability, agricultural sustainability, and protection of the environment. PF involves using cultural practices specific to the needs of crops in small areas or parts of a field.

PF is based on spatial variability. **Spatial variability** is a term that defines the differences found in a field or other area. An example is the soil variation within a field. All areas of a field are not identical. Applying identical treatments to an entire field is wasteful and can be environmentally damaging.

Precision farming is beneficial because large fields can be divided into smaller areas (mini-fields) based on important cropping information, such as water-holding capacity and fertility of the soil. Fields can be divided into areas or grids a few feet (meters) square for analysis. Application of fertilizer and other crop-supporting inputs can be precisely made for each grid. Equipment is controlled by computers to provide varying amounts of fertilizer, irrigation water, and other inputs based on the small areas within a field.

2-11. Microcomputers are used to study soil fertility information collected by remote sensing. (Courtesy, Agricultural Research Service, USDA)

2-12. Using a global positioning unit to study soil salinity. (Courtesy, Agricultural Research Service, USDA)

2-13. Dividing a large field into grids, or mini-fields, is a part of mapping in a geographic information system. (This shows the use of information on nitrate leaching.) (Courtesy, Agricultural Research Service, USDA)

Precision farming is used to manage the following crop production activities: fertilizing, planting, tilling, controlling pests, and harvesting, including yield mapping.

*T*ECHNOLOGY USED

Four areas of technology are used in precision farming:

- **Microcomputers**—Microcomputers in offices and on equipment are used to process information about fields. Information about the crop requirements is used to provide instructions for precise areas in fields.

- **Geographic information systems (GIS)**—*Geographic information systems* are used to map fields in small areas, known as grids. The grid maps provide information on soil fertility, texture, moisture-holding capacity, pesticide residues, and crop yields in previous years. The grid map is used with global positioning to locate and control the equipment operating in a field.

CONNECTION

USING TRIANGULATION TO MEASURE DISTANCES ON THE EARTH

Global positioning involves measuring distances between a point on the earth and a group of satellites. The distances are measured by timing how long it takes for a radio signal to reach the point on the earth from each of the satellites. Three satellites are used in making the measurement with a fourth satellite used to verify accuracy. The satellites serve as faraway reference points.

Triangulation, also known as satellite ranging, is used to identify a single point on the earth. Distances between the satellites and the earth are quickly handled by computers. The information is used by machinery controllers to vary rates of materials' application or record harvest yields.

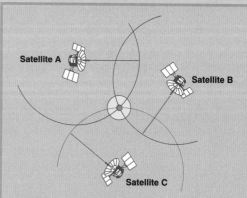

GLOBAL POSITIONING SYSTEM (GPS)

- Global positioning systems (GPS)—A **global positioning system** connects satellites orbiting the earth and a ground receiver that may be near or in a field or on equipment. The system can locate exact points in fields. These points are used with the GIS grid maps to help control machinery operating in the field.

- Machinery controllers—**Machinery controllers** are used on applicators and other implements to apply varying amounts of fertilizer and other crop inputs in fields using GIS and GPS information. Also known as variable rate technology (VRT), each grid, or mini-field, receives the exact amount of input needed.

Much of the precision farming technology is undergoing additional research and development. Costs of the equipment needed have decreased so the technology is more affordable. Individuals who use the equipment need training in its operation.

2-14. Checking readings on a digital yield monitor in a combine. (Courtesy, Agricultural Research Service, USDA)

PRESCRIPTION FARMING

Prescription farming is similar to precision farming. It involves mapping fields and applying fertilizer and other inputs based on what is needed in particular areas of fields. Water is often used to apply fertilizer and other inputs. This works well with dryland farming

2-15. Variable rate technology requires receivers and controllers that operate properly. (This shows potash being applied by an applicator responding to a yield map.) (Courtesy, Agricultural Research Service, USDA)

2-16. Black and white aerial photographs are simple forms of remote sensing that show important land features. (Courtesy, U.S. Department of Agriculture)

where irrigation is frequently used. Prescription farming does not use all of the global positioning techniques of precision farming.

REMOTE SENSING

Remote sensing is collecting information about something without actually contacting what is being studied. It can be used to collect information about crop fields without going into the fields. Most remote sensing is with satellites and airplanes. The information is used to create maps that provide details for variable rate technology.

Ground truthing is used to verify the accuracy of remote sensing information. It involves actual field investigation.

Aerial photography has been a widely used form of remote sensing for many years. Aerial photographs are useful in determining the shapes, elevations, and sizes of fields and other land and water areas. Most aerial photographs used in agriculture have been in black and white.

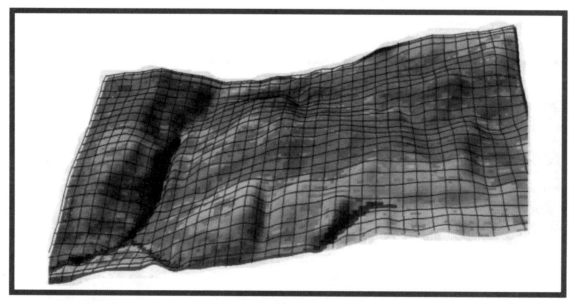

2-17. A vegetative index map of a farm prepared using remote sensing information. (Courtesy, Montana State University)

A **sensor** is a device used to remotely collect information. Sensors are sensitive to light, temperature, radiation, and other characteristics. The sensor transmits information to a control instrument or computer for processing. Most satellite-based sensors use electro-optical sensors. These devices create electrical signals proportional to the amount of electromagnetic energy that is detected. The signals are used to create images that provide useful information. An example is a map that shows the amount and kind of vegetation growing on an area of land.

Infrared, multispectral imagery, microwave radar, and high-frequency radar are used in remote sensing. Infrared mapping has been particularly useful in assessing cropping conditions on a large scale.

REVIEWING

MAIN IDEAS

Meeting the needs of people requires sustaining plant and soil productivity indefinitely. Resources must be used so they last a long time. This especially includes the nonrenewable natural resources.

Sustainable agriculture is using practices that assure the ability to produce crops and livestock into the future. Rotating crops, using biological pest control, preventing disease, and using improved crop plants are four important approaches in sustainable agriculture.

Conservation of soil and plants is important. It involves using them so some will be available in the future. Both soil and water conservation are vital in cropping.

Quality of life is important to all people. This requires providing a good environment and protecting it from damage. Adequate wholesome food plus clothing and shelter must be provided.

New technology continually emerges in agriculture. Environmentally sound production practices are being used. These involve various applications of precision farming. Four technologies are used: computer applications, geographic information systems, global positioning systems, and machinery controllers.

QUESTIONS

Answer the following questions using complete sentences and correct spelling.

1. What is sustainable resource use? How does it relate to renewable and nonrenewable natural resources?
2. What is sustainable agriculture?
3. What are the four components in a sustainable agriculture system?
4. What two areas of conservation are important in agriculture? Briefly describe each.
5. What is quality of life?
6. Distinguish between science and technology.

7. What is precision farming? Why is it beneficial?

8. What is remote sensing?

EVALUATING

Match the term with the correct definition. Write the letter by the term in the blank that is provided.

a. machinery controller
b. sustainable resource use
c. global positioning system (GPS)
d. conservation
e. geographic information system (GIS)

f. technology
g. precision farming (PF)
h. remote sensing
i. erosion
j. irrigation

1. _____ loss of topsoil

2. _____ using resources so they last a long time

3. _____ artificial application of water to encourage plant growth

4. _____ collecting information about something without contact

5. _____ information- and technology-based crop management

6. _____ mapping fields by forming them into small grids

7. _____ using satellites and a ground receiver to locate a point on the earth

8. _____ regulates the rate of application of crop inputs

9. _____ using resources to assure that they will be available in the future

10. _____ practical use of science

EXPLORING

1. Study the use of global positioning systems. Arrange for a demonstration of a global positioning system at your school. For assistance, contact the dean of agriculture at the land-grant university nearest you, the local forest service, or a National Guard or Army Reserves Unit. Try to locate different points on the school grounds using the system. Write a report on your observations.

2. Invite a local soil and water conservationist to serve as a resource person in class and discuss the importance of soil and water conservation. Contact the local office of the Natural Resource Conservation Service of the U.S. Department of Agriculture for assistance.

3. Arrange to make a field trip to a farm or other place that uses precision technology. Observe the equipment needed and how it is used to promote sustainable resource use.

Part Two

Fundamentals of Plant Science

PLANTS are fascinating. There is so much to know about them. The more we know, the better able we are to efficiently grow them in ways that meet our needs. We need to remember that plants are living organisms. As such, they carry out life processes. Any attempt to culture plants must be based on promoting these processes. How plants grow and reproduce influence how they are cultured.

CHAPTER 3

Plants as Living Organisms

OBJECTIVES

This chapter highlights the different characteristics of plants and their contribution to human life. It has the following objectives:

1 Identify the major characteristics of plant life.

2 Explain life cycles.

3 Describe the structure of the parts of plants.

4 Describe the functions of the major parts of plants.

5 Identify the parts of plants used for food and fiber.

TERMS

adventitious root	dicot	leaves	root cap
angiosperm	dioecious	life cycle	root hair
annual	epidermis	mesophyll	secondary root
anther	evergreen	monecious	sepal
apical dominance	fibrous root system	monocot	simple leaf
apical meristem	filament	narrowleaf plant	spongy layer
biennial	flower	ovary	stamen
broadleaf plant	guard cell	palisade layer	stigma
buds	gymnosperm	perennial	stomata
bud scale	hardy	perfect flower	style
calyx	herbaceous	petal	tap root system
cambium	imperfect flower	petiole	terminal bud
complete flower	incomplete flower	phloem	transpiration
compound leaf	lateral bud	pistil	vascular cambium
cuticle	leaf blade	pollen	woody
deciduous	leaflet	primary root	xylem

3-1. Plants are complex organisms that come in many shapes, sizes, and colors.

ROOTED in one place for their entire lives, plants are thought of as simple organisms, if they are thought of at all. Most people find it easier to relate to animals. Why? Is it because animals have the ability to move about? Or, because animals have features similar to ours? You might have more in common with plants than you think.

Plants are complex multicellular organisms. Groups of plant cells form tissues and organs that perform specialized functions. Plants also have Deoxyribonucleic acid (DNA) molecules that contain genetic information to regulate life processes just as people have. This genetic material is passed to future generations through reproduction.

Plants make their food; animals cannot. It is because of this ability to make food that animals, including humans, depend on plants. All the food we eat can be traced back to plants. Of course, plants provide people with more than food. Fibers, such as cotton, are used for clothing. Wood from trees is used for houses, furniture, and fuel. Chemicals taken from plants are used for medicine.

THE PLANT KINGDOM

The plant kingdom consists of thousands of plant species. Plants come in all shapes and sizes from giant redwood trees to mosses growing on the surface of the ground. Not only do they look different from each other, but they are nearly everywhere. Plants are found in great variety in the tropical rainforests. They are also found in harsh climates, such as the arctic tundra or the desert.

There are four major groups of plants living today: mosses, ferns, gymnosperms, and angiosperms. The most successful of the four groups are the seed producing gymnosperms and angiosperms.

The word, **gymnosperm**, comes from the Greek language and means "naked seed." These plants produce seeds that lay naked on the scales of cones. Examples of gymnosperms include pine, spruce, fir, ginkgo, and redwood.

Angiosperms are called flowering plants. The name, angiosperm, is also derived from Greek. It means "seed enclosed in a vessel."

3-2. Plants are found in rainforests.

3-3. Plants are found in arctic tundra.

3-4. Plants are found in volcanic craters.

3-5. Seeds are a significant source of quality nutrition. (Courtesy, Agricultural Resource Service, USDA)

The seeds of flowering plants develop within a fruit. Examples of angiosperms include corn, apples, wheat, petunias, and oaks.

Flowering plants gave humans the opportunity to practice agriculture and to become civilized. In fact, all of our major crops are flowering plants. Seeds of flowering plants have a concentration of proteins, carbohydrates, oils, and vitamins that are good for human nutrition. More important, seeds can be collected in abundance, saved for extended periods of time, and planted to produce new crops.

TWO CLASSES OF FLOWERING PLANTS

There are two major classes of flowering plants. They are the monocots and the dicots. **Monocots** include lilies, grasses, corn, and palms. **Dicots** include oaks, cacti, roses, and soybeans. Some of the differences of the two types of plants are shown in Table 3-1.

Table 3-1. Comparison of Monocots and Dicots

	Monocots		Dicots	
Leaves:	Long, narrow blades with parallel veins		Broad to narrow leaves with netted veins	
Flowers:	Flower parts in multiples of three		Flower parts in multiples of four or five	
Stems:	Vascular bundles scattered		Vascular bundles arranged in circle	
Seedlings:	One seed leaf (cotyledon)		Two seed leaves (cotyledons)	
Seeds:	Contain one cotyledon		Contain two cotyledons	

3-6. Corn is a monocot.

3-7. Soybeans are dicots.

PLANT LIFE CYCLES

A plant *life cycle* is the time from when a plant begins growth until the time it dies. Some individual plants, like the Bristlecone pine in the western United States, have been alive for more than five thousand years. Many other types of plants live for less than one year. Knowing this, we can begin to classify plants by their life cycle. The three major life cycle groups in which we can place plants are annuals, biennials, and perennials.

ANNUALS

Plants that complete their life cycle within one year are called *annuals*. Annuals germinate, grow leaves and roots, flower, produce seed, and die, all within one year.

Many of our crops and garden plants are annuals. Corn, soybeans, rice, wheat, potatoes, and tomatoes are examples of annual food crops. Petunias, impatiens, marigolds, and zinnias are examples of garden annuals. Many plants that are considered weeds, such as ragweed, pigweed, lambsquarter, and crabgrass, are also annuals.

BIENNIALS

Biennials are plants that normally require two growing seasons to produce flowers and fruit before dying. Plants in this group germinate and grow leaves and roots during the first

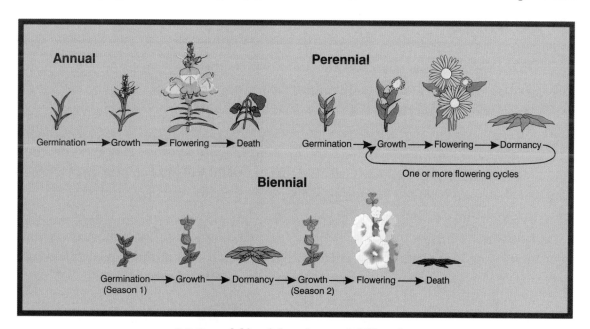

3-8. Annual, biennial, and perennial life cycles.

CAREER PROFILE

BOTANIST

A botanist is a specialist in the science of plants or botany. Botanists may specialize in plant physiology, breeding, classification, or other area. The work may be in laboratories or in the outdoors or forests. Botanists may also teach and carry out research programs.

Botanists have college degrees in botany or biology. Most have masters degrees and many have doctoral degrees in botany. Begin preparation in high school by taking biology, agriculture, and horticulture classes.

Jobs for botanists are with colleges, research stations, arboretums, and other places where plants are studied. This photo shows a botanist working with an olericulturist in solving a problem with tomatoes. (Courtesy, Agricultural Research Service, USDA)

growing season. In the fall, they go dormant and rest until the following spring. Growth is resumed in the second season. They then flower, produce fruit, and die.

This group of plants is fewer in number than the other two groups. Some examples include hollyhock, Sweet William, parsley, beets, and carrots.

PERENNIALS

Perennials are plants that have a life cycle of more than two years. Once these plants reach maturity they may flower and produce seeds every year for many years. During the winter months, they go dormant or stop any growth. Perennials are classified as herbaceous or woody.

Herbaceous perennials have soft shoots that are killed by frost. However, the roots and the crown of the plant remain alive. In spring, the plant will send up new shoots. Examples of herbaceous perennials include daylilies, peonies, asparagus, tulips, and Oriental poppies. (Annuals and biennials are herbaceous, too.)

Woody perennials are trees, shrubs, or vines that have wood and buds above the ground that survive winter. In spring, new shoots or flowers emerge from the buds. Some woody perennials are maple, honeysuckle, grape, blueberry, and English ivy.

Perennials are adapted to withstand climatic conditions in a certain area. Plants that are tolerant to cold temperature in an area are said to be **hardy**. For example, the Southern

3-9. Trees are woody perennial plants.

magnolia is hardy in Atlanta, Georgia, but not in Naperville, Illinois, where it cannot tolerate the cold winters.

PLANT STRUCTURES AND THEIR FUNCTIONS

Plants are complex organisms. They are made of organs or structures that carry out specific jobs. Each is important to the plant's survival. The major organs are the roots, stems, leaves, and flowers.

ROOTS

Very often, the health of a plant can be linked to its root system. If the soil conditions or environmental conditions put stress on the root system, the whole plant will suffer. That is because the roots perform important functions for the plant:

- Roots absorb water and minerals from the soil.

- Roots anchor the plant so it can grow upright.

- Roots store food manufactured in the leaves.

3-10. Young, healthy roots are creamy white. (Courtesy, Agricultural Research Service, USDA)

3-11. Root hairs develop on young roots. (Courtesy, Agricultural Research Service, USDA)

The vast majority of roots are found in the top two feet of soil. This is because roots need air to live. Topsoil has a better exchange of gases with the atmosphere than does subsoil. Compacted, heavy clay, or waterlogged soils have poor air exchange. Only plants adapted to those extreme conditions can survive. In general, roots will grow best in soil that is moist, yet is loose enough for air to reach the roots.

If you were to take a close look at a root, you would notice certain features. The first thing you might notice is that most healthy roots are white or cream in color. Another feature is the abundance of tiny **root hairs** found near the growing tip of the root. Root hairs greatly increase the surface area of the root allowing more water and minerals to be absorbed into the plant.

When a seed begins to grow, it sends out a root, called the **primary root**, to absorb water and nutrients and to anchor the seedling. The primary root is a single, main root. As the primary root grows, it produces many smaller root branches called **secondary roots**. It is on the young, developing roots of a plant that root hairs can be seen. As roots grow and push through the soil, their tips are protected from coarse soil. This is accomplished by a mass of cells called the **root cap**.

Not all roots begin growth from root tissue. Some roots begin growth from the stem of a plant or a leaf. These roots are called **adventitious roots**. Philodendrons, orchids, and corn plants have adventitious roots that grow from their stems. In addition, methods of propagating (starting) new plants rely on the development of adventitious roots from the stem or leaf.

Root Systems

Not surprisingly, different plants have different types of root systems. Two major categories of root systems are the tap root and fibrous root systems.

Tap root systems are characterized by a thick main root that grows straight down. As the plant develops, smaller roots branch from the main root. Plants with tap roots often tolerate dry periods because the primary root extends deep into the soil. Oak trees and dandelions have tap roots. Carrots, beets, and radishes are vegetables that store food in their tap root.

Fibrous root systems consist of numerous slender roots. A mature plant with a fibrous root system has many roots of about equal size. They are usually located near the soil surface. Because the roots of these plants are thin and located in the upper part of the soil, they are often less tolerant of dry conditions. Examples of plants with fibrous roots are grasses, soybeans, impatiens, rhododendrons, and magnolia trees.

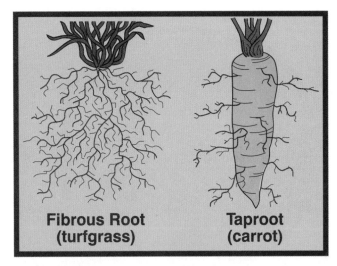

Fibrous Root (turfgrass)   **Taproot (carrot)**

3-12. Plants tend to have either fibrous or tap root systems.

STEMS

As with roots, stems serve important functions for the plant. Some of these include:

- Stems hold leaves in a position to take the best advantage of sunlight.

- Stems conduct water and minerals absorbed by the roots to the leaves.

- Stems conduct food made in the leaves to the rest of the plant.

- Stems store food and water in their tissues.

- Stems produce new living stem tissues.

3-13. The woody stems of cherry trees support flowers, leaves, and fruit. (Courtesy, Agricultural Research Service, USDA)

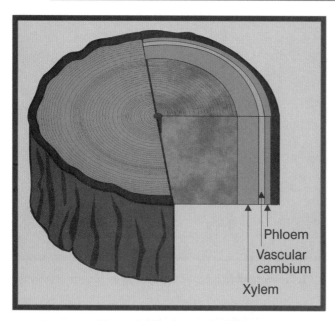

3-14. Wood is made of xylem and phloem tissues.

Phloem
Vascular cambium
Xylem

Conductive Tissues

The life flow of a plant is found in its stems. Just as we need to continually pump blood through our bodies, a plant needs to conduct water, minerals, and food to all of its parts. The plant can accomplish this with special conductive tissues called xylem and phloem. Xylem and phloem tissues in the root are continuous with the xylem and phloem tissues of the stem, which are continuous with the xylem and phloem in the leaves.

Water and minerals are transported from the roots to the leaves in a tissue called the *xylem*. Xylem is located in the roots, stem, leaves, and flowers. Xylem cells also have stiff walls that provide structural support for the plant. Food made in the leaves is transported to the rest of the plant through the *phloem* tissue. Phloem, like the xylem, is found throughout the plant.

The location of xylem and phloem within a stem differs with the type of plant. Monocot plants have xylem and phloem scattered throughout the stem. Dicot plants have xylem and phloem in a ring within the stem. Dicots also have a layer of cells called the *cambium*

CONNECTION

DIRECTING GROWTH

Orange producers need trees that are nicely shaped and strong enough to support large crops of fruit. Trees sometimes do not naturally grow as desired. Strong branching structure can be established while the trees are still young.

Young orange trees growing in this nursery are being straightened, staked, and pruned. Workers ride through the field on a slow-moving tractor. Note that the implement mounted on the tractor has been specially designed for the work.

After another year or so of growth, the trees will be ready for transplanting to an orchard. (Courtesy, U. S. Department of Agriculture)

where cell division takes place. These dividing cells become either xylem or phloem cells depending on which side of the cambium they are located.

Trees have a layer of **vascular cambium** beneath the bark of the stem or trunk. To the outside of the cambium, phloem cells are produced, and to the inside of the cambium, xylem cells are produced. Xylem cells are larger and make up the bulk of the wood. Each spring, when growth starts, a new layer of conductive tissues is made. It is interesting to know that you can count the layers of growth or rings in tree wood to determine the age of the tree.

Buds

Stems of woody perennial plants have structures called **buds** that contain undeveloped leaves, stem, or flowers. **Bud scales** cover and protect these undeveloped parts. Buds are most visible in the fall, winter, and spring. When the buds begin to grow and develop, they form the stems on which flowers or leaves are held.

The large bud at the tip of a twig is referred to as the **terminal bud**. The terminal bud is important because it contains the **apical meristem** or the primary growing point. Buds located along the sides of a stem are **lateral buds**. Lateral buds can be found where the leaves are attached to the stem. Lateral buds are prevented from developing by hormones produced in the apical meristem. This is called **apical dominance**. As a plant grows, the influence of the hormones on the older lateral buds is lessened. The lateral buds can then develop into secondary branches. The lateral buds also will develop if for any reason the terminal bud is removed.

Stems of plants provide us with many valuable products. The most obvious is wood, which is used for construction, furniture, and fuel. We harvest and eat sugar from the stems of sugar cane. Another common food is the Irish potato, which is actually an underground stem. Some plants, like hemp and flax, provide us with fibers for rope or fabric. The bark of the pacific yew contains a cancer-fighting chemical called taxol. Stems of the kenaf plant provide us with the fibers to make paper.

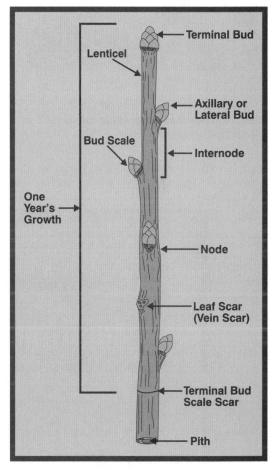

3-15. A typical stem.

LEAVES

Leaves are plant organs responsible for the production of food (sugars) for the plant. In fact, the most important chemical reaction on Earth is this food-producing reaction that takes place in the leaves. It is called photosynthesis.

Leaves are thin organs composed of specialized cells. There is an upper and lower **epidermis** or protective layer of cells. The epidermis cells have a waxy coating, called a **cuticle**, that serves to prevent excessive water loss. Within the lower epidermis there are pores or openings in the leaf called **stomata** (singular: stoma) that allow the exchange of oxygen, carbon dioxide, and water vapor. Most stomata can be found on the underside of leaves. The movement of water vapor through stomata is referred to as **transpiration**. The opening and closing of stomata are controlled by a pair of **guard cells**. When water is plentiful and light is shining, the guard cells are pumped full of water. This causes the cells to push apart, creating an opening to allow an exchange of gases. The guard cells close at night and also when the plant is experiencing water stress. This happens when water leaves the

3-16. Leaves are the plant organs responsible for the production of food. Leaves of the paulownia tree may reach 30–36 inches (75–91 cm) in diameter.

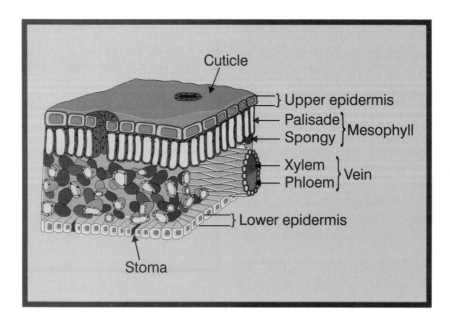

3-17. This cross section of a leaf shows how specialized cells are arranged.

guard cells, the cells collapse, and the opening closes. By closing the stomata, the plant can conserve moisture.

Sandwiched between the epidermal layers is the photosynthetic tissue of the leaf known as the **mesophyll.** Mesophyll is a Greek word meaning "the middle of the leaf." The bulk of photosynthetic activity for a plant takes place in the mesophyll cells. Just below the upper epidermis is a stacked layer of mesophyll cells called the **palisade layer**. Underneath the palisade layer is a loosely arranged layer of mesophyll cells called the **spongy layer**.

Throughout the mesophyll there is a network of veins. Each vein contains xylem and phloem tissues. The veins are numerous enough so every cell is reached for exchange of materials. The xylem conducts water and minerals to the mesophyll cells. Those materials are used in the cells to make food. The food is then transported to the rest of the plant through the phloem.

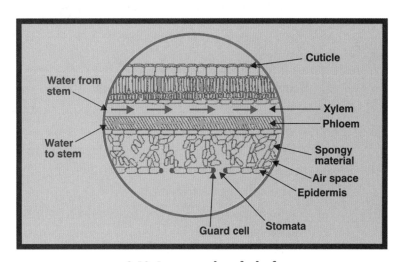

3-18. A cross section of a leaf.

Types of Leaves

Leaves come in a great variety of shapes, sizes, and colors. As a result, leaves are useful when trying to identify plants. We can start to recognize plants by first determining if they have broad leaves or narrow leaves.

Broadleaf plants have wide, flat leaves. The large broad part of a leaf is called the **leaf blade**. The leaf blade provides a large surface area that increases the amount of solar energy absorbed for photosynthesis. The leaf blade is connected to the stem by the **petiole** or leaf stalk. Water and minerals flow through the xylem in the petiole and

3-19. Leaves may be simple or compound.

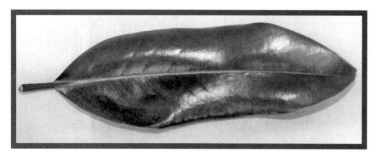

3-20. This is a simple leaf consisting of a leaf blade and a petiole.

through the veins of the leaf to the individual cells.

Some broadleaf plants have simple leaves. **Simple leaves** consist of a single leaf blade and a petiole. Other broadleaf plants have compound leaves. **Compound leaves** are made of a petiole and two or more leaf blades called **leaflets**. Examples of broadleaf plants are maple, soybean, honeylocust, and elm.

Narrowleaf plants have needle shaped or scale shaped leaves. The long, narrow leaves of corn, Kentucky bluegrass, iris, pine, and spruce place them in the narrowleaf plant group.

When a woody perennial plant loses its leaves in the fall, it is said to be **deciduous**. This happens with both broadleaf and narrowleaf plants. The plants that keep their leaves year round are called **evergreen**.

People eat plant leaves, such as lettuce, spinach, and mustard greens. Some leaves like chives, oregano, thyme, and basil are used to flavor foods. Leaves of trees provide us with shade. Because of transpiration from the leaves, they also have a cooling effect on hot summer days. One of the greatest benefits is the oxygen we breathe that is produced and released by leaves during photosynthesis.

3-21. Deciduous trees lose their leaves in the fall and produce new leaves in the spring.

FLOWERS

Flowers are reproductive organs. They can be extremely colorful and beautiful to look at. They can take on very unusual shapes and can smell like perfume. On the other hand, they may be very plain and escape our notice. However they appear to us, keep in mind that they have one function. The function of the flower is to produce seeds that will become the next generation of plants.

3-22. The function of a flower is to produce seed.

Parts of a Flower

Flowers have different parts that perform different functions. Until a flower opens, it is protected by green, leaf-like structures called **sepals**. Sepals are located at the base of a flower. The sepals fold back as the flower opens. All the sepals of a flower are called the **calyx**.

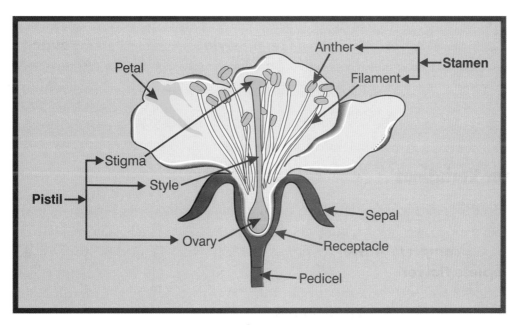

3-23. Flower parts.

3-24. Flower parts can be easily identified on this lily flower.

The petals just inside the sepals are usually thought of first when discussing flowers. **Petals** appear leaf-like and are often very colorful. Brightly colored petals attract pollinators, such as insects and birds. The petals surround the center part of the flower.

Stamens are located to the inside of the petals. **Stamens** are the male reproductive parts of a flower. They consist of a stalk called the **filament** and an anther. The **anther** produces and holds the pollen. The **pollen** grains contain the male sex cells.

At the center of the flower is the **pistil** or the female part of the flower. It is sometimes called the carpel. The pistil has three main parts: the stigma, the style, and the ovary. The **stigma** is found at the end of the pistil and it has a sticky surface on which pollen can be caught. The neck of the pistil is referred to as the **style**. The third part is the **ovary**, which contains one or more ovules where eggs are produced and seeds develop. As the seeds form, the ovary becomes a fruit. Apples are ripened ovaries or fruit. Some people are surprised to learn that tomatoes, green peppers, cucumbers, and watermelons are considered to be fruits too.

Flower Variations

There is a great variety to the basic structure of a flower involving sepals, petals, stamen, and pistil. If the flower has the four parts including the sepals, petals, stamen, and pistil, it is said to be a **complete flower**. A flower that lacks any one of these parts is called an **incomplete flower**.

If a flower has both stamen and pistil, the two parts involved in fertilization, it is referred to as **perfect flower**. An **imperfect flower** is one that lacks either stamen or pistil. A flower that has only stamen is a male flower, while those with only pistil are female flowers. Some plant species with imperfect flowers, such as corn, cucumbers and oaks, have both male and female flowers on the same plant. They are said to be **monecious**. Many other plant species with imperfect flowers produce only male flowers or female flowers on a single plant. These plant species, which include willows, ginkgo, and holly, are called **dioecious**. Pollination between a male plant and a female plant must take place for seed to form.

Table 3-2. Some Perfect, Monecious, and Dioecious Plants

Perfect	Monecious	Dioecious
Apple	Birch	Ash
Pear	Corn	Asparagus
Raspberry	Cucumber	Date palm
Cranberry	Cantaloupe	Hop
Citrus fruits	Fir	Juniper
Tomato	Hickory	Mulberry
Pepper	Larch	Poplar
Bean	Oak	Willow
Carrot	Pine	Spinach
Sweet potato	Pumpkin	Ginkgo
Rose	Watermelon	Holly

Humans Benefit from Flowers

The reproductive part of plants provides us with an abundance of food. Broccoli and cauliflower are flowers that we eat. Pears, peaches, and tomatoes are edible fruits. Wheat, corn, soybean, coconut, peanut, sesame, and rice seeds are extremely nutritious. Seeds are also used to feed livestock.

Seeds are used for countless nonfood products today and research continues for new uses. Soybeans are used to make diesel fuel and printer's ink. Corn is used to make ethanol fuel and biodegradable plastics. Linseed oil is used in paints.

REVIEWING

MAIN IDEAS

Animal life, including human life, is dependent upon plant life. Plants have the ability to make their own food. We eat plants that have stored this food or animals that have fed on plants. We also benefit from fibers used for clothing, wood used in construction and fuel, chemicals used for medicine, and numerous other applications.

There is great variety in the plant kingdom. Because there are so many types of plants, there is a need to group them according to characteristics. The flowering plants are the most important plants to us. Monocots and dicots make up the two classes of flowering plants. Annuals, biennials, and perennials are three common life cycles of plants. Plants can be categorized further by their structures.

Plants are complex living organisms. They have four main structures including the roots, stem, leaves, and flowers. Each structure performs functions that are critical for the survival of the plant and species. If any single part fails its duty or is damaged, the future of the plant or species is at risk. Roots absorb water and minerals and anchor the plant. Stems hold the leaves and flowers. Leaves are designed to make food. The function of flowers is to reproduce the species.

QUESTIONS

Answer the following questions using complete sentences and correct spelling.

1. How do plants improve our lives?

2. What are the differences between monocot and dicot plants?

3. What are the three common plant life cycles? Describe each.

4. What are the functions of a root?

5. What are the functions of a stem?

6. What does the xylem do for a plant?

7. What is the function of the leaf?

8. What is the function of the flower?

9. What parts make up a complete flower?

EVALUATING

Match the term with the correct definition. Write the letter by the term in the blank that is provided.

a. angiosperm   e. stamen   i. phloem
b. stigma   f. perennial   j. bud
c. root hairs   g. hardy
d. deciduous   h. stomata

1. _____ plant that loses its leaves in fall
2. _____ plant with a life cycle of more than two years
3. _____ plant that can withstand cold temperature in an area
4. _____ male part of a flower
5. _____ thin roots that absorb water and minerals
6. _____ flowering plant
7. _____ pore or opening in the leaf
8. _____ conductive tissue through which food moves
9. _____ contain undeveloped leaves, flowers, or stem
10. _____ sticky surface at the end of the pistil

EXPLORING

1. Look at the fruits and vegetables in your home. Identify the different plant structures that we eat. Go to a grocery store where there is a wide selection of vegetables and do the same.

2. Identify your favorite fruit. Find out which part of the world it came from. List interesting facts about this fruit.

3. Look at a flower in the garden. See if you can identify the major parts of the flower. Remove the sepals and petals to get a better view of the pistil.

CHAPTER 4

Plant Reproduction

OBJECTIVES

In this chapter, the various ways plants reproduce are introduced. It has the following objectives:

1 Name and describe the ways plants reproduce.

2 Explain the formation and structures of seeds.

3 Describe the conditions for seed germination.

4 Explain methods of asexual reproduction.

5 Apply genetics and breeding in plant reproduction.

6 Explain basic concepts of plant genetic engineering.

TERMS

asexual reproduction
callous
chromosome
clone
cotyledon
cross-pollination
cultivar
cuttings
Deoxyribonucleic acid (DNA)
diploid
disseminate
division

double fertilization
embryo
endosperm
epicotyl
explant
fruit
gene
genetic engineering
genotype
germination
grafting
haploid
hybrid

hybridization
hybrid vigor
hypocotyl
layering
mutation
phenotype
plant breeding
plantlet
plumule
pollination
radicle
rootstock
scarification

scion
seed
seed coat
selection
self-pollination
separation
sexual reproduction
stratification
tissue culture
transgenic plants
viability
vigor
zygote

4-1. Harvested wheat is being loaded for delivery to a grain elevator. Each individual grain is a seed containing a living embryo. (Courtesy, Agricultural Research Service, USDA)

WHAT would happen if plants did not reproduce? The plants we see today would grow older, they would die, and, eventually, they would cease to exist. Plants would be extinct. The earth would be rocky and sandy. In fact, life as we know it would end. Fortunately, through evolution, plants have devised efficient ways to reproduce.

Throughout history, people have relied on seeds and plant parts to grow new plants for food and fiber. Plant breeding, along with developments in technology, is credited for the production of an adequate food supply to feed the world's 6 billion people. Studies have shown that the world's population is expected to increase considerably by the year 2050. Will current crops and practices produce enough food and plant products to meet the needs of a much larger population? Many informed people do not think so.

SEXUAL REPRODUCTION

Sexual reproduction in flowering plants involves flowers, fruits, and seeds. In **sexual reproduction**, sperm carried in the pollen from the male part of a flower fuses with the egg in the female part of the flower. Both the sperm and the egg contribute genetic material to the new life within the seed.

What advantage does sexual reproduction provide to a plant? For survival, a plant species must live to reproduce in its environment. However, the earth and the environment have been changing since the beginning of time. Plants adapted to conditions thousands of years ago may not be able to survive in today's conditions. Sexual reproduction gives the plant species the means to change with a changing environment. Every time sexual reproduction occurs there is a recombining of genetic material. Some genetic changes are beneficial. Plants receiving genes that enable them to adapt to a changed environment are more likely to survive to pass genes onto their offspring. Thus, species evolve in order to survive.

4-2. Insect pollinators, such as butterflies, are attracted to flowers by color and nectar.

POLLINATION

Flowering plants have evolved with a number of methods to accomplish pollination. **Pollination** is the transfer of pollen from the male to the female part of a plant. Colorful, scented flowers attract birds, insects, bats, and other animals. These creatures unknowingly pick up pollen from the anthers and, when they visit another flower, deposit the pollen on the stigma. Pollinators are rewarded by the plant with food. They feed on the nectar (a sugary, energy-rich food made in the flower) or on the pollen itself, which is rich in protein.

Other plants rely on the wind to transfer pollen to the stigma. The force of the wind physically moves pollen from one flower to another. Since there is no need to attract pollinators, these plants do not produce colorful flowers with large petals, scents, or nectar. Plants that depend on wind for pollination actually rely on luck for the pollen of a plant species to blow onto the stigma of the same species. Plants that depend on wind for pollination produce a great abundance of pollen to improve the odds that pollination will occur. Ragweed, many trees, and grasses are wind-pollinated plants. Hay fever sufferers are well aware of when certain plants are producing pollen.

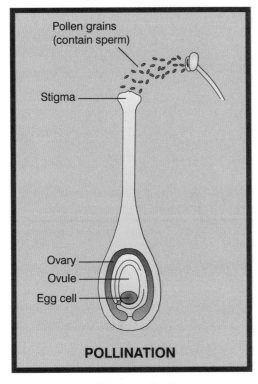

4-3. Pollination of a flower.

4-4. Some plants, such as wheat, depend on wind for pollination and lack showy flowers. Wheat flowers appear as a head begins to form.

When the pollen of a plant pollinates a flower on the same plant, it is called **self-pollination**. Some plants have this ability; others do not. When the pollen of a plant pollinates the flower of another plant, it is said to be **cross-pollination**. For example, cross-pollination is required for most varieties of apple trees. Also, it is important to understand that plants must be closely related for cross-pollination to occur. For example, an oak tree cannot pollinate a soybean plant.

Once the pollen lands on the stigma, it grows a thin pollen tube down the style to the ovary. The cell within the grain of pollen divides to form two sperm nuclei. These travel down the pollen tube to the embryo sac that holds the egg.

FERTILIZATION

Fertilization in flowering plants is unlike fertilization in any other living organism. This is because both sperm nuclei in the pollen grain are involved in fertilization. Flowering plants actually have a **double fertilization**.

One fertilization occurs when one sperm fuses with the egg. In this process, the sperm carries genetic material from the male part of the flower. The egg contains genetic material

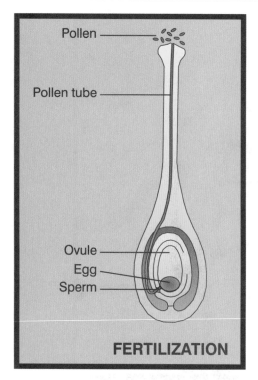

FERTILIZATION

4-5. Fertilization of a flower.

4-6. Hybridization improves plant vigor, uniformity, and insect and disease resistance.

from the female part of the flower. The cell that results from the combining of the sperm and egg is called a **zygote**. As the zygote cell divides, it develops into the seed. The result is a seed that contains a combination of genetic material received from the male and female parts of a flower.

The other fertilization has the second sperm nuclei fusing with two nuclei in the embryo sac. It develops into the endosperm. The ovule of the flower becomes the seed.

OFFSPRING AND GENETICS

The offspring of genetically different parents is said to be a **hybrid**. People have greatly improved agricultural crops and animals through hundreds of years of hybridization. The first farmers saved seed from their best plants for planting the following year. Over many generations of saving the best seed and discarding the less valuable seed, crops changed. In many cases, you would find it difficult to recognize the distant ancestors of today's crops and animals. For example, ancient corn looked more like a grass than today's hybrids. The advantage of hybridizing is the best traits of each parent may be expressed in the offspring. (More information on hybridization is later in the chapter.)

An understanding of genetics is necessary when hybridizing plants. Genetic information is stored in every cell of a plant or animal in long molecular chains made of **Deoxyribonucleic acid (DNA)**. Segments of the DNA, called **genes**, code for life processes and the appearance of a plant. For instance, genes in a petunia plant may code for a red flower. Genes in that petunia plant also tell it when to flower and how many petals to make. An individual plant may have 100,000 genes informing it what to do and how to look.

The genes are arranged in a set of **chromosomes**. Normal cells contain a double set of chromosomes and are said to be **diploid**. Reproductive cells, sperm and egg, have a single set of chromosomes and are said to be **haploid**. When fertilization occurs, both the sperm and the egg contribute a single set of chromosomes. Each cell of the resulting seed ends up with a normal double set. In this way, traits from each of the parent plants may be passed on to the offspring.

SEEDS

The largest seed in the world is the double coconut found in the Indian Ocean's Seychelles archipelago. It can weigh up to fifty pounds. Other seeds, like those of the begonia, are as small as dust. However they may appear, their function is to grow and develop into a mature plant that will produce more seeds. Seeds are truly a wonder of nature.

FRUITS

After fertilization, the ovary wall enlarges and forms the *fruit*. Seeds are formed in the fruit. The fruit may be either fleshy or dry. Fleshy fruits, like the tomato, are juicy to prevent the seeds from drying until they are mature. They also serve to help disperse the seeds. Animals are at-

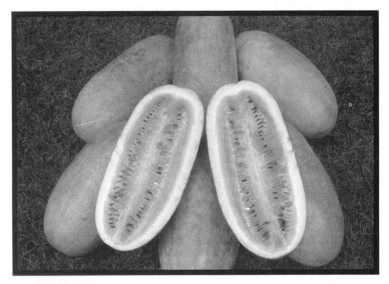

4-7. Watermelons are fleshy fruits. (Courtesy, Agricultural Research Service, USDA)

CAREER PROFILE

PLANT TECHNICIAN

A plant technician studies ways to improve plant growth, plant yields, and plant quality. Much of the work is performed in a laboratory. Genetic engineering and biotechnology skills are often needed.

There are different levels of plant technicians. The extent of education and work experience dictate the level for which a person is qualified. A high school diploma is sufficient for a few entry-level jobs. Advanced levels require a college degree in biology, agronomy, or horticulture and laboratory experience. Knowledge of botany and genetics is essential. Practical experience in agriculture is helpful.

Jobs for plant technicians are with large agricultural companies, government agencies, experiment stations, and colleges. (Courtesy, Cargill Hybrid Seeds)

4-8. Soybeans have dry fruit.

tracted to the nutritious fruit, eat it with the seeds, and disperse or **disseminate** the seeds away from the parent plant.

The dissemination of seeds is an important process for the survival of a plant species. Fruit forms, sizes, and shapes have evolved to support continuation of a species. Not all fruits are fat and juicy like the tomato. The dandelion has evolved a dry, feathery fruit to take advantage of the wind for dissemination. Soybean pods, cotton bolls, and the helicopters (samaras) of maple trees are technically dry fruits, too.

SEED STRUCTURES

Seeds of flowering plants have a seed coat, an embryo or young plant, and a source of stored food. The **seed coat** is a protective shell surrounding the embryo and the endosperm. It protects the seed from drying and from physical injury. The seed coat also plays an important role in determining when conditions for **germination** or the beginning of growth are right.

The **embryo** is a little plant that eventually grows and develops into the mature plant. It is in a dormant or resting phase inside the seed. It has a stem, root, and one or two seed leaves called **cotyledons**. The first bud of the embryo consists of miniature leaves enclosing a growing point and is called the **plumule**. Embryos of monocot plants have one cotyledon while those of dicot plants have two cotyledons. In dicot plants, the portion of the em-

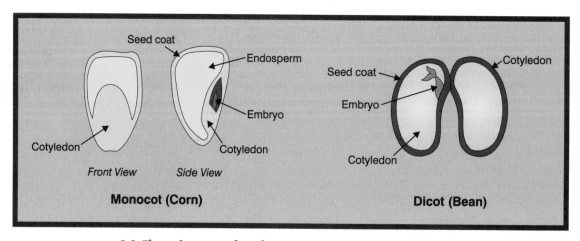

4-9. The major parts of seeds are common to dicots and monocots.

bryo above the cotyledons is referred to as the **epicotyl**. The portion of the embryo below the cotyledons is the **hypocotyl**.

Food in the form of starch and protein is stored in the seed for the embryo. When germination begins, the embryo draws energy from the food to emerge from the soil and to develop leaves so it can begin to manufacture its own food. With monocot plants, much of this energy source is found in the **endosperm**. Dicots store all their food in the two cotyledons.

SEED GERMINATION

Seeds are designed to wait for favorable conditions to begin growth. The wait for the favorable conditions may take many years. In some cases, the embryo's life within the seed dies before the arrival of conditions favorable for the plant's survival. It is fascinating to think about how seeds have the ability to lay dormant for many years and then to begin growth when given the right conditions for survival.

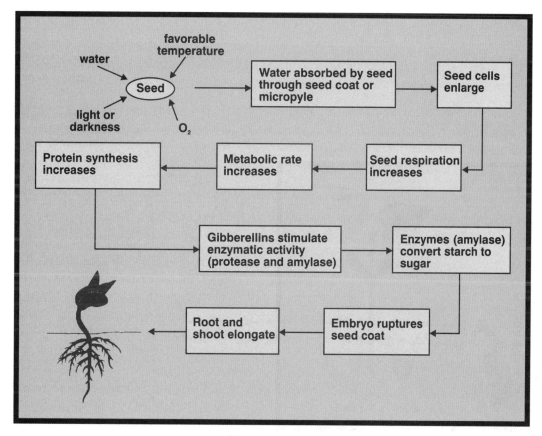

4-10. The germination process.

DORMANCY MECHANISMS

One mechanism seeds have to help ensure survival is called stratification. **Stratification** is the process whereby a seed must go through a period of cold temperatures before it will germinate. This mechanism makes sense. Without this waiting period, a seed could germinate during a warm spell in December and die with the arrival of freezing temperatures.

Another mechanism is **scarification**, or the breaking down of the seed coat. Some seeds have very hard, thick seed coats that provide excellent protection for the seed. However, they prevent the absorption of water and germination under normal conditions. These seeds must pass through the acid stomach of an animal to wear down the seed coat or lay in the soil where microorganisms can eat away the seed coat before they can germinate. Some examples of plant seeds that require scarification are geranium, lupine, honeylocust, and Kentucky coffeetree.

ENVIRONMENTAL FACTORS

What are the right conditions for seed germination? Environmental factors play key roles in seed germination.

- Water—Water is necessary for a seed to germinate. Water stimulates chemical reactions within the seed.

- Oxygen—Along with moisture, seeds need air. Oxygen is particularly important as starches stored in the seed are converted to energy through cellular respiration.

- Temperature—Seeds need the right temperatures to germinate. The optimal temperature varies with the plant species. Most plant seeds germinate at 60 to 80°F (15.5 to 26.7°C), but different plant species are adapted to germinate at 40 to 104°F (3.5 to 40°C). Naturally, plants from the tropics will need a warmer temperature than those found in the Arctic.

- Light—Some seeds have certain light requirements to germinate. They might need total darkness or exposure to light. Lettuce seeds, for exam-

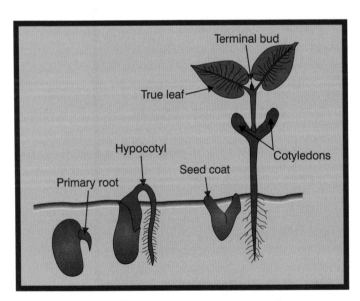

4-11. The germination of a bean seed.

Terminal bud

True leaf

Hypocotyl

Seed coat

Cotyledons

Primary root

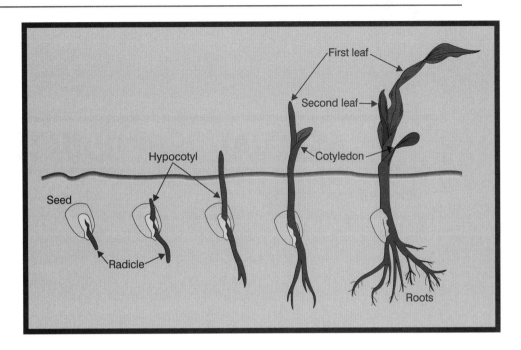

4-12. The germination of a corn seed.

ple, must be exposed to light before they will germinate. If any one of these environmental factors is not favorable, the seed will not germinate.

The first step in the germination process is the absorption of water, which triggers a number of activities within the seed. The seed swells with moisture. The embryo changes from a dormant state to an actively growing plant. During these early stages, the embryo draws energy from starches stored in the endosperm or the cotyledons. The embryo's root or **radicle** emerges from the seed and develops into the primary root. Then, the stem of the embryo sprouts upward. After germination, the seedling needs light and minerals to begin producing its own food.

Many of our most important crops are grown from seed. Corn, soybeans, wheat, and vegetables are just a few crops grown for food. Bedding plants and turfgrass grown in the horticulture industry are started from seeds. Trees for reforestation begin as seeds. Fiber crops, such as cotton and hemp, are grown from seeds, too.

Viability and Vigor

The quality of seed used is very important to those in production agriculture. Seed quality refers to two factors, viability and vigor. **Viability** is the ability of a seed to germinate under optimum conditions. **Vigor** is the ability of a seed to germinate under different conditions and still produce healthy seedlings. Seeds that are viable and have vigor are valued when planting time comes around because the producer wants to see a high percentage of seed germination. Seed companies run tests to determine what percentage of the seeds will germinate. That information must be printed on the seed bag. A high percentage of viable seeds is the goal. Proper humidity and temperature during storage of the seeds from one growing season to the next help to maintain a high proportion of viable seeds.

ASEXUAL REPRODUCTION

Asexual reproduction is the reproduction of a plant using leaves, stems, or roots. You may have seen your parents or grandparents start new plants by placing a leaf or stem in a glass of water. Plants like African violets and philodendrons have the natural ability to begin new plants from plant parts. This type of reproduction produces new plants that are the genetic duplicates of the parent plant. The new plants are said to be **clones** of the parent plant.

There are certain advantages associated with asexual reproduction:

1. Plants with outstanding characteristics can be produced without the risk of losing the desired characteristic through the recombining of genes that would occur in sexual reproduction.

2. Some plants are difficult to reproduce sexually. They may produce very few viable seeds.

CONNECTION

HARDY OATS

Winter oats are sown in the fall in southern areas of the United States. They grow through the winter and are harvested in late spring. Of the grain crops, oats are more susceptible to cold damage. The term used to describe resistance to cold is "hardiness." Varieties that resist damage by cold weather are said to be hardy.

Understanding how grain plants resist cold weather may help develop new cold-tolerant oat varieties. Plant breeders are interested in carbohydrate storage and biochemical changes in plants that make them hardy. Speculation is that the presence of certain sugars in plants lowers the freezing point.

The winter oat seedlings shown here are being prepared for a hardening experiment. They will be exposed to temperatures just above freezing. Changes occurring in the plants will be studied. (Courtesy, Agricultural Research Service, USDA)

3. Huge numbers of genetically identical plants (clones) can be produced.

4. Plants can be grown that are free of viral diseases.

CUTTINGS

There are many ways in which plants are reproduced asexually. **Cuttings** of stems, leaves, and roots are some of the most common and simplest methods of asexual reproduction. The time of year, the stage of plant growth, and the type of cutting being made are all important factors to consider. Once taken, they may be treated with a plant hormone to stimulate the formation of adventitious roots. Special care is given to providing proper environmental conditions to reduce stress on the cutting and to speed root growth. In general, cuttings require humid conditions, slightly lower light intensity, and warmer temperatures. Many trees, shrubs, and greenhouse plants are started from cuttings.

GRAFTING

A method of asexual reproduction common to the orchard and nursery industries is grafting. The Red Delicious apples we eat and the Marshall Seedless ash shade tree used in landscaping are propagated by grafting. **Grafting** is the process in which the stem of one plant is made to grow on the roots of another plant. The portion of the graft that is to become the stem is the **scion**. The lower portion of the plant that includes the root system is called the **rootstock** or the understock.

Grafting may involve placing individual buds into a stem or a stem being placed onto another stem. In either case, it is important that the cambium wood of both the scion and the rootstock line up. Once placed together, the union is protected from moisture loss. The scion and

4-13. Banana varieties grown commercially produce fruit without pollination and fertilization. As a result, they must be propagated asexually.

4-14. Cuttings are the most common method of asexual reproduction.

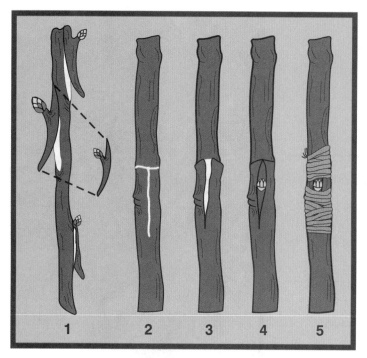

4-15. Budding is a method of grafting in which the scion consists of a bud and a small piece of stem.

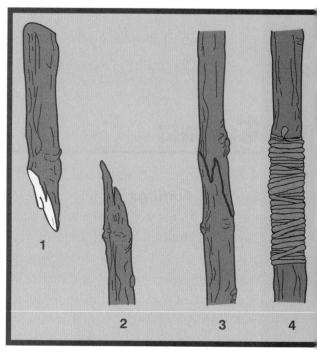

4-16. Whip-and-tongue grafting.

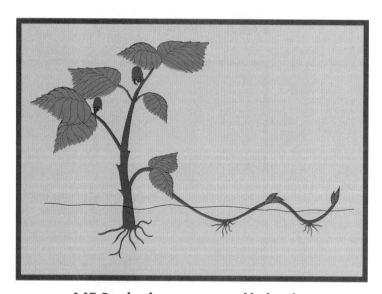

4-17. Raspberries are propagated by layering.

the rootstock materials must also be capable of growing together for a successful graft. For instance, a magnolia will not grow on an elm.

LAYERING

Layering is a method of asexual reproduction in which roots form on a stem while the stem is still attached to the parent plant. This occurs in nature when branches of some woody plants are in close contact with or buried by soil. The advantage to layering is the parent plant provides the new plants with water and minerals until it produces its own roots. Fewer plants can be produced through this method as compared to other methods.

SEPARATION AND DIVISION

Many garden plants can be propagated by separating or dividing vegetative plant structures from the parent plant. In **separation**, the vegetative structure is removed and planted. With **division,** the plant roots or the entire plant may be cut to make two or more plants from the original plant. Daylilies can be divided by digging a plant and cutting it into smaller portions.

TISSUE CULTURE

Another method of asexual reproduction developed recently is tissue culture. **Tissue culture** is a very technical form of reproduction. As the name implies, small pieces of plant tissue are cultured or grown on an artificial medium under sterile conditions. One advantage to tissue culture is thousands of identical plants can be produced from a small piece of plant tissue. The tissue culture propagation process can be defined in four main stages.

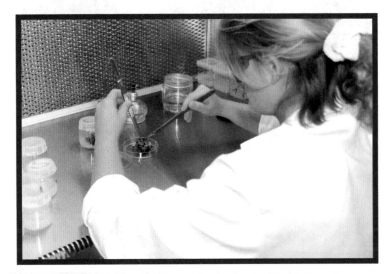

4-18. Tissue culture is carried out in a work area that is free of contamination.

First, small pieces of plant material, called **explants**, are carefully removed from the parent plant. The explants are cleaned of bacteria or fungal spores before being placed on an agar media in glass bottles or test tubes. The agar media is a gel that contains water, sugars, nutrients, and plant hormones to support and promote plant growth. It is critical that the first three stages take place under sterile conditions to reduce the chance of contamination from microorganisms.

In stage two, the cells of the explants multiply in one of two ways. They may form a **callous**, which is a group of cells with no particular function. Even-

4-19. Tissue culture is a highly technical method of asexual reproduction. (Courtesy, Agricultural Research Service, USDA)

4-20. Explants are placed on an agar gel in glass containers. (Courtesy, Agricultural Research Service, USDA)

tually, these callous cells develop into a normal plant. The second possibility is branching that occurs from existing buds on the explant.

When the shoots have developed, they are ready for the third stage. Shoots are transplanted to another medium containing hormones that induce the growth of roots. The *plantlets* or small plants are also given higher light intensity in preparation for stage four.

In stage four, the plantlets are removed from the glass container. They are divided, planted in pots, and placed in a greenhouse. The exciting thing about tissue culture is huge numbers of genetically identical plantlets can be grown from the initial explant.

Tissue culture is used with a number of herbaceous perennial plants including daylilies, hostas, chrysanthemums, and orchids. Some woody plants, including fruit trees, are propagated by tissue culture, too. The importance of tissue culture in plant breeding continues to grow as the field of genetic engineering expands.

GENETICS AND PLANT BREEDING

Plant breeding is the systematic process of improving plants using scientific methods. The methods used may include selection, inbreeding, hybridization, and genetic engineering. Plant breeders use methods to prevent the unwanted breeding (pollination) of plants.

Goals of plant breeding vary with the desired improvements to be made in plants. Plant breeding may be used to:

- gain disease resistance
- gain insect resistance
- improve environmental adaptation
- improve productivity
- make a species more suited to cultural practices
- obtain a more desirable product from plants

Plant breeders look for new genotypes that will result in plants with different and desired phenotypes. **Genotype** is the genetic makeup of a plant that is not readily observable though DNA analysis may provide the information. **Phenotype** is the outward or physical appearance of a plant. These are the changes developed by plant breeders that are most obvious to the human eye.

The work of plant breeders leads to varieties of crop plants that have desired qualities. These varieties are popular with producers and are known as cultivars. A **cultivar** is a cultivated plant that has specific and distinguishable characteristics. The plants retain the characteristics when reproduced either sexually or asexually. In the United States, the term cultivar is considered synonymous with variety.

Selection

Selection is the process of breeding plants selected for a particular characteristic. This leads to the dominance of certain genetic traits. Most often, these are the traits that are most desired by plant producers or consumers of products.

Selection is often used to obtained desired colors in ornamental plants. Through mutation a variation in color may appear. A **mutation** is a genetic

4-21. Plant breeders prevent unwanted pollination by covering the forming ears and silks on corn plants.

variation that naturally occurs. Mutations cannot be predicted. Mutations are often used in breeding programs to obtain desired plants. The goal is to have a cultivar that has the desired qualities of the mutant plant.

Hybridization

Hybridization is the process of breeding individuals from distinctly different varieties. The goal is to gain a superior characteristic in the offspring. The offspring may have a trait known as hybrid vigor. **Hybrid vigor** is a condition where the offspring may have greater yield, height, disease resistance, or other traits than either of the parents.

Many crop and ornamental plants are hybrids. Corn was the first crop hybridized and planted throughout the United States. The seeds of hybrid plants are not usually planted because the desired trait may not be evident in the offspring.

GENETIC ENGINEERING

Plant breeding and reproduction play a critical role in supplying the plant materials we need for our survival. Scientists suggest the answer to this increased demand for food and fiber may lie in genetically engineered plants. In fact, plant breeding practices are being dramatically changed with new discoveries in biotechnology and genetic engineering. Recent discoveries in science have enabled scientists to select and move genetic material from one living organism to another.

Plants into which genes from another organism have been incorporated are said to be **transgenic plants**. This process, which involves the alteration of an organism by deliberately changing its DNA, is called **genetic engineering**. Genetic engineering holds great potential for improving agricultural production.

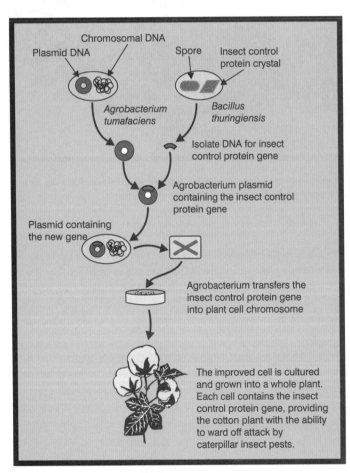

Chromosomal DNA
Plasmid DNA
Spore
Insect control protein crystal

Agrobacterium tumafaciens
Bacillus thuringiensis

Isolate DNA for insect control protein gene

Agrobacterium plasmid containing the insect control protein gene

Plasmid containing the new gene

Agrobacterium transfers the insect control protein gene into plant cell chromosome

The improved cell is cultured and grown into a whole plant. Each cell contains the insect control protein gene, providing the cotton plant with the ability to ward off attack by caterpillar insect pests.

4-22. Cotton has been genetically engineered to resist certain insect pests.

Actually, in a crude way, farmers and plant breeders have been practicing genetic engineering for thousands of years. By selectively breeding the most productive, disease-resistant plants, they have created better hybrids. Of course, it takes many years to produce improved hybrids through traditional breeding. The results are hard to predict, and the offspring may or may not show improvement over the parent plant. Another disadvantage of traditional breeding is the available genetic material is limited to that found in a certain species. Only a corn plant can be crossed with a corn plant!

With genetic engineering, the available genetic material is expanded to include all organisms. For example, genes from bacteria (*Bacillus thuringiensis*) that are deadly to caterpillars have been removed and spliced into the DNA of a cotton plant. The new genes tell the plant to make the poison that kills caterpillars. Then, when the caterpillar feeds on the plant, it eats the poison and dies. Money is saved because damage to the crop is lessened, and the need for applying pesticides is reduced.

ADVANTAGES OF GENETIC ENGINEERING

There are other advantages to genetic engineering, aside from the increased number of genes available. Desired results can be obtained much more quickly than with traditional breeding methods. There is also greater control over which characteristics will be expressed in the offspring.

Researchers have targeted a number of areas for improved performance in plants. They are developing plants that will be resistant to diseases, herbicides, insects, and viruses. There is an approved, genetically engineered soybean that is resistant to the herbicide glyphosate so the field can be sprayed with little damage to the soybean plants. Work is being done to develop genetically engineered plants that will be tolerant to cold, drought, and salty soils. The nutritional properties of foods may be improved through genetic engineering. Storage properties may be improved to reduce losses to spoilage before the food can be sold. Also, the appearance of fruits and vegetables may be improved, resulting in increased consumer appeal.

4-23. Reading DNA sequencing gels helps researchers make decisions in genetic engineering. (Courtesy, Agricultural Research Service, USDA)

APPROVAL OF NEW PLANTS

Agricultural biotechnology companies invest millions of dollars and many years to produce genetically engineered plants. They then seek the approval from the United States Department of Agriculture (USDA), the Food and Drug Administration (FDA), and the Environmental Protection Agency (EPA) before the engineered plants can be put into production. The time and money involved in the process are considered well worth the effort as the results can be extremely profitable.

An example of an approved genetically engineered plant is the squash. Viral infections of squash over the years have cost growers 60 percent of the crops. With the introduction of a squash genetically engineered for viral resistance, yields and profits have been greatly improved.

REVIEWING

MAIN IDEAS

Plants can reproduce in two ways. They may reproduce sexually or they may reproduce asexually. Sexual reproduction in flowering plants involves flowers, fruits, and seeds. Pollination of a flower occurs when pollen is transferred to a stigma. Fertilization of the egg held in the ovule takes place when it fuses with the sperm carried in the pollen. The result of sexual reproduction is a seed. The offspring is a hybrid of the parent plants. The significance of sexual reproduction is the combining of genetic material from two parents.

Seeds produced through sexual reproduction contain living embryos and food to help them begin growth. Germination or the start of growth happens when the seed receives the right conditions. Those conditions include water, oxygen, proper temperature, and possibly light.

Asexual reproduction involves the leaves, stems, or roots. There are many methods of asexual reproduction. Traditional methods of asexual reproduction include cuttings, grafting, layering, separation, and division. A newer, highly technical method of asexual reproduction is tissue culture. Tissue culture consists of placing explants on an agar gel under sterile conditions. The offspring resulting from asexual reproduction are genetically identical or clones of the parent plant.

Recent research in biotechnology has given scientists and plant breeders the ability to improve plants through genetic engineering. Genetic material in the cell of an organism can be removed and placed into another organism through genetic engineering. This technique has tremendous potential for improving agricultural crops and practices.

QUESTIONS

Answer the following questions using complete sentences and correct spelling.

1. What is the major difference between sexual and asexual reproduction?
2. How does pollination occur?
3. What is the function of a fruit?
4. What are the advantages of hybridization?
5. How is genetic information stored in a cell?
6. What are the main parts of a seed and their functions?
7. What are stratification and scarification?
8. What are the advantages of asexual reproduction?
9. What are the major methods of asexual reproduction?
10. What are the advantages of genetic engineering?

EVALUATING

Match the term with the correct definition. Write the letter by the term in the blank that is provided.

a. sexual reproduction   e. grafting   i. explant
b. germination   f. fertilization   j. pollination
c. cotyledon   g. stratification   k. clone
d. tissue culture   h. DNA   l. fruit

1. _____ molecular chain that stores genetic information in all living cells
2. _____ ripened ovary
3. _____ beginning of growth from a seed
4. _____ transfer of pollen from the male part of a flower to the stigma
5. _____ when sperm carried in the pollen fuses with the egg held in the ovule
6. _____ mechanism that requires a seed to go through a cold period before it will germinate
7. _____ offspring obtained through asexual reproduction that is genetically identical to the parent plant
8. _____ seed leaf that contains starch and protein for the embryo
9. _____ asexual method of reproduction in which the scion is fused with the rootstock
10. _____ small portion of a plant used in tissue culture
11. _____ form of reproduction involving the combining of genetic material from two parents
12. _____ highly technical method of asexual reproduction

EXPLORING

1. In late summer, walk through your neighborhood and collect seeds from plants. Examine how they differ in their structure. See if you can determine how they are disseminated.

2. Collect some honeylocust seeds. Plant some in a pot. Plant a second group in a similar way after using a file to cut just through the seed coat. See whether you have fulfilled the scarification requirement.

3. Place 25 bean seeds on a moist paper towel and put it in a plastic bag at room temperature. Open the bag after five days. Count the number of germinated seeds and calculate the germination rate. Open a seed and identify the major seed parts.

4. Obtain cuttings from various house plants. Place them in a glass of water so their lower stems extend into water. Set the glass by a window. Observe the cuttings for several weeks to determine if roots begin to grow.

CHAPTER 5

Plant Growth

OBJECTIVES

This chapter focuses on plant growth and development. It has the following objectives:

1 Explain the cellular structure of plants.

2 Explain how plants grow.

3 Describe the role of plant hormones.

4 Explain photosynthesis.

5 Describe the nutrient needs of plants.

6 Identify signs of nutrient deficiencies.

TERMS

abscisic acid
auxin
cell
cell wall
cellular respiration
cellulose
chloroplast
cytokinin
cytoplasm
enzyme

ethylene
gibberellin
golgi body
gravitropism
hormone
macronutrient
meiosis
micronutrient
mitochondria
mitosis

nucleus
organelle
photosynthesis
phototropism
rough endoplasmic reticulum
smooth endoplasmic reticulum
soluble salt
thigmotropism
tropism
vacuole

5-1. It is amazing how Douglas fir in the Pacific Northwest can grow so large and live for hundreds of years.

Most of us are familiar with the story about Jack and the beanstalk. In that story, a boy named Jack trades a cow for some magic bean seeds. In anger, his mother throws the seeds out the window. The following morning, Jack finds that a huge beanstalk has grown from the seeds. He climbs the beanstalk and encounters a goose that lays golden eggs and a giant.

We know that the "Jack and the Beanstalk" story is just a fable. Plants do not grow by magic, but how do they grow? How do leaves, stems, and roots form? How can they reach such great sizes?

People who work with plants are better able to do their jobs if they know the basics of plant growth. They must know how growth takes place and what plants need to grow. This knowledge will help them become successful in their plant-oriented careers.

PLANT LIFE

Plant life begins at the cellular level. Individual cells are very small. A microscope is needed to see a single cell. Flowering plants are a collection of thousands, even millions, of these tiny cells. The cells of a plant are organized into tissues that perform specialized functions. The tissues are organized further into organs, such as roots, stems, leaves, and reproductive parts.

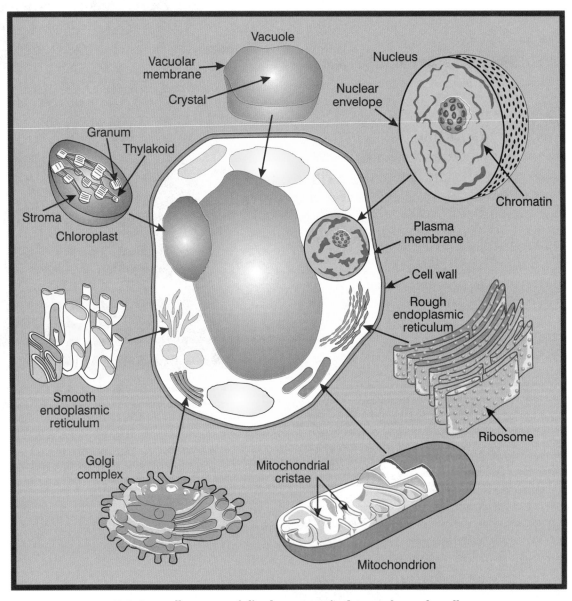

5-2. Organelles are specialized structures in the cytoplasm of a cell.

Life functions of plants from the cell level to an organ are very complex. Photosynthesis, growth, and development are controlled by genetic material found in the chromosomes. The life processes are regulated by hormones produced in the cells. The survival of a plant is based on the coordination of life processes in the cells, tissues, and organs and how the plant adapts to environmental conditions.

CELLS

The **cell** is the basic unit of life. It consists of living material bound by a membrane or wall. The living material inside a cell is referred to as the **cytoplasm**. Cytoplasm is the home to a number of specialized structures called **organelles**. Each organelle carries out an important duty for the cell. The major organelles are:

- **Nucleus**—The command center of the cell is the **nucleus**. Almost all of the DNA in a cell is located in the nucleus. The DNA is arranged in structures called chromosomes. Segments of chromosomes, or genes, contain chemically coded instructions for life processes that take place in the cell.

- **Vacuole**—Plant cells have a structure called the vacuole. The **vacuole** is a large sac bound by a membrane. It may occupy up to 90 percent of the cell. It contains water, stored foods, salts, pigments, and wastes.

- **Mitochondria and chloroplasts**—The life of the cell and the plant is dependent upon energy. Two structures found inside a cell that convert energy are the mitochondria and the chloroplasts. **Mitochondria** convert food into energy through cellular respiration. **Chloroplasts** contain green pigments called chlorophyll that trap light energy for photosynthesis. Chloroplasts are produced in cells exposed to light and are abundant in leaves. It is in the chloroplasts that food is made.

- **Endoplasmic reticulum and golgi bodies**—Other structures in the cell include the smooth and the rough endoplasmic reticulum. The **smooth endoplasmic reticulum** is the site for the production of lipids (fats) and hormones. The **rough endoplasmic reticulum** produces the proteins for the cell. Proteins are processed, sorted, or modified in the **golgi body**. These processes result in the complex molecules needed for plant growth.

- **Cell wall**—The **cell wall** itself is made of multiple layers of cellulose. **Cellulose** is a poly-saccharide or complex sugar molecule. The layers of cellulose offer great strength. Once a cell has stopped growing, the cell wall thickens and becomes rigid. The paper industry utilizes cellulose fibers from wood and fibrous crops in making paper.

CAREER PROFILE

PLANT PHYSIOLOGIST

A plant physiologist studies the cells, tissues, organs, and systems of plants. This includes the processes that occur to support plant life. They also study the uses of physiological processes in plants. The plant physiologists shown here are investigating the ability of sorghum plants to remove toxic materials from the soil—a physiological process. (Courtesy, Agricultural Research Service, USDA)

Plant physiologists need college degrees in botany, agronomy, or a closely related area. Most all have masters or doctors degrees in the area of plant physiology. Begin preparation in high school by taking science classes and agriculture classes that focus on plants.

Jobs for plant physiologists are with colleges, experiment stations, environmental research organizations, and private industries that deal with plant physiology.

CELL DIVISION

Flowering plants have two major types of cell division: meiosis and mitosis. Meiosis is associated with sexual reproduction and the production of sex cells or gametes. Mitosis involves the division of cells that make up the plant body, which consists of the roots, stem, leaves, and flower parts.

Meiosis

In **meiosis**, cells divide twice to produce gametes, which are the sperm and egg. The gametes contain a haploid number of chromosomes (n) or half that of a typical cell. A typical cell has two sets of chromosomes (2n). Through this division, each gamete is unique. No two gametes have the same genetic content.

When fertilization takes place during sexual reproduction, the haploid gametes fuse to form a diploid cell called the zygote. The zygote will develop into the embryo. It inherits a single set of chromosomes from each gamete providing it with a full set of chromosomes. Each offspring differs genetically from all others.

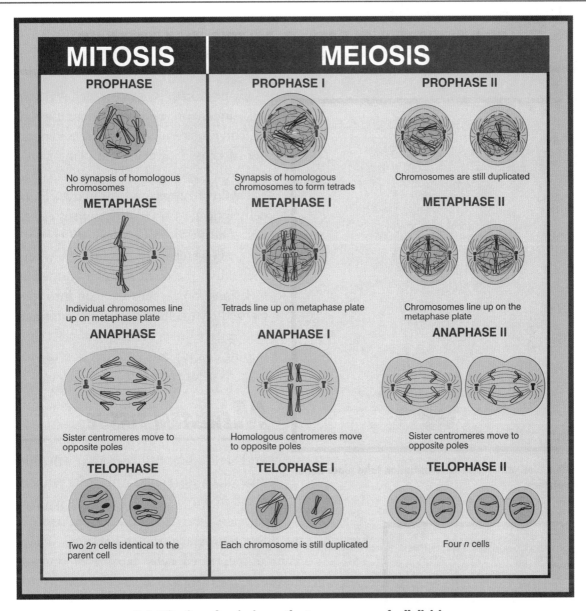

5-3. Mitosis and meiosis are the two processes of cell division.

Mitosis

After fertilization, the zygote grows by cell division. This complex process in which the cell multiplies is called *mitosis*. The first division of the diploid zygote results in two diploid cells. Those two cells divide to make four. Those four divide to make eight, and so on. Each new cell inherits a nucleus containing the same number and type of chromosomes as the original nucleus. The cytoplasm also divides providing each new cell with the essential organelles.

GROWTH AND DEVELOPMENT

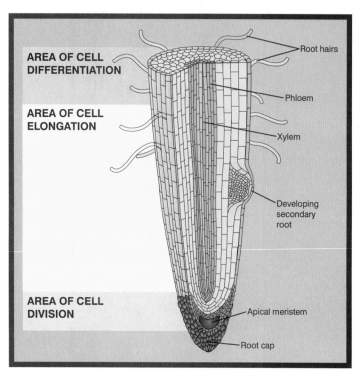

5-4. Cell division, elongation, and differentiation take place in the root apical meristem.

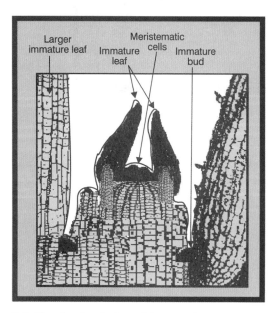

5-5. The shoot apical meristem produces increased length of the stem.

Plant growth is based on cell division, cell elongation, and cell differentiation. Cell division results in an increased number of cells. However, the plant would not grow in size unless the cells also had the ability to elongate and enlarge. Thus, cell elongation or enlargement is essential. The third process, cell differentiation, allows cells, depending on where they are in the plant, to take on specific functions. Some cells develop into specialized root cells that absorb water and nutrients. Others become leaf cells containing chloroplasts and function to produce food for the plant.

MERISTEM TISSUE

There are two types of growth that take place in plants. The first, primary plant growth, occurs in areas called apical meristems. Apical meristem tissue is found at the tips of roots and at the end of stems. These areas contain unspecialized cells capable of division through mitosis. Growth at the apical meristem produces increased length to the plant. Primary growth occurs in both herbaceous and woody plants. Secondary growth takes place only in woody plants. It involves cell division in layers ringing the stem that result in an increase in the width of the stem or trunk of the plant.

Primary Plant Growth

The meristem of a root is located right behind the root cap. The root cap is a layer of cells that protect the meristem as the root grows through the soil. Just behind the root cap is the area of cell division. Further back is an area of cell elongation. It is the pressure from these elongating cells that pushes the root through the soil. As the cells in the root mature, they differentiate to perform specific functions. They may become root hairs, xylem, phloem, etc.

Like the root meristem, the meristem of the shoot has areas of cell division, cell elongation, and cell maturation. One way in which the meristem of a shoot differs from that found in the root is it has immature leaves and immature buds. As the cells in these immature structures divide, elongate, and mature, they become the leaves.

Secondary Plant Growth

Secondary growth in woody plants takes place in the cambium. There are two layers of cambium. Secondary growth results in increased thickness of a plant.

The vascular cambium is a layer of meristematic tissue found between the wood and the bark. As the cells in the vascular cambium divide, they become either xylem or phloem. Xylem forms to the inside of the vascular cambium and phloem to the outside. Cell division occurs only when the plant is actively growing. Most plant growth takes place during the spring and summer. Each year, layers of new xylem and phloem cells are produced in the cambium rings of a tree.

The other meristematic tissue in the stem is cork cambium. It is located in the outer bark region, and the cells produced there form the outer bark. The bark protects the stem from insect and disease attack, excessive heat and cold, and other injuries. The bark in some plants, like the redwood trees of California, is very thick and fire resistant.

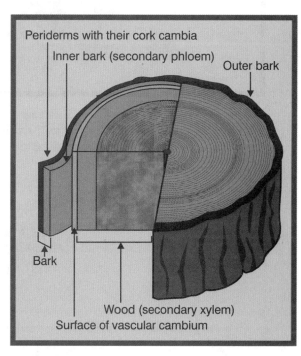

5-6. Secondary growth occurs in the vascular cambium of a tree.

TROPISMS

External stimuli, or environmental factors, cause plants to respond in certain ways. These responses are termed **tropisms**.

Plants will grow toward the source of light. In a forest situation, the tree best able to compete for light by growing into holes in the branches is more likely to survive to reproduce. This movement toward light is called **phototropism**.

5-7. Auxins move to the lower side of a horizontal stem causing the lower cells to grow faster than those on top. The response of stems curving up is gravitropism.

5-8. This potato plant exhibits the effects of gravitropism within a few days of being laid on its side.

Gravity of the earth affects plant growth. Stems respond by growing away from the gravitational forces, while roots grow toward the force, or down. This tropism is referred to as **gravitropism** (traditionally called geotropism).

Another tropism is called **thigmotropism**. Thigmotropism is the response of a plant to a mechanical stimulus, like a solid object. For instance, tendrils of cucumber vines will wrap around twigs they touch. The tendril is then able to help support the plant.

PLANT HORMONES

Plant growth is influenced by different forces. External or environmental factors have an important role in seed germination, flowering, dormancy, and other processes. Internal signals are equally important. Plants are complex organisms. All aspects of their growth are affected by chemicals called **hormones**. Hormones are produced in a part of a plant and transported to another where they regulate plant processes.

Five different hormones are produced in plants: auxins, gibberellins, cytokinins, ethylene, and abscisic acid. Each interacts with the others in complex ways to produce plant responses. Each hormone promotes many different responses, and each is effective in very low concentrations.

AUXINS

Auxins are produced in the apical meristem of a plant's stem and migrate down the stem. Auxins cause cells to elongate. **Auxins** are responsible for phototropic responses, or the ability of a plant to bend toward a light source. The way this works is the auxins move down the shaded side of a stem causing those cells to elongate more than the cells on the bright side of the stem. The effect is a stem bending toward the light.

5-9. Auxins caused this phototropic response in which the plant leans toward the light.

Auxins play a role in apical dominance. They move down the stem from the apical meristem and inhibit growth of side shoots. Pinching off the apical meristem stops the flow of auxins down the stem and side shoots are free to develop. This pinching practice is commonly used in the greenhouse industry to produce bushy, well-branched crops.

Auxins are also used in the horticulture industry to promote rooting of cuttings. The more quickly a cutting roots, the better chance of survival. Plant propagators apply certain types of auxins to the base of the cuttings before placing them in a rooting media.

GIBBERELLINS

Gibberellins are produced in stem and root apical meristems, in seed embryos, and in young

5-10. Treating vines with gibberellins results in larger, more well-spaced fruit.

5-11. Cytokinins are added to tissue culture media to promote cell division. (Courtesy, Agricultural Research Service, USDA)

leaves. They induce stem cell elongation and cell division. Gibberellins have been linked to hybrid vigor in corn plants. Larger growing hybrid plants contain higher concentrations of gibberellins than inbred plants. In addition, gibberellins play a key role in stimulating the development of flowers.

CYTOKININS

Cytokinins are responsible for cell division and differentiation. In tissue culture, cytokinins must be added to the media for cell division to occur. Cytokinins also have the effect of delaying the aging process. Cut flowers that lose their source of cytokinins, produced in the roots, are sometimes sprayed with cytokinins to extend their vase life.

ETHYLENE

Ethylene is a colorless gas produced in the nodes of stems, ripening fruits, and dying leaves. It plays a role in the aging of plant parts, particularly fruit.

Ripening fruit produces ethylene, which triggers the production of more ethylene, speeding the ripen-

5-12. Ripe apples release ethylene that causes other apples to ripen.

ing process. It's true that a rotten apple in a bushel of fresh apples will cause the fresh apples to ripen and rot. Another example involves bananas. Bananas are picked green for shipment to reduce bruising. Prior to arriving at the store, they are treated with ethylene to promote the ripening response. Cut flowers are never stored with ripe fruit or decaying leaves that might give off ethylene and shorten the life of the flowers.

ABSCISIC ACID

Abscisic acid is most apparent when a plant is under stress. It is abscisic acid that causes stomata to close, thus reducing water loss during periods of drought. **Abscisic acid** is largely responsible for seed dormancy and the loss of leaves by deciduous plants in the fall. Before the leaves have turned color and dropped, abscisic acid prompts the formation of bud scales to protect the buds during the winter.

5-13. Abscisic acid plays a role in the falling of leaves in deciduous plants.

PHOTOSYNTHESIS

Green plants have the ability to convert solar energy into a stored chemical energy. This process is known as **photosynthesis**. It is the most important chemical reaction on Earth. The chloroplasts serve as the "factory" where photosynthesis takes place. Green pigments, called chlorophyll, are found in the chloroplasts. It is the chlorophyll that gives plants their green color.

REQUIREMENTS FOR PHOTOSYNTHESIS

For photosynthesis to take place, a plant must receive light energy. Chlorophyll traps the light energy and converts it to stored chemical energy. Water absorbed by the roots and carbon dioxide, which enters through stomata, are used by the plant to make food. The food, or stored chemical energy, is produced in the form of sugar (glucose). Oxygen and water are other products of photosynthesis. A simple chemical equation for photosynthesis follows:

$$6CO_2 + 12H_2O \xrightarrow[Chlorophyll]{Light} C_6H_{12}O_6 + 6O_2 + 6H_2O$$

(Carbon dioxide) (Water) (Sugar) (Oxygen) (Water)

The sugars that are produced fuel plant growth. They provide the energy for all the plant's life processes. Sugars move down the phloem to the rest of the plant. They may be converted to carbohydrates (complex sugars) and stored in the roots or stem for future use. Potatoes, for example, can store carbohydrates in the form of starch for the plant.

CELLULAR RESPIRATION

How do plants use the sugars and carbohydrates they have made? Sugars provide the energy used to run life processes in the cells. The energy is released in the mitochondria. This process is known as **cellular respiration**. Cellular respiration is the opposite of photosynthesis. It requires oxygen to break

5-14. Potatoes store carbohydrates for the plant.

CONNECTION

AUTUMN GOLD

A golden-leafed hickory stands out on a clear day in the Blue Ridge Mountains of Georgia. From spring until autumn, the leaves are green. The brilliant gold lasts only a week or so in late October. Why is it so colorful?

Green leaves contain green, yellow, orange, and red pigments. The green of the chlorophyl usually dominates leaf color. Shorter days and cooler nights of autumn cause the chlorophyl to break down and the other colors begin to manifest themselves.

The yellow of the hickory is caused by pigment known as xanthophyll. Other trees may be orange (caused by carotene pigments) and red or purple (caused by anthocyanins). The changes occur as the leaves age and prepare to die. Abscisic acid produces an abscission layer of cells that separate the leaf from the stem so it falls to the ground. The tree is dormant until warm weather of the following spring.

Table 5-1. A comparison of photosynthesis and respiration.

Photosynthesis	Respiration
CO_2 and H_2O are used.	O_2 and food are used.
Food and O_2 are produced.	CO_2 and H_2O are produced.
Energy from light is trapped by chlorophyll and food.	Every living cell carries out respiration.
Only cells containing chlorophyll carry out photosynthesis.	Respiration occurs in both light and dark.
Photosynthesis occurs only in light.	

down food and produces carbon dioxide as a by-product. All plant and animal cells respire, or use energy to live. The embryos in seeds survive by respiration. They break down food stored in the seed. A simple equation for cellular respiration follows:

$$C_6H_{12}O_6 + 6O_2 + 6H_2O \longrightarrow 12H_2O + 6CO_2 + Energy$$

(Sugar)   (Oxygen)   (Water)   (Water)   (Carbon dioxide)

Utilizing the carbohydrates is extremely important in the growth and development of cells in plants. Plant respiration and actual plant growth take place primarily at night when photosynthesis is shut down. With signals from hormones, **enzymes** (chemical activators) are produced. All enzymes have specific jobs. With split-second timing they break down sugars and recombine them with nitrogen and other minerals. The complex molecules produced are starches, lipids (fats), cellulose, and proteins.

NUTRIENT NEEDS

Plants have needs that go beyond carbon dioxide, water, and light. The life of a plant is also dependant upon chemical elements. Most elements needed by a plant are obtained from the soil. Water in the soil dissolves the elements. Then, as the plant roots absorb the water, they take up these elements.

There are seventeen essential elements for plant growth. The elements used by plants for growth and development are often referred to as nutrients. Some nutrients are required in greater quantity than others. They are called **macronutrients**. Other nutrients that are needed, but in low

5-15. Nutrient deficiencies influence trees, such as this red maple showing signs of an iron and/or a manganese

Table 5-2. Seventeen Essential Plant Nutrients

Nutrient	Influence/Function
Non fertilizers	
Carbon	Building block for carbohydrates, proteins, fats, nucleic acids
Hydrogen	Building block for carbohydrates, proteins, fats, nucleic acids
Oxygen	Building block for carbohydrates, proteins, fats, nucleic acids
Primary macronutrients	
Nitrogen	Produces stem and leaf growth; gives plants dark green color
Phosphorus	Stimulates root development and growth; aids in cell division; encourages flower bud formation; improves winter hardiness; helps plants to a vigorous and rapid start
Potassium	Increases plant vigor and disease resistance; aids in the transport of foods through the phloem; has key role in opening and closing stomata; thickens cell walls
Secondary macronutrients	
Calcium	Maintains strength of cell walls; promotes early root growth
Magnesium	Essential for chlorophyll and photosynthesis; activator for many plant enzymes
Sulfur	Stimulates root growth; needed for protein formation
Micronutrients	
Boron	Essential for pollination and reproduction
Copper	Regulates several chemical processes
Chlorine	Involved in photosynthesis
Iron	Important in chlorophyll formation
Manganese	Important in chlorophyll formation; part of enzymes involved in respiration and nitrogen metabolism
Molybdenum	Part of enzymes involved in nitrogen metabolism
Nickel	Converts urea to ammonia in plant tissue
Zinc	Important in chlorophyll formation; part of enzymes involved in respiration

amounts, are called ***micronutrients***. People involved in production agriculture add nutrients in the form of fertilizers to growing media and soil to ensure that the crops will have the nutrients they need. Close attention is given to providing the optimal amount of nutrients. Plants obtain non-fertilizer elements—carbon, hydrogen, and oxygen—from air and water.

MACRONUTRIENTS

There are six macronutrients. They are nitrogen (N), phosphorus (P), potassium (K), calcium (Ca), magnesium (Mg), and sulfur (S). Nitrogen, phosphorus, and potassium are considered to be primary nutrients because they are used by plants in large amounts. Calcium, magnesium, and sulfur are said to be secondary nutrients because moderate amounts are needed.

Table 5-3. Selected Nutrient Deficiency Symptoms

Nutrient	Deficiency Symptom
Nitrogen	Entire plant lighter green, lower leaves yellowing; slow or dwarfed growth
Phosphorus	Purplish coloration of leaves and stems; stunted growth
Potassium	Yellowing or death of tissues at tips and outer edges of older leaves
Calcium	Short, much branched roots; young leaves at growing points die back
Magnesium	Loss of green leaf color starting with bottom leaves
Sulfur	Young leaves light green with veins being lighter
Boron	Young leaves yellow and thick
Copper	Yellowing of leaves, younger leaves affected first
Chlorine	Symptoms have not been recognized
Iron	Young leaves yellow first; veins remain green
Manganese	Young leaves yellow first; veins remain green
Molybdenum	Older leaves yellow; stunted growth
Zinc	Older leaves yellow; stunted growth

5-16. Examples of nutrient deficiencies are potassium deficiency in alfalfa (top left), nitrogen deficiency in corn (top right), iron deficiency in a tomato plant (bottom left), and magnesium deficiency in a grape plant (bottom right). (Courtesy, Potash and Phosphate Institute, Norcross, Georgia)

MICRONUTRIENTS

Another eight nutrients are required in small quantities. They are called micronutrients, or trace elements. They are boron (B), copper (Cu), chlorine (Cl), iron (Fe), manganese (Mn), molybdenum (Mo), nickel (Ni), and zinc (Zn).

A little phrase can be used to help memorize the 17 essential elements for plant growth. It is "C. B. Hopkins Café Mighty Good Closed Monday Morning See You Zen." It represents the following: Carbon (**C**), Boron (**B**), Hydrogen (**H**opkins), Oxygen (**HO**pkins), Phosphorus (**HoP**kins), Potassium (**HopK**ins), Nitrogen (**HopkiN**s), Sulfur (**HopkinS**), Calcium (**Café**), Iron (**caFé**) Magnesium (**M**ighty good), Chlorine (**Cl**osed), Manganese (**Mo**nday), Molybdenum (**Mo**rning), Nickel (**morN**ing), Copper (See you = **Cu**), Zinc (**Zen**).

NUTRIENT DEFICIENCY

If a plant fails to receive the needed amount of nutrients, it will show signs of deficiency. Nutrient deficiencies most often result in an unhealthy plant appearance. Symptoms vary with the nutrient that is in short supply. Common symptoms of deficiencies include discoloration of the leaves, death of leaf tissue, and stunted growth. Because of the complex interactions of nutrients in plant processes, deficiency symptoms for different nutrients are often very similar. Laboratory tests can be used to determine which nutrient is lacking.

NUTRIENT EXCESS

Sometimes, in an effort to provide plants with nutrients through fertilization, too much fertilizer is applied. Nutrient fertilizers that dissolve in water are referred to as ***soluble salts***. High levels of soluble salts are harmful to plants. The primary damage caused by soluble salts is the burning of the roots. The injury or death of the roots then results in wilting and the death of leaf tissues. The buildup of soluble salts in soil is a major problem in agricultural practices in some parts of the western United States where crops must be irrigated.

5-17. Soluble salts built up in the soil on this land making it useless for crops. (Courtesy, Agricultural Research Service, USDA)

HOW A FLOWERING PLANT GROWS

LIGHT

Light energy from the sun is converted to chemical energy in green pigments called chlorophyll. This process involving the manufacture of sugars is known as **photosynthesis**.

AIR

The atmosphere or air provides **carbon dioxide** that is needed for photosynthesis. It also provides oxygen for the release of stored energy. This second process is called **respiration**.

APICAL MERISTEM

The growth of a plant in height or length occurs in the **apical meristem**. In this area cells divide, elongate, and differentiate to take on specific tasks.

ENZYMES

With signals from hormones, **enzymes** or chemical activators are made to perform specific jobs. They break down sugars and combine them with nutrients to make starches, fats, cellulose, proteins, etc.

FLOWERS

The reproductive organs of a plant are called **flowers**. A complete flower has **sepals**, **petals**, **stamens**, and a **pistil**. Pollination of the flower, followed by fertilization of the egg, results in seed and a fruit.

TEMPERATURE

Plants have evolved to grow in different conditions. The speed of photosynthesis, respiration, and enzyme activities are affected by temperature.

HORMONES

Chemicals, called hormones, produced within the cells of a plant regulate life within the plant. Five different hormones are produced in plants: auxins, gibberellins, cytokinins, ethylene, and abscisic acid.

WATER

The life of a plant depends upon **water**. Water is required for both photosynthesis and respiration to occur.

LEAVES

Leaves are plant organs that function to manufacture food. They combine water and carbon dioxide in the presence of light to make sugars.

SOIL

Soil serves to hold water and nutrients for the plant. It also functions as a base to help plants grow.

ROOTS

Roots function to absorb water and nutrients from the soil. They also anchor the plant and store food.

NUTRIENTS

Sixteen **essential elements** are used as building blocks for plant growth and development.

STEMS

Stems serve as the transport system for plants. **Xylem** tissues transport water and minerals from the roots to the leaves. **Phloem** tissues transport sugars from the leaves to the rest of the plant.

CAMBIUM

Growth in the thickness of a stem occurs in a meristem tissue called the **cambium**. Cells in the cambium layer ringing a stem divide and become xylem, phloem, or bark.

5-18. Needs and processes in the growth of a flowering plant.

REVIEWING

MAIN IDEAS

Plant growth and development begins at the cellular level. Cells contain organelles that perform specific life functions. The cells are organized into tissues and organs including roots, stems, leaves, and flowers. There are two types of cell division: meiosis, which produces sex cells, and mitosis, in which each daughter cell is an exact duplicate of the parent cell.

Plant growth and development is a result of cell division, cell elongation, and cell differentiation. Primary growth occurs in the apical meristem tissues located at the tips of roots and stems. Secondary growth takes place only in woody plants. It occurs in the vascular and cork cambiums and results in an increased width of a stem. External factors trigger plant responses known as tropisms.

Life processes in plants are regulated by hormones produced in the cells. The five hormones regulating plant growth are auxins, gibberellins, cytokinins, ethylene, and abscisic acid. The most important chemical reaction of plants is photosynthesis. Photosynthesis is the process where light energy is converted to stored chemical energy. Water and carbon dioxide are used, while sugars and oxygen are produced. Another reaction, respiration, uses stored chemical energy to fuel life processes in the cells.

Essential elements are needed for plant growth. Hydrogen, carbon, and oxygen serve as basic building blocks for the carbohydrates, lipids, proteins, and nucleic acids manufactured in cells. Macronutrients are required by the plant in large quantities. They include nitrogen, phosphorus, potassium, calcium, magnesium, and sulfur. Micronutrients are needed in lesser quantities. They are boron, copper, chlorine, iron, manganese, molybdenum, nickel, and zinc.

QUESTIONS

Answer the following questions using complete sentences and correct spelling.

1. What are the major cell organelles? Explain the major function of each.
2. How do meiosis and mitosis differ?
3. In what tissues does cell division resulting in plant growth take place?
4. What are phototropism and gravitropism?
5. What are the five hormones that regulate plant growth? What is the major function of each?
6. How does photosynthesis take place?
7. Why is respiration important in plant growth?
8. What is the difference between macronutrients and micronutrients?
9. What are the essential nutrients for plant growth?

EVALUATING

Match the term with the correct definition. Write the letter by the term in the blank that is provided.

a. hormone   e. cellular respiration   i. photosynthesis

b. apical meristem   f. cell   j. macronutrient

c. chloroplast   g. mitosis

d. auxin   h. nucleus

1. _____ command center of a cell

2. _____ responsible for phototropic responses in plants

3. _____ essential nutrient needed in large quantities

4. _____ most basic unit of life

5. _____ chemicals affecting all aspects of plant growth

6. _____ growth occurs in this area at the tips of roots and stems

7. _____ organelle that contains chlorophyll

8. _____ process in which light energy is converted to stored chemical energy

9. _____ cell division in which the daughter cells are exactly like the parent cell

10. _____ cells convert carbohydrates to a form of energy to run life processes

EXPLORING

1. Plant seeds in a pot and place it by a window. Observe how the plants will lean toward the light source. Try to picture in your mind how auxins stimulate cell elongation on the shaded side of the stem.

2. Place a bean seed on a moistened paper towel. Wrap the paper towel and the seed in clear plastic wrap. Tape the package onto a window. The seed will germinate in several days. Rotate the package on the window 90 degrees every day for a week. How has the plant responded?

CHAPTER 6

Cultural Practices in Producing Plants

This chapter explores the environmental and cultural factors that play a large role in the health of a plant. It has the following objectives:

1 Explain the conditions needed for plant growth.

2 Explain land preparation and tillage.

3 Describe the use of fertilizer and other soil amendments.

4 Describe water management and the use of irrigation.

5 Describe the role of pest management.

TERMS

bark
biological control
chemical control
compost
conservation tillage
cultural control
cultural practices
day-neutral plant
drought
electromagnetic spectrum
etiolation

fertilizer
fungicide
herbicide
inorganic fertilizer
insecticide
integrated pest management
   (IPM)
leaf mold
long-day plant
miticide
natural organic fertilizer

peat moss
perlite
pest
photoperiodism
sand
short-day plant
soil amendment
synthetic organic fertilizer
turgidity
vermiculite
wilting

6-1. Farming in Pennsylvania. (Courtesy, Agricultural Research Service, USDA)

DROUGHT HITS THE MIDWEST!
Crops dry up and die in the fields for lack of water.

FREEZING TEMPERATURES DESCEND ON FLORIDA!
The citrus crop is severely damaged.

THE MED FLY INVADES CALIFORNIA!
Agricultural yields greatly reduced.

These and other newspaper headlines about crop failures have appeared in recent years. Crops have been ruined by natural conditions that are next to impossible to control. Even with careful selection of plant varieties, hybridization, and genetic engineering, plant productivity is not guaranteed. Success with plants or crops often rests on uncontrollable environmental conditions.

People are smart. They have learned how to take advantage of natural resources to help attain a needed level of production and offset natural problems. This involves using important cultural practices.

ENVIRONMENTAL CONDITIONS NEEDED FOR PLANT GROWTH

Evolution has allowed plants to adapt to certain environmental conditions. That is why cacti grow in the desert and Douglas fir in the Pacific Northwest and the Rocky Mountains.

6-2. Chrysanthemums have been carefully spaced to grow in a controlled environment.

In agriculture, humans have taken plants from natural settings for production purposes. The plants are often forced to grow in situations different from the situations in which they evolved in nature. Where in nature do you see acres and acres of corn, neatly manicured grasses, or evenly planted rows of pistachio trees? Of course, the benefits of agriculture include higher-quality foods, more food, useful fibers, and ornamental crops.

Environmental conditions have always been a concern to the agricultural producer. Sometimes, extreme conditions occur and the plants are damaged or killed. Since people are dependent on plants for food and clothing, such damage can result in suffering or death. It can also result in the loss of money. Environmental conditions, including light, temperature, air, and water, greatly influence plant growth. Where possible, agricultural producers control these conditions to improve plant growth. The control or management of light, temperature, air, nutrients, pests, and water is often referred to as *cultural practices*.

LIGHT

Plant life depends on light energy for food production through the process of photosynthesis. Light itself is the transfer of energy in the form of radiation. Plants convert this radiant energy to a form of chemical energy. There are three key aspects of light affecting plant growth. They are color, duration, and intensity.

Color

Visible light is a small segment of all the radiant energy given off by the Sun. X-rays, gamma rays, ultraviolet rays, microwaves, and radio waves are some other forms of radiant energy. The different rays of radiant energy have been measured on the basis of their wavelengths and placed on an ***electromagnetic spectrum***.

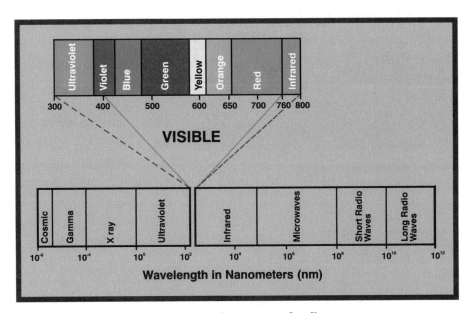

6-3. The electromagnetic spectrum of radiant energy.

Sunlight contains a blend of visible colors. We can see the colors when light passes through a prism. The prism separates the colors. This phenomenon of separating the colors of visible light occurs naturally with rainbows. Each color has a slightly different wavelength.

Light is absorbed or reflected by objects. Objects that appear black absorb all the colors of visible light. Therefore, no colors are reflected into our eyes. White objects, on the other hand, reflect all the colors of visible light. If an object absorbs all the colors but green, it will appear green to us because green wavelengths have been reflected from the object into our eyes. This is the case with green leaf plants. Because green wavelengths are reflected from the leaves, they have little effect on plant growth. The colors that have the greatest influence on plant growth are blue and red. In general, blue wavelengths affect photosynthesis, and red wavelengths influence flowering and growth.

Light Duration

Plants are responsive to the length of exposure to light. A mechanism within plants detects day lengths. The length of the days is known to play a role in different phases of plant

6-4. Chrysanthemums are short-day plants.

6-5. These soybean seedlings, which were grown in low light, exhibit etiolation or stretching toward the source of light.

growth. Some of these phases are seed germination, enlargement of leaves, development of buds, and flowering. A plant's response to light duration is called **photoperiodism**.

Different plants respond differently to light duration. Chrysanthemums and poinsettias are two plants that will begin to flower in the fall when the day lengths naturally get shorter. They are said to be **short-day plants**. Radish and lettuce plants flower as days lengthen in the summer. They are termed **long-day plants**. A third group of plants are unaffected by day length and are classified as **day-neutral plants**.

Horticulturists will control the length of days in a greenhouse to bring on a desired plant response. Greenhouse crops may be covered with black cloths during times of the year when days are long to give the illusion of short days. Other crops might be lighted at night during times of the year when days are short to simulate long days. By changing the periods of daylight, growers can force plants to flower at unnatural times of the year. This is why you can find flowering chrysanthemums in the stores at any time of the year.

Light Intensity

The third effect light has on plants involves intensity or brightness. Intensity of light depends largely on the angle of the Sun, clouds, and dust in the atmosphere. Light intensity is greater in the summer months when the Sun is higher in the sky. Where artificial light is provided, the source of light and the distance from the light source to the plants are important.

Different plant species have evolved to survive in conditions where the intensity of light varies. The major agricultural crops produce the highest yields when they receive full sunlight. It stands to reason that crops receiving bright sunlight will have an elevated rate of photosynthesis and will produce more food. Plants like impatiens, African violets, ferns, and many house plants are adapted to shady conditions and suffer if placed in full sun. In fact, most house plants have been selected because of their ability to tolerate low levels of light.

Plants will show symptoms of having inadequate light or too much light. Normally, a plant receiving the proper level of light will be compact and have good leaf color. One that is experiencing low light levels will stretch for the light source, a condition known as **etiolation**. The leaves are also smaller than normal and lighter in color. Plants adapted to low light situations can be damaged by high light intensities. Those receiving too high of light intensity may take on a bleached look as chloroplasts are killed. Excessive light intensity may even cause death of leaf tissue.

TEMPERATURE

Temperature is an important factor governing plant growth. Temperature is heat energy. It is measured in degrees. Life on Earth depends on heat energy from the Sun.

Plants grow best in the temperature range to which they have adapted. In fact, cold hardiness or heat tolerance are major factors in determining where native plants grow.

Two examples to illustrate how temperature influences plant growth are:

- Corn Hybrids—The development of corn hybrids has increased the range in which corn is grown. Today, corn is grown in all 50 of the United States including Minnesota, North Dakota, and Montana. Before hybridization corn was found only in warmer climates that have longer summers.

- Bluegrass—Kentucky bluegrass is native to the northern United States. It is commonly grown as a lawn grass. It grows best when temperatures are cool in the spring and fall. If temperatures soar, as they often do in the summer, the plant goes dormant. Kentucky bluegrass does not grow well in the southern United States, where temperatures are hotter for a longer period of the year.

6-6. Leaves on this apple tree were damaged by an early frost.

6-7. Grapes in Napa Valley, California, can be protected from frost damage by using large fans to create air movement.

Important biochemical reactions, including photosynthesis and respiration, are affected by temperature. Life processes occur more quickly as temperature rises. They slow down as temperatures become cooler. This is because the enzymes that drive those reactions are sensitive to temperature. If the temperature becomes too warm or too cold, the enzymes cannot carry out their functions.

Freeze Damage

Freezing can cause physical damage to plant cells. This varies greatly with the plant species and the climate to which it has adapted. Freeze damage occurs when the cell walls of plants break as the contents of the cells freeze. Hardier plants are better able to withstand lower temperatures than less hardy plants.

The orchard industry is very concerned about heavy freezes when fruit trees are in flower. Flower cells are easily damaged by cold and entire crops can be lost in a spring freeze. To prevent serious damage to their crops, growers may choose to spray their trees with water. The sprayed water offers some protection to the flowers, leaves, and stems. A continuous application of water prevents the temperature of the plant tissues from dropping below 32 degrees.

Landscape plants that are marginally hardy can be damaged by cold. The stems, buds, flowers, and leaves can be damaged or killed. However, the greatest concern for the survival of the plant is with the roots. If soil temperatures drop too low, roots can be killed and the plant will die. One reason landscapers and gardeners place a layer of mulch (wood chips, stone, peanut hulls, etc.) on the soil is to insulate the roots from cold.

6-8. Using mulch around landscape plants helps to maintain uniform soil temperatures.

Most of our agricultural crops are herbaceous annual plants. Herbaceous plants are more easily damaged by cold temperatures. The concern of freeze damage is usually the greatest during the planting season or before harvest.

Heat Damage

Heat can also damage plants. This is particularly true when it is both hot and dry. Plants have some ability to cool themselves when it is hot. Have you ever noticed how cool it is under a large shade tree or in the woods on a hot summer day? Plants transpire water through their stomata. As this water evaporates, it has a cooling effect. The stomata close if the plant roots cannot supply water quickly enough to replace water lost from transpiration. This helps the plant to conserve water. Further water

6-9. Wilting is caused by inadequate moisture.

loss causes wilting. **Wilting** is a drooping condition and a lack of firmness to the plant tissues caused by inadequate water. Wilting is a result of a loss of **turgidity**, or water pressure, in the plant cells. Severe heat conditions can cause plant tissues to dry up and die.

Air

Air has carbon dioxide and oxygen that are needed for photosynthesis and respiration. Exchange of oxygen and carbon dioxide through stomata in the leaves keeps photosynthesis operating at peak efficiency. The ability of air to move in and out of the soil is important in providing oxygen for healthy root growth. Root cells must have oxygen to undergo the vital life process of respiration.

The gaseous form of water is water vapor. Humidity, a term used to express water vapor in the air, affects plant growth. The growth rate of plants increases under conditions of high humidity. Lush, tropical forests exist, in part, because of frequent rain and high humidity. High humidity reduces water stress of a plant so photosynthesis can function smoothly. If the humidity is low, the dryness of the air can put stress on the plant. This is especially true if soil moisture is inadequate and wilting occurs.

Air pollution can be damaging to plants. Dust in the air can reduce light intensity, slowing photosynthesis. Chemical pollutants, such as sulfur dioxide, can actually kill plant cells

6-10. Drought may cause death of leaf tissues, stunted growth, and reduced yields. (Courtesy, Illinois Farm Bureau)

or the entire plant. Although plants in urban settings often suffer from air pollution, they provide a benefit to people by cleaning the air of pollutants.

WATER

Life processes of a plant depend on water. Photosynthesis and respiration require water. Roots can only absorb minerals if the minerals are dissolved in water. Water is the carrier of materials through the xylem and phloem. Water, in the form of rain, aids pollination of corn during tasseling. Water also makes up a large percentage of the plant cells, tissues, and organs.

The amount of water required by plants varies with the type of plant. Some plants, like wheat, produce well with low amounts of rainfall. Other plants, including many fruits and vegetables, will not produce high yields unless they receive high levels of rainfall.

Drought occurs when there are periods of insufficient rainfall. Normal plant growth is slowed during peri-

CAREER PROFILE

GREENHOUSE MANAGER

A greenhouse manager oversees the production of plants in a greenhouse. Managers are often responsible for several greenhouses. They plan crops to be grown, obtain materials, organize facilities, supervise workers, oversee greenhouse conditions, and prepare plants for marketing.

Greenhouse managers often have college degrees in horticulture. Some may have postsecondary degrees from specialized horticulture programs in community colleges or technical schools. Practical experience in greenhouse work is essential.

Jobs are found where greenhouse facilities are operated. Some are entrepreneurs and set up their own greenhouses. This photo shows a greenhouse manager checking a timed heater that burns elemental sulfur at night (when no people are present) to control pests in a greenhouse.

ods of drought. Stomata close to conserve water. When closed, the stomata do not permit an exchange of carbon dioxide and oxygen. Without carbon dioxide, photosynthesis, or the manufacture of sugars, cannot occur.

Excessive rainfall can be harmful to plants. Roots of plants may die if they do not receive oxygen for respiration. This is common if soils stay wet for extended periods. Wet soils, coupled with cool soil temperatures can also prove damaging to newly planted seeds. Seeds will often rot in the ground before germinating under wet, cold conditions.

6-11. Standing water and saturated soil have killed corn in the foreground of this field.

Water Management

Sometimes, normal precipitation is not enough to grow healthy crops or to produce high yields. In these cases, it is beneficial to provide additional water to the plants. Furnishing plants with water by flooding or sprinkling is known as irrigation. Water for irrigation purposes can be obtained from lakes, rivers, reservoirs, and underground sources.

As is the case with many of our natural resources, water must be managed wisely. Population growth in cities and towns has increased the demand for clean drinking water. There is also a need to increase agricultural production to feed, clothe, and shelter the people of the world. This amounts to greater demands for clean water to grow healthy plants. Sometimes, the demands for water create conflict. Legal battles are being fought in the courtrooms of the United States over the rights to water. The important thing is that water is both conserved and kept clean.

6-12. These potatoes are irrigated to boost yields.

6-13. In drier regions, irrigation is required to raise crops. The large circles represent areas of land that are irrigated using pivot systems. (Courtesy, Agricultural Research Service, USDA)

With so much demand for clean water, it is important that water be conserved. Overuse of water can deplete sources. Care must be taken not to exceed the amount needed for optimal plant growth. In many cases, water is recycled to extend its use. Pollution in water can prevent its use for people or crops. Water pollution also damages the environment, putting stress on native plants and animals. Therefore, efforts should be maintained to keep surface and ground water free of pollutants.

CONNECTION

ULTIMATE MANAGEMENT

What would happen if crops were given ideal growing conditions? One would expect the crops to grow vigorously and produce high yields. Unfortunately, nature seldom provides crops with optimal conditions on a day-to-day basis.

However, advances in engineering and research have resulted in greenhouse facilities that give people control over temperature, light, air, and water. Greenhouses are now widely used to provide optimum growing conditions for specialty crops and for research purposes.

Many schools have greenhouses where students can learn how to care for plants. These Texas students are applying what they have studied in the classroom in their school's greenhouse.

SOIL PREPARATION

Soil is important to plant growth. It provides the plant with minerals. It serves to hold water for the roots. Soil also provides solid footing so plants can grow upright. Care must be taken in preparing soils for crop production.

Soil is prepared for plants or seeds before planting. This typically involves loosening the soil. Plows, disks, rototillers, and spades are all used to break up the soil. Seeds are more easily planted and air enters the soil. Also, the plant roots can grow and establish themselves much more quickly in loosened soil. When working the soil, large clumps need to be broken up.

Soil preparation needs to be done at the right time. If the soil is either too moist or too dry, damage can be done to the soil structure. Plowing wet soils produces large, hard clods and causes compaction of subsoils. The weight of the machinery squeezes air pockets from between the soil particles. When the soil is dry, it is often difficult to pull the machinery through the field. There is a simple way to determine when the soil has the right moisture level for working. Pick up a handful of soil. Squeeze it together. It should form a ball. Then, it should crumble apart easily using your fingers. If it cannot be formed into a ball, it is too dry. If it will not crumble, it is too wet.

There are some drawbacks to working the soil. The soil is exposed to water and wind that can carry soil particles away from the field. This

6-14. A disk is used to prepare the soil on this sod farm for planting grass seed.

6-15. Erosion can remove soil needed for productive plant growth. (Courtesy, United States Conservation Service)

6-16. With conservation tillage, planting may be done without tilling the soil. (Courtesy, United States Soil Conservation Service)

loss of soil is called erosion. Eventually, erosion may reduce the depth of the soil and cause a decline in productivity. Also, populations of beneficial animals, such as the earthworm, are reduced when soils are worked.

New practices are being used to reduce soil loss from erosion. **Conservation tillage** includes several farming practices that eliminate traditional methods of turning the soil. Seed planting, fertilization, and weed control are conducted while keeping soil disturbance to a minimum. The aim of conservation tillage is to reduce soil loss. It also conserves fuels by eliminating some machinery use.

FERTILIZERS AND SOIL AMENDMENTS

Materials are added to soils to improve productivity. The most common additive is fertilizer. **Fertilizer** is any material added to soil to provide nutrients that will increase growth, yield, or nutritional value of the plants. Sometimes materials are needed to improve other characteristics of the soil. The materials added to soils to improve water drainage, moisture holding ability, and aeration (the exchange of gases) are called **soil amendments**.

FERTILIZER

Fertilizers may be inorganic, natural organic, or synthetic organic materials. **Inorganic fertilizers** originate from nonliving sources, such as mineral deposits or rocks. They may also be chemically made. **Natural organic fertilizers** are of plant or animal origin. For instance, animal manure, cottonseed meal, and dried blood are natural organic fertilizers. **Synthetic organic fertilizers** are artificial, carbon-based materials.

In commercial production, soils or plant tissues are tested to determine if nutrients are lacking. If nutrients are found lacking, fertilizer can be applied to raise the nutrient level of the soil. It is to the advantage of the grower to provide the crops with an optimum level of nutrients. Too much fertilizer may damage the plants, and in other cases, pollute surface water and ground water.

6-17. Fertilizers are used to add nutrients to the soil.

SOIL AMENDMENTS

Soil conditions can be improved, if necessary, by adding certain materials. Water drains slowly through tight soils or soils with high clay content. These types of soils also have a poor exchange of gases. Sandy soils, on the other hand, lack the ability to hold water and minerals, but have very good aeration. Both of these conditions make it difficult for plant growth and can be improved by mixing soil amendments with the soil. Plant growth can be greatly enhanced with the addition of soil amendments to certain soils.

Soil amendments have an organic or inorganic origin. Organic soil amendments originate from living organisms, plants, or animals. Inorganic soil amendments have their origin from nonliving sources, such as rocks. A brief description of common soil amendments follows.

Organic Soil Amendments

Bark—Bark from trees obtained through the timber industry can be added to soil to improve moisture holding ability and aeration. Bark ground into fine pieces is the most useful.

Compost—This is a soil amendment that can be made at home. It may consist of garden wastes, such as leaves, grass clippings, weeds, and vegetable scraps from the kitchen. The materials are placed in a pile and allowed to decompose before being added to the soil. The soil's water holding ability and aeration are improved.

6-18. Peat moss is a popular organic soil amendment.

Leaf mold—Leaf mold is similar to compost. The difference is it is derived entirely from leaves. Leaf mold improves moisture holding ability and aeration of the soil.

Peat moss—One of the most popular soil amendments is peat moss. It is an organic material dug from peat bogs. When mixed with soil, it improves moisture holding ability and it opens spaces for the exchange of gases.

Inorganic Soil Amendments

Perlite—This amendment originates from volcanic rock. The rock is crushed, then quickly heated to temperatures of 1800°F. The heat causes it to pop like popcorn. This white soil amendment is very light. Perlite improves soil drainage and aeration.

Sand—Sand is a widely used amendment. It is found naturally and is a result of the wearing of rock. Sand improves soil drainage and aeration.

Vermiculite—The origin of vermiculite is a mineral called mica. Mica is heated to 1800°F, which causes it to expand like an accordion. Vermiculite improves moisture holding ability and aeration of the soil.

MANAGEMENT OF PLANT PESTS

A plant **pest** is any living organism that causes injury, loss, or irritation to a plant. A plant pest can take many forms. A pest could be an insect, a mite, a mouse, a deer, a fungus, a bacterium, a virus, a nematode, or another plant. Pests may eat the plant, compete with the plant, or cause disease of the plant. Producers cannot totally eliminate most pests. They can, however, manage the pests to reduce the damage they might cause.

Recommended pest management programs involve a strategy known as **integrated pest management (IPM)**. IPM uses a combination of methods to control a pest with care taken to limit damage to the environment. Of course, the most important step in pest management is the correct diagnosis of the problem. Once identified, knowledge of the pest's biology can be applied in selection of treatments.

Integrated pest management involves three primary methods for management of the pest problem. They are biological, cultural, and chemical controls. The three pest control methods may be used together or independently in an IPM program. Integrated pest management is effective when practiced by trained individuals. Those individuals follow certain guidelines when designing a program:

1. The control selected will cause the least amount of injury to the pest's natural enemies.

2. The controls used will have the least potential to cause harm to humans.

3. The controls used will cause the least damage to the environment.

4. The controls used will be the most cost-effective.

5. The controls used will be the easiest to carry out effectively.

Biological control consists of using a pest's natural enemies to control the pest population. A good example involves the white fly, which is a common pest of greenhouse crops. The white fly has a tiny wasp as a natural enemy. Eggs of the wasp can be placed in the greenhouse populated with white flies. Wasps emerge from the eggs and seek white fly larvae to attack. Other good examples are ladybugs that eat aphids, and a bacterium (*Bacillus thuringiensis*) that is deadly to caterpillars.

Cultural control consists of many practices. The most important cultural practice is plant selection. Choosing plants that are resistant to diseases and insects is important. They have natural defenses. Biotechnology has opened the door for the development of genetically enhanced agronomic crops. Some of these transgenic plants have been given greater defenses against pests and diseases.

Plants should be selected that are hardy enough for the climate and tolerant of the growing conditions. Plants adapted to cer-

6-19. This ladybug is eating an aphid, which is a form of biological plant pest control. (Courtesy, Agricultural Research Service, USDA)

6-20. Biological insecticides are being sprayed on this field of peppers in Florida. (Courtesy, Abbott Labs)

tain growing conditions experience less stress. Many pests become a problem only when the plant is under stress.

Cultural control may also involve physically destroying the pest, disrupting the ability of the pest to reproduce, removal of food used by the pest, or altering the pest's environment. Examples are rotating crops, weeding by hand or with a cultivator, watering properly and at the right time, hand removal of an insect pest, and pruning a tree to allow more light to pass.

Chemical control utilizes chemicals to reduce pest populations. These include chemical pesticides, traps with sex hormones as attractants, and plant growth regulators. The most common chemical control utilizes pesticides. Different types of pesticides are used depending on the type of pest. *Insecticides* are used to control insects, *miticides* for mites, *herbicides* for plants, and *fungicides* for fungi, to name a few.

Pesticides are the most dangerous means of pest control in terms of the environment and the risk to humans. Even highly trained applicators must be extremely cautious when using pesticides.

REVIEWING

MAIN IDEAS

Plants have evolved to grow and reproduce. They have done this by adapting to the earth's environment. Environmental or cultural factors play a large role in the health of a plant. The major factors affecting plant growth are light, temperature, air, and water. Other important environmental factors are soils, nutrients, and pests.

Three characteristics of light affect plant growth. They are color, duration, and intensity. Red and blue are the two colors of light from the electromagnetic spectrum most important for plant growth. The duration of light triggers photoperiodic responses in plants, such as flowering. Bright light helps most plants produce an abundance of food through photosynthesis.

The life processes of plants are influenced by temperature. Life processes, such as respiration, are driven by enzymes. Enzyme activity speeds up as temperatures rise. The same processes slow down as temperatures drop. Plants are adapted to temperatures in the particular area of the world where they evolved. Plant cells may freeze and die if temperatures drop below a level to which they are hardy. Plants also suffer when temperatures rise too high.

The air or atmosphere around a plant is important. Plants need oxygen to carry out photosynthesis and carbon dioxide for respiration. Plants benefit from a good exchange of gases through stomata. Well-aerated soil is critical for healthy root growth. Plants are harmed by air pollutants like sulfur dioxide and ozone.

Water is a resource without which plant life could not exist. Plant tissues wilt if there is a lack of water. A shortage of precipitation may cause drought conditions resulting in injury to plants. Growers can promote growth and improve yields by irrigating crops. Water is a natural resource that should be conserved.

Soil provides water, minerals, and a firm medium in which the roots can anchor the plant. Plants get off to a quicker start if the soil has been worked or loosened before planting. The drawback to working the soil is soil erosion. Conservation tillage has grown in popularity, as it is a way to protect soil from erosion. Poor soils can be improved with the addition of fertilizers and soil amendments.

Pests are organisms that may cause damage or injury to plants. Integrated pest management is a strategy that uses several methods to control pests. Three pest control methods that can be employed independently or together are biological controls, cultural controls, and chemical controls. The aim is to provide control of the pest with the least damage to the environment and risk to humans.

QUESTIONS

Answer the following questions using complete sentences and correct spelling.

1. How does light affect plant growth?
2. What effect does temperature have on plants?
3. Why is water so important to plant growth?
4. What is irrigation? Are there any concerns with irrigation?
5. What are some advantages and disadvantages to working the soil?
6. How can soils be improved?
7. What is integrated pest management? Explain each component.

EVALUATING

Match the term with the correct definition. Write the letter by the term in the blank that is provided.

a. pest
b. soil amendment
c. photoperiodism
d. biological control

e. etiolation
f. irrigation
g. wilting
h. drought

i. herbicide
j. integrated pest management

1. _____ combination of pest control practices
2. _____ drooping of plant tissues caused by lack of water
3. _____ stretching of a plant caused by low light levels
4. _____ method of pest control utilizing the natural enemies of the pest
5. _____ plant's response to light duration
6. _____ period of insufficient precipitation
7. _____ living organism that causes injury, loss, or irritation to a plant
8. _____ furnishing plants with water by flooding or sprinkling
9. _____ materials added to soil to improve drainage, moisture holding ability, or aeration
10. _____ pesticide used to control plants

EXPLORING

1. Sow tomato or bean seeds in several pots. Place the pots in areas of different light intensity. Observe how the plants respond.

2. Obtain two packages of well-sealed lettuce seeds, moistened paper towels, and zipper bags. In a lighted room, place 10 lettuce seeds from one package on a moist paper towel. Place the paper towel in the zipper bag and zip. Wrap the zipper bag in aluminum foil so no light can enter. Undertake the same procedure with the other package of seeds in nearly *complete darkness*. This can be done in a closet. Be extremely careful not to allow light to strike the second batch of seeds. In five days, open the foil and observe what has happened. (Note: Lettuce seeds require light to germinate.)

3. Obtain two mason jars, paper towels, steel wool, and corn seeds. Line the inside of both jars with paper towels. In one jar, place steel wool in the center. In the other, fill the space with paper towels. Wet the paper towels and the steel wool in the jars. Now, plant 10 corn seeds in each jar between the paper towels and the glass about half way down. Seal the jars tightly. Make observations every day for about two weeks. What has happened? Why?

4. Obtain moist paper towels, soybean seeds, zipper bags, and a thermometer. Lay 25 soybean seeds on three different paper towels. Roll the paper towels up and place each in a different zipper bag. Locate one bag in a very warm area, one in a cold area, and one at room temperature. Record the temperature in each area. In one week, open the bags. Has temperature affected seed germination?

5. In your garden, see if you can identify some natural biological controls in action. Some things to look for might be ladybugs feeding on aphids, paper wasps snatching cabbage loopers, and parasitic wasp eggs on the back of a tomato hornworm.

Part Three

Fundamentals of Soil Science

PLANTS require nutrients for growth. Without sufficient nutrients, all or some productivity is lost. In extreme cases, the plants will die. Most nutrients are obtained from the soil. Plant roots remove water, minerals, and other substances along with the nutrients needed for growth. The ability of soil to support plant growth is its fertility. Understanding the nature of soil and how it is used and protected are important. Good producers know how to add needed nutrients to soil. Some growers attempt to produce plants without soil. They use a liquid as a source of needed nutrients.

CHAPTER 7

Soil Materials and Formation

OBJECTIVES

This chapter introduces the formation and materials of soil. It has the following objectives:

1 Explain the meaning and nature of soil.

2 Describe how soil is formed.

3 Explain the structure of soil.

4 Assess soil profiles.

5 Describe water relationships in the soil.

TERMS

aggregates
alluvial soil
capillary water
clay
decomposition
glacial soil
gravitational water
horizon
humus

hygroscopic water
infiltration
leaching
loess
mineral matter
organic matter
peds
percolation
permeable

pore spaces
silt
soil profile
soil structure
soil texture
subsoil
tilth
topsoil
weathering

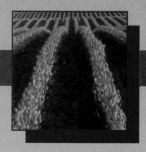

7-1. The fertile soil and favorable climate of Maui make it a productive island.

What you might know about soil is you can get it on your shoes, you might dig in it, agriculture producers raise crops in it, but is it important? It's just dirt, right? Wrong.

Soil is an important part of the environment, nature, and the agricultural industry upon which we depend. President John F. Kennedy once said, "Our entire society rests upon and is dependent upon our water, our land, our forests, and our minerals. How we use these resources influences our health, security, economy, and well-being."

Everyone has a role in maintaining soil. Those who farm the land have major roles. Homeowners with lawns and flower beds use soil. All of us can work to reduce pollution and damage to the soil. Our future depends on it!

WHAT IS SOIL?

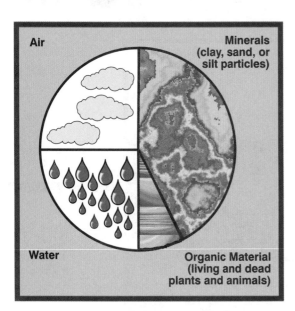

7-2. Contents of an average soil.

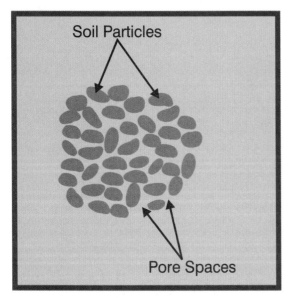

7-3. Pore spaces are the openings between solid soil particles.

Soil is the outer layer of the earth's surface. Depending on where you are on Earth, the layer of soil may be several inches thick or many feet deep. Soil is a complex material of great value. It is a renewable natural resource that supports life. It may take a thousand years for just one inch of soil to form. Therefore, it is critical that we manage our soils properly for the benefit of future generations.

Some people refer to soil as dirt. That is not really accurate. Dirt is soil that is out of place. You might find dirt under your fingernail or in the corner of the garage. Soil, on the other hand, is made of solids, liquids, gases, and living organisms. It provides food, water, air, and support for plant life.

Soil is made of mineral matter, water, air, and organic matter. The solid mineral matter and organic matter materials of soil make up about 50 percent of the soil. The average soil contains about 45 percent of mineral matter, and about 5 percent organic matter. These solids are not tightly packed. Instead, there are spaces between the solid particles. These are called **pore spaces**. Pore spaces are filled with either water or air. In good soil, water makes up about 25 percent of the soil and air makes up 25 percent of the soil.

MINERAL MATTER

Mineral matter is inorganic material, meaning it did not originate as a plant or animal. Mineral matter began as rock. Soils generally have mineral particles of different sizes. These particles are labeled sand, silt, or clay, based on their size.

Sand—Sand is the largest size particle. Sand particles create large pore spaces that improve aeration. Water flows through the large pore spaces quickly. They are, therefore, considered to be well-drained. Sandy soils lack the ability to hold nutrients and are not fertile. Also, soils that contain sand particles feel gritty to the touch.

Silt—Silt is the mid-size soil particles. It has good water-holding ability. It also has the ability to retain nutrients for plants. Silt feels like flour when dry and smooth like velvet when moist.

Clay—Clay is the smallest size soil particle. Clay has the ability to hold both nutrients and water that can be used by plants. It has very small pore size, resulting in poor aeration and poor water drainage. Clay forms hard clumps when dry and is sticky when wet.

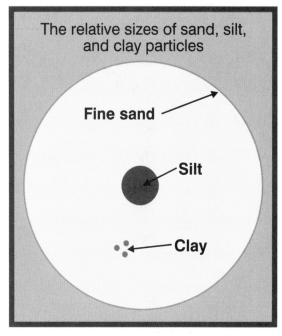

The relative sizes of sand, silt, and clay particles

Fine sand

Silt

Clay

7-4. Relative sizes of sand, silt, and clay particles.

Table 7-1. Characteristics of the Various Soil Materials

Characteristics	Sand	Silt	Clay
Looseness	Good	Fair	Poor
Air Space	Good	Fair to Good	Poor
Drainage	Good	Fair to Good	Poor
Tendency to Form Clods	Poor	Fair	Good
Ease of Working	Good	Fair to Good	Poor
Moisture Holding Ability	Poor	Fair to Good	Good
Fertility	Poor	Fair to Good	Fair to Good

Soils are classified by the size of the mineral particles within that soil. The proportions of the three soil particles determine the **soil texture**. Soils with different amounts of sand, silt, and clay are given different names. For instance, a soil containing 40 percent sand, 20 percent clay, and 40 percent silt is called loam soil. The textural triangle is used to classify soils based on their texture.

Soil texture may limit which crops can be grown. For example, root crops, such as carrots and onions, perform best in a sandy soil because it is loose and allows the plant to expand.

Table 7-2. Soil Textural Classes

Sand	Dry—Loose and single grained; feels gritty. Moist—Will form very easily crumbled ball Sand: 85–100%, Silt: 0–15%, Clay: 0–10%
Loamy Sand	Dry—Silt and clay may mask sand; feels loose, gritty. Moist—Feels gritty; forms easily crumbled ball; stains fingers slightly. Sand: 70–90%, Silt: 0–30%, Clay: 0–15%
Sandy Loam	Dry—Clods easily broken; sand can be seen and felt. Moist—Moderately gritty; forms ball that can stand careful handling; definitely stains fingers. Sand: 43–85%, Silt: 0–50%, Clay: 0–20%
Loam	Dry—Clods moderately difficult to break; somewhat gritty. Moist—Neither very gritty nor very smooth; forms a ball; stains fingers. Sand: 23–52%, Silt: 28–50%, Clay: 7–27%
Silt Loam	Dry—Clods difficult to break; when pulverized feels smooth, soft and floury, shows fingerprints. Moist—Has smooth or slick, buttery feel; stains fingers. Sand: 0–50%, Silt: 50–88%, Clay: 0–27%
Clay Loam	Dry—Clods very difficult to break with fingers. Moist—Has slight gritty feel; stains fingers; ribbons fairly well. Sand: 20–45%, Silt: 15–53%, Clay: 27–40%
Silty Clay Loam	Same as above, but very smooth. Sand: 0–20%, Silt: 40–73%, Clay: 27–40%
Sandy Clay Loam	Same as for Clay Loam. Sand: 45–80%, Silt: 0–28%, Clay: 20–35%
Clay	Dry—Clods cannot be broken with fingers without extreme pressure. Moist—Quite plastic and usually sticky when wet; stains fingers. (A silty clay feels smooth, a sandy clay feels gritty.) Sand: 0–45%, Silt: 0–40%, Clay: 40–100%

On the other hand, yields from crops like corn are lower in sandy soils because they lack the water and nutrient holding ability required of corn.

The texture of a soil also influences the ease with which a soil can be worked and the timing of the work. Soils with a larger percentage of sand are easier to work than soils with a large percentage of clay. Clay soils tend to be tighter, making it more difficult to break up or cultivate, whereas, sandy soils are looser. It also takes longer for a clay soil to dry after a rain

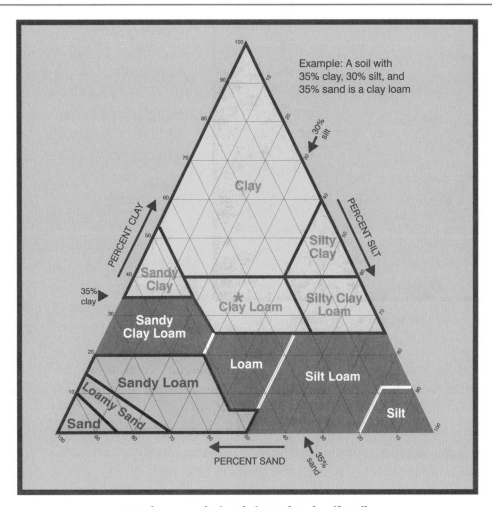

7-5. The textural triangle is used to classify soils.

than a sandy soil. Because of better drainage, a sandy soil can be worked sooner. In the case of wet clay soils, the producer or gardener must wait longer for the soil to dry sufficiently.

ORGANIC MATTER

Organic matter is the accumulation of decayed or partially decayed plants and animals. Most organic matter is from plant leaves, roots, and stems. Soils with a high organic matter content tend to be dark in color. Organic soils are more productive than soils with low organic matter content. This is because the organic matter contributes to soil fertility. Additional advantages of organic matter are improved aeration and water holding capacity.

For years, agriculture producers have improved their soils by adding organic matter to their fields. Manure from livestock is spread in the fields adding to the levels of organic matter. Another practice is to grow a green manure, such as alfalfa, and plow the plants into the

7-6. Crop residue adds to the organic matter content of a soil.

soil. Gardeners also recognize the importance of organic matter for plant health. It is a common practice to add peat moss or composted landscape and kitchen wastes (grass clippings, leaves, potato peels, etc.) to the garden.

WATER

When there is rainfall, water either enters the soil or flows off the soil as surface drainage. A soil with good texture and structure will absorb a great deal of rainfall. The process of the water soaking into the soil is known as *infiltration*. Once in the soil, the water moves downward. The downward movement of water through the soil is known as *percolation*. In a heavy rainfall, the pore spaces of soil fill up more quickly than water can drain through the soil. When this occurs, and all pore spaces are filled with water, the soil is considered saturated. Additional rainwater forms puddles or flows downhill on the surface of the soil. A quality soil that allows water movement by infiltration and percolation is said to be *permeable*.

CAREER PROFILE

SOIL CONSERVATION TECHNICIAN

A soil conservation technician studies, designs, and promotes proper soil management practices. They often work with land owners to implement approved practices. This photo shows a soil conservation technician using a transit to mark a field for the construction of a terrace.

Soil conservation technicians need a college degree in agronomy, soils, or closely related area. Practical experience farming, gardening, and exploring nature are beneficial. In high school, take science and agriculture classes.

Most jobs for soil conservation technicians are with government agencies or local soil and water conservation districts. (Courtesy, U.S. Department of Agriculture)

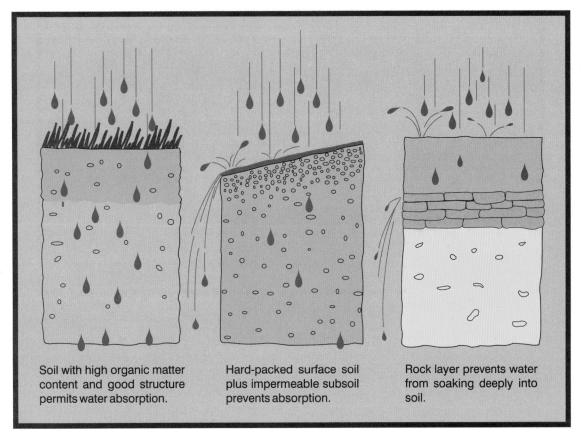

Soil with high organic matter content and good structure permits water absorption.

Hard-packed surface soil plus impermeable subsoil prevents absorption.

Rock layer prevents water from soaking deeply into soil.

7-7. Permeable soil allows water to infiltrate and percolate.

Water found in soils is placed into one of three categories. The categories are gravitational water, capillary water, and hygroscopic water.

After a rain, much of the water drains down through the soil through the pore spaces. This is called ***gravitational water*** because gravity pulls the water down. The water ends up in the ground water below the soil surface. Gravitational water flows quickly through sandy soils that have large pore spaces and more slowly through clay soils with tiny pore spaces.

As gravitational water moves through the soil, it carries with it dissolved minerals, chemicals, and salts. This process is known as ***leaching***. The soil texture will determine how quickly minerals, chemicals, and salts are leached, or washed, through a soil. Because of leaching, crops grown on sandy soils need more frequent fertilizer applications.

Capillary water is a second type of water. It is also the water that plants are most able to use. ***Capillary water*** is the water held between the soil particles against the force of gravity. This water can move through soil upwards or sideways by capillary action. As surface soil dries out, some water moves up into the open pore spaces by capillary action. Clay soils have many more pore spaces than sandy soils. Therefore, they hold more capillary water.

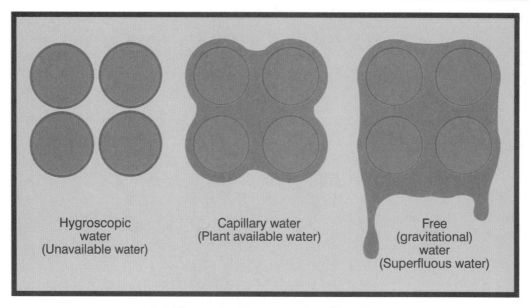

| Hygroscopic water (Unavailable water) | Capillary water (Plant available water) | Free (gravitational) water (Superfluous water) |

7-8. Three types of water are found in soils.

A third type of soil water is called hygroscopic water. **Hygroscopic water** is the water that forms a thin film around individual soil particles. Even the driest soils have hygroscopic water. Plants are unable to absorb and use this type of water.

CONNECTION

MUDDY WATER

Muddy water is not very appealing in a stream. We like to see crystal clear water. Fish and other animals in streams usually prefer clear water. Muddy water can often harm these animals.

Why is water sometimes muddy? It is because the water contains suspended particles of soil and other materials. The soil particles have been washed from land that has been loosened and left unprotected. Plowed fields, excavation sites for homes and businesses, and unprotected areas around golf courses and athletic fields are sources of particles. Loosening the soil makes it easy for flowing water to pick up and carry the particles away.

As soil is carried away, fertility is being lost. The ability to produce crops is reduced. The soil particles also may clog streams or create large delta areas in oceans and lakes. We can all help keep our streams clear by protecting the soil.

Air

Air is also found in the pore spaces. Roughly 25 percent of a soil will be air. After a rain, many of the pore spaces are filled with water. As that water drains away, air fills the pore spaces. In dry soils, air may take up the majority of the pore spaces. Sandy soils with large pore spaces tend to be well-aerated. Clay soils with their tiny pore spaces often are not well-drained and have a smaller percentage of air.

Most plants perform best when there is a balance between water and air in a soil. Sandy soils are loose and well-aerated, but do not hold water very well. Clay soils hold water, but are not well-aerated. Soils that have a mix of sand, silt, and clay tend to be both well-drained and well-aerated, which is good for plant growth.

Life in the Soil

There is a tremendous amount of life in soil. We cannot see most of the life forms with the naked eye. Soil is the home to earthworms, insects, bacteria, fungi, and other organisms. It is estimated that one teaspoon of soil contains more bacteria than there are people on the earth. The bacteria and fungi play an important role. They break down organic matter and release nutrients into the soil.

Earthworms, ants, crawfish, moles, and other organisms improve soil tilth. **Tilth** is the ease at which soil can be worked. These animals tunnel through the soil. Their openings help to loosen the soil. Water drains more quickly, and air exchange is improved. The castings, or waste, left on the soil surface by earthworms are fertile and beneficial to plant growth.

7-9. Earthworm activity improves soil tilth.

SOIL FORMATION

The formation of soils takes hundreds to thousands of years. It is a complex process that requires time. Initially, parent material, or rock, is weathered.

WEATHERING

Weathering is a process in which the climate literally causes rock to break into small particles. The resulting small particles are mineral matter found in soils.

Weathering is caused by a number of forces. The repetition of freezing and thawing temperatures causes the rock to expand and contract. This action forces parts of the rock to break away. Water that fills small cracks in rock and then freezes forces the rock apart. Rainfall wears the rock away a little at a time. Wind also wears the rock away. The force of water in streams, rivers, and ocean waves causes rocks to hit one another and break into smaller pieces.

Plants also play a role in breaking parent material into smaller pieces. Lichens secrete acids that help to break down the rock. Fine particles of rock break off, contributing to the

7-10. The force of waves helps to make soil.

7-11. Rainfall, and the streams it creates, slowly wears down mountains.

7-12. Lichens, mosses, and the roots of plants help develop soil.

soil mass. Roots of larger plants grow in cracks of the rocks. As the roots expand, they force the rocks apart.

TRANSPORT OF SOIL MATERIALS

The small particles of rock that have broken off are usually transported. Water washes the particles down the stream. The particles settle in flat areas, such as deltas and flood plains. The deposits left from the action of water are known as *alluvial soil*.

Glaciers, or huge sheets of ice, that have moved across much of the earth during colder periods of time created and transported soil. These massive sheets of ice, a mile or more thick, ground rock into sand, silt, and clay as they moved across the landscape. When warmer periods came, the glaciers melted. The soil created, moved, and deposited by glaciers is known as *glacial soil*.

A third way soil is transported is by the wind. The force of the wind can pick up soil particles and carry them

7-13. This river exiting a glacier is milky white with silt ground by the force of the glacier.

7-14. This midwestern soil is dark because of an accumulation of organic matter.

great distances. These soils are referred to as *loess*. Large deposits of soil in China and in the Midwest and Northwest of the United States are loess.

As the mineral matter from rocks accumulates, greater numbers of plant species can be supported. The plants grow and eventually die. New plants take their place. Fungi, bacteria, insects, and other organisms feed upon the dead plant tissues. This process, in which organic matter is broken down, is known as *decomposition*. The plant material or organic matter that reaches the advanced stage of decomposition is called *humus*. Soils with high levels of humus are dark in color, loose, and fertile.

SOIL STRUCTURE

Sand, silt, clay, and organic matter particles in a soil combine with one another to form larger particles of various shapes and sizes. These larger particles or clusters are often referred to as *aggregates*. The clusters of soil particles are called *peds*. In some soils, the peds are distinct and difficult to crush between the fingers. In other soils, the peds are diffi-

Granular

Crumb

Platy

Prismatic

Massive

Columnar

Blocky

Single grain

7-15. Soil structure categories.

cult to identify and fall apart easily when handled. The way in which aggregates or clusters are arranged is referred to as **soil structure**.

There are eight types of soil structure, including blocky, columnar, crumb, granular, platy, prismatic, single grain, and massive. Granular is the most desirable structure type, because it has the greatest proportion of large openings between the individual aggregates. Another desirable type of soil structure is crumb.

Soil structure affects water and air movement in a soil, nutrient availability for plants, root growth, and microorganism activity. The pore spaces created by the larger soil particles are larger than those in between individual particles of sand, silt, or clay. This allows for greater air and water movement, better root growth, and open passageways for small animals. The aggregates are also better able to hold water and nutrients.

Soil structure can be destroyed. A major cause of damage is driving heavy equipment over wet soil. Damage is also caused by working soil when it is excessively wet. These conditions lead to the clay particles clogging up the pore spaces. The soil becomes compacted, very dense, and when it dries, it becomes very hard. It is very difficult for most plants to survive in a soil whose structure has been destroyed.

7-16. Heavy equipment on construction sites often destroys soil structure. The result in this case is poor water drainage.

SOIL PROFILES

A **soil profile** is a view of a cross-section of soil. Much like you would view a person's profile from the side, the soil is viewed from the side. This view reveals the layers of soil, or **horizons**. The depth of the soil and the thickness of each horizon varies with the location. Soils typically have four main horizons:

- *O horizon*—Not considered one of the four main horizons, this layer consists of leaves, roots, limbs, and decaying matter laying on the surface of the soil. It is common in forests and swamps.

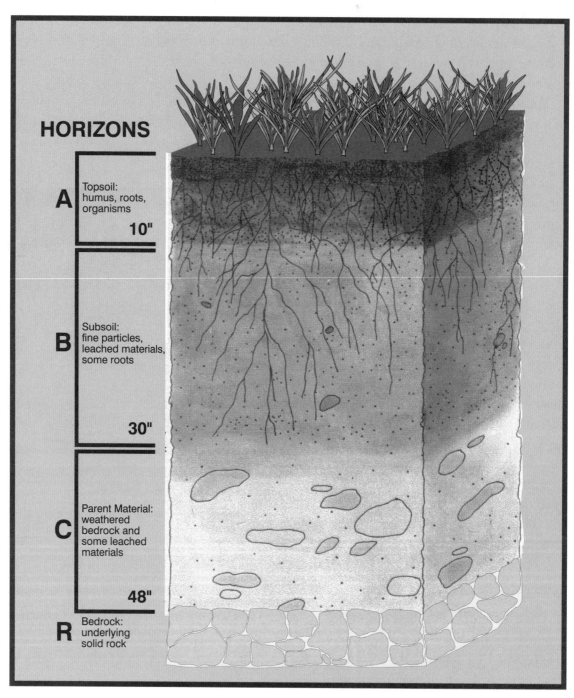

HORIZONS

A Topsoil: humus, roots, organisms
10"

B Subsoil: fine particles, leached materials, some roots
30"

C Parent Material: weathered bedrock and some leached materials
48"

R Bedrock: underlying solid rock

7-17. A diagram of a soil profile.

- *A horizon*—The A horizon is the top layer of soil or the **topsoil**. This layer contains the greatest amount of decomposed organic matter. It is fertile and supports healthy plant growth. The thickness of the A horizon is anywhere from nonexistent to about 10 inches deep.

- *B horizon*—The B horizon is also known as the **subsoil**. It is located beneath the A horizon. Some nutrients leach into the B horizon from above. The amount of organic matter is very low. Some plant roots grow into the subsoil and obtain some nutrients and anchorage.

- *C horizon*—This layer of soil contains parent material, weathered bedrock, and leached materials. There is no organic matter or root growth found in the C horizon.

- *R horizon*—Bedrock or solid rock located below the C horizon.

Observation of a profile provides useful information. The depth of the topsoil is obvious in most profiles. The presence of hardpans and other structures that restrict root growth and water movement are easily seen. The activities of soil organisms, such as earthworms, may also be evident.

REVIEWING

MAIN IDEAS

Soil is the outer layer of the earth's surface. It is a complex material made of mineral matter, organic matter, air, and water. There are also many living organisms in soil. Soils support plant life by providing nutrients, water, air, and support to plants. It is said that an ideal soil would be about 45 percent mineral matter, 5 percent organic matter, 25 percent water, and 25 percent air.

Weathering of rock produces small particles of mineral matter. The particles of mineral matter in a soil are sand, silt, and clay. Sand is the largest particle, while clay is the smallest. Soil texture is determined by the percentage of sand, silt, and clay in a soil. Soil particles may be transported by water to become alluvial soil, glaciers to become glacial soil, or wind to become loess. Decomposed plants and animals contribute organic matter to the soil.

Pore spaces are the openings between solid soil particles. Pore spaces are filled with air or water. A permeable soil allows the movement of water into the soil by infiltration and downward through the soil by percolation. Three types of water are in soil. Gravitational water is pulled downward by gravity. Capillary water moves upwards or sideways and is most available to plants. Hygroscopic water forms a film around soil particles and is not available to plants.

Soil particles combine to form aggregates. The aggregates are better able to hold nutrients and water than individual particles. Aggregates also create larger pore spaces, improving aeration. The way in which the aggregates are arranged is the soil structure. Working soil when it is either too wet or too dry can destroy soil structure and make it difficult for plants to grow.

Four main layers, or horizons, are in a soil profile. The A horizon is the topsoil and has the greatest amount of organic matter, making it the most fertile horizon. The B horizon is the subsoil. Plant roots can be found growing in both the A and B horizons. The C horizon is below the B horizon. The R horizon is the bedrock.

QUESTIONS

Answer the following questions using complete sentences and correct spelling.

1. What are the major components of soil?

2. How does organic matter improve soil?

3. How do the types of soil water differ?

4. What are the advantages of sand? Clay?

5. How are living organisms associated with tilth?

6. How is soil formed?

7. What are three ways soil is transported and deposited?

8. What is a soil profile?

9. What are the horizons of a soil profile?

10. What is soil structure?

EVALUATING

Match the term with the correct definition. Write the letter by the term in the blank that is provided.

a. loess   e. horizon   i. tilth
b. topsoil   f. pore spaces   j. capillary water
c. clay   g. organic matter
d. soil texture   h. soil structure

1. _____ water held between the soil particles against the force of gravity

2. _____ ease at which soil can be worked

3. _____ soil deposited by wind

4. _____ finest particle mineral matter in soil

5. _____ accumulation of decayed or partially decayed plants and animals

6. _____ openings between the solid particles of soil

7. _____ proportions of sand, silt, and clay in a soil

8. _____ layer of soil seen in a soil profile

9. _____ soil found in the A horizon

10. _____ way in which soil aggregates are arranged

EXPLORING

1. Visit a construction site that has exposed a soil profile. See if you can determine the different horizons. What are some differences between the layers of soil?

2. Take an empty, narrow, tall jar and fill it about half full with your garden soil. Add a couple of drops of dishwashing detergent and enough water to fill the jar to within about one-half inch from the top. Shake the jar vigorously. Set the jar down and watch the soil settle. Which particles settle first? Let the jar set undisturbed for a day. Can you determine layers of sand, silt, clay, and organic matter? Measure the thickness of the layers, calculate the percentage of sand, silt, and clay. Then, determine the textural class of your soil.

3. Determine the texture of a soil by following these steps:

a. Moisten a sample of soil to the consistency of putty.

b. From this sample, make a ball about ½ inch in diameter.

c. Hold the ball between the thumb and forefinger, and gradually press the thumb forward, forming the soil into a ribbon.

d. If the ribbon forms easily, and is long and pliable, the soil is fine textured.

e. If the ribbon forms, but breaks into pieces ¾ to 1 inch long, the soil is moderately textured.

f. If no ribbon is formed, and the soil feels very gritty, the soil is moderately coarse textured.

g. If the sample consists almost entirely of gritty material and leaves little or no stain on the hand, it is coarse textured.

Land Classification and Use

This chapter explains the characteristics of land and describes land classification. It has the following objectives:

1 Explain the features and uses of land.

2 Describe land capability and list capability factors.

3 Explain land capability classification.

TERMS

arable land	hardpan	slope
best land use	land	soil depth
capability factors	land capability	surface runoff
climate	land capability classification	topography
cropland	land forming	

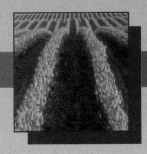

8-1. Strip cropping of corn and alfalfa is used on this farm to protect the land. (Courtesy, U.S. Department of Agriculture)

WHY do some places produce certain crops and others do not? We all know that the kinds of crops produced and the methods of production vary from one location to another. These depend on important capabilities of the land.

Making decisions about buying cropland and selecting the crops to grow require careful study. Soil characteristics are important. The overall environment in which the soil supports plant life is critical to success.

Land is more than soil. Successful crop production considers the nature of the land. Some land can be easily cultivated; other land cannot be cultivated at all. What makes the land different and why are differences important?

LAND IS MORE THAN SOIL

Land is more than soil fertility, though fertility is important in some uses. **Land** is all of the natural and artificial characteristics of an area to be used for agricultural or other purposes. It includes both renewable and nonrenewable resources plus improvements. The concept of land is not always easy to understand.

Most of the time, land refers to the surface of the earth that is not permanently covered with water. Crop producers often refer to their land and its condition. Sometimes, land can be temporarily or permanently covered with water. Land near a creek may be briefly flooded following a heavy rain. This may interfere with crop terrestrial production. Land is also used for roads, building sites, fence rows, and other essential purposes.

8-2. This photograph shows several features of land. Which can you identify? (Courtesy, U.S. Department of Agriculture)

Land used for aquaculture may be covered with water on a long-term basis, if not permanently. A pond used for aquaculture is land. The area of an ocean near the shore where oysters are farmed is land. The location of roads, towns, and other features also influences land and its use. For example, proximity of land to markets and consumers may influence its use.

PRODUCTIVE CROPLAND

Cropland is land used for growing crops. Typically, the crops grown involve tilling the land. Increasingly, cultural practices that do not require as much tilling are being used.

Land that is productive for growing crops has certain characteristics. It provides a good environment for the crop to grow. The soil provides the nutrients that the crop needs to grow (often with a little fertilizer added). Other features also influence the crops that can be grown. All of the features combine to set the value of land.

The major characteristics of cropland include:

- Soil—The nature of the soil has a big impact on productivity. Soil texture, nutrients, and internal structure are major areas that influence soil productivity. (The area of soil fertility is covered in detail in other chapters.)

- Climate—The crops that can be grown on land are determined by the climate. **Climate** is the average of the weather conditions over a long time. Some locations have tropical climates, where the weather is always warm. Other locations have temperate climates, where the weather varies between warm and cool. Arctic climates have cool or cold weather conditions that may severely limit the crops that can be grown.

- Topography—**Topography** is the form or outline of the surface of the earth. Some land may have a nearly level or gently rolling topography. Other land may have steep hills and mountains. Topography is important in the use of land. Obviously, row crops cannot be planted on the sides of steep mountains.

8-3. A wide area was destroyed by the eruption of Mt. St. Helens in Washington. Twenty years after the destruction, the area remains barren. (The crater is in the center of the photo.)

- Water supply—Water supply is the amount of water available for crops. The water may be from precipitation, moisture within the soil, or applied artificially by irrigation. Land in dry areas may have water rights that allow the owner to use water from a canal or well. Without the water, the land may be of little value for crops.

- Subsurface conditions—Conditions found below the surface are important in crop production. Soil texture may influence internal drainage. Some areas may have shallow rock or hardpans. **Hardpan** is a tightly compacted layer of soil that interferes with water movement and root growth. Hardpans may be naturally present or result from driving heavy equipment over fields. Land in a location with a good climate and fertile soil may

8-4. The topography of the land used for a playground often results in temporary flooding by overflow water.

not be productive because of subsurface conditions. Cultural practices can be used to help alleviate subsurface problems. However, these add to the costs of production.

- Pollution—Land may be polluted and unsuited for crop production though all other factors are favorable. Sometimes, the pollution is in the soil and results from materials dumped on the land. At other times, the pollution may be in the air or water. Sites near large cities and highways may be polluted by emissions from cars, factories, or homes. Plants in dryland areas can be damaged by the tiny particles that settle onto leaves from the air. (These particles can stay on the leaves for a long time because precipitation does not fall and wash it off the leaves. This has increasingly become a concern in Arizona and other areas of the southwestern United States.)

ALTERNATIVE USES

Some land has many uses; other land has few potential uses. The use of land for crops must sometimes be based on other potential uses. People are frequently faced with determining the best use of land.

Best land use is the use of land that produces the most benefits to society. Crop producers view land from the standpoint of cropping. They view the best land use as the use that will result in the highest returns. Land varies in the kinds of crops to which it is best suited. Some land is suited for row crops, while other land is suited for pasture and forests.

8-5. Quality cropland now in soybeans is offered for sale for industrial development. Is a national land use policy needed to protect cropland from being diverted to other uses?

Sometimes, the best use is for factories, homes, shopping malls, or other purposes. Land near major highways and cities may have a higher value for uses other than crops. The economic return to the owner may be greater if not used as cropland.

Some locations are too valuable for crop production. A land owner cannot buy the land and pay for it with the returns from crops. This is especially true around cities with urban sprawl.

Each year thousands of acres of good cropland are lost in the United States. The land is taken out of crops. It is used for factories, homes, shopping centers, golf courses, and other facilities. Productive soybean land may be advertised for sale as industrial property. Once sold, those acres will no longer be planted to soybeans.

8-6. Streets, hotels, shopping centers, and other structures take up much of the land in Honolulu.

Fortunately, the United States has sufficient land for pastures, crops, forests, and other agricultural uses at the current level of demand. Some people are concerned. They feel that the best agricultural land should be reserved for agriculture. They feel that we will not be able to provide for the needs of future generations. Others feel that land should be used for its highest-value use. What do you think?

CAREER PROFILE

SOIL SCIENTIST

A soil scientist studies the soil and prepares maps to depict features of the land. They may go into fields, collect samples, plot land features, and do other activities. Aerial photographs and global positioning systems may be used.

Soil scientists need degrees in soils, agronomy, horticulture, or a closely related area. Baccalaureate degrees are typically required. Many have masters degrees and some have doctorates in the area. Considerable skill is needed in the use of computers and other equipment.

This photograph shows a soil scientist collecting a soil sample for use in mapping a field. (Courtesy, Potash and Phosphate Institute)

Using good cropland for non-crop purposes often involves making tough decisions. The earth's population is increasing at a rapid rate. These people need food. What will happen if too much productive cropland is used for non-crop purposes?

LAND CAPABILITY

8-7. A California farm has arable land well suited for lettuce.

Land capability is the suitability of land for agricultural uses. The uses should not result in damage to the land, though nutrients may be removed from the soil. Cropping includes good fertility management so the productivity of the land is sustained.

ARABLE LAND

Arable land is land that can be used for row crops. These are crops that typically require some tillage. Arable land can be tilled and is where crop production is practical. Some land should not be used for crops that require plowing. Alternatives include pasture and forest crops.

Land Improvement

Various practices can be used to improve arable land. Four common practices with arable land are:

- Irrigation—Irrigation is the artificial application of water to soil or a growing medium to assure adequate moisture for plant growth. It is essential to crops in the dryland areas of the western United States. Irrigation is often used on a supplemental basis in other areas where seasonal shortages of water may reduce crop yields. To a crop producer, the availability of water for irrigation may be the most important factor in the productive use of arable land. (Chapter 10 covers irrigation in more detail.)

- Erosion control—The long-term productivity of land can be assured by controlling soil erosion. Excessive erosion may result in land that is no longer fertile. A crop will not grow on it. Using good cultural practices can reduce erosion.

8-8. Clear-cutting in a forest has resulted in unprotected land subject to severe erosion.

- Drainage—Land sometimes needs surface or internal drainage. Surface drainage is used to remove water from the surface of the land. Ditches and terraces are often used for surface drainage. Internal drainage may be improved with drain tiles or tubes installed below the surface. The depth is below the normal plow depth.

8-9. An Indiana farm has used a grassy strip between corn fields to control runoff.

- Forming—**Land forming** is used to have a good surface for cropping. The surface is smoothed or reshaped to enhance the use of the land. Small dips are filled and high places are taken down. How the surface is formed depends on the crop to be grown. Rice, for example, requires a different surface from corn or wheat. Land forming involves using laser-guided equipment to assure a good surface.

8-10. A laser-guided system is being used to form this land before turf is installed in Georgia.

8-11. Contrasting soil tilth is quite obvious. The top photo shows garden soil with good tilth. The bottom photo shows soil with poor tilth caused in part by plowing when the soil was too wet.

Soil Tilth

Soil tilth is the physical condition of soil that makes it easy or difficult to work. Soil with good tilth is loose, has adequate moisture, and forms a good seedbed. Soil with poor tilth has hard clods, may be very dry or wet, and does not form a good seedbed.

Tilth is a product of the texture and structure of soil. It is an important part of raising crops on land. Plows, harrows, disks, and other implements more easily prepare a good seedbed in soil with good tilth. This feature is often associated with crop land.

CAPABILITY FACTORS

Capability factors are the characteristics of land that determine its best crop use. These factors include both surface and subsurface characteristics. Six common factors are:

- Surface texture—Surface texture is the proportion of sand, silt, and clay in the soil to the typical depth of plowing—about 7 inches (17.8 cm). Soil can be classified into three major areas: sandy, loamy, and clayey soils. Sandy soils have a coarse, gritty feel when rubbed between the fingers. Loamy soils are high in silt and have a feeling of flour when rubbed between the fingers. Soils high in clay have a sticky feeling when moist and will form pieces that will stick together. Each of these textures is further divided based on the soil triangle shown in Chapter 7.

- Internal drainage—Internal drainage is known as permeability. Soil permeability is the movement of air and water through the soil. It is determined by the texture and structure of the soil. When water leaves soil, it is replaced by air. Soil permeability is classified as very slow, slow, moderate, and rapid. Sometimes, parent material influences permeability. Layers of shale or rock restrict the movement of water.

  Water quickly soaks into sandy soil with high permeability. Because of this movement, the water is lost from high permeability soil by evaporation and movement to lower levels in the earth. Soil high in clay has slow permeability. It holds water a long time. Loamy soil high in silt usually has moderate permeability. With crops, sandy soils hold moisture poorly and require more precipitation or irrigation than soils high in clay. In wet areas, sandy soils dry out faster than those high in clay or silt.

  The nutrient content of soil is related to permeability. Water movement carries dissolved nutrients. Soils high in sand are often low in nitrogen and other nutrients.

8-12. Water standing on this turf is a sign of soil with poor internal drainage.

- Soil depth—**Soil depth** is the thickness of the soil layers that are important in crop production. This varies somewhat with the crop. Both topsoil and subsoil are included. Some crops, such as sugar beets, have roots that go fairly deep into the soil. Other crops have shallow roots, with corn an example. Rock, shale, or other underlying materials affect soil depth. Four soil depths are used:

  — very shallow soil—less than 10 inches [25 cm] deep
  — shallow soil (10 to 20 inches [25 to 61 cm] deep)
  — moderately deep soil (20 to 36 inches [61 to 91 cm] deep)
  — deep soil (more than 36 inches [91 cm] deep).

Shallow and very shallow soils often result on eroded land, or are over rock or shale formations. Deep and moderately deep soils are the most productive.

- Erosion—Erosion is the loss of topsoil by water, wind, or other forces. Much of the fertility of land is in the topsoil. When it is lost, the productivity of land is gone. Four categories of erosion are used: very severe erosion—75 percent or more of the surface soil has eroded, and large gullies are present; severe erosion—75 percent of the surface soil has eroded, but no large gullies are present; moderate erosion—25 to 75 percent of the surface soil has been lost with small gullies present; and none to slight erosion—less than 25 percent of the topsoil has been lost and no gullies are present.

8-13. Small rills are forming on this hill as evidence of severe erosion.

- Slope—**Slope** is the rise and fall in the elevation of land. It is commonly measured in percent or the number of feet of rise and fall in 100 feet (metric conversions can also be used). For example, a hill that drops 5 feet in 100 feet has a 5 percent slope. Land slope is associated with hills and mountains.

Six classes of land slope are commonly used: very steep—more than 12 percent, steep—8 to 12 percent, strongly sloping—5 to 8 percent, moderately sloping—3 to 5 percent, gently sloping—1 to 3 percent, and nearly level—less than 1 percent slope.

Land slope is important in determining the best use of the land. The more slope land has, the less suited it is to crops that require tillage. Sloping land is also more susceptible to erosion, particularly the formation of gullies.

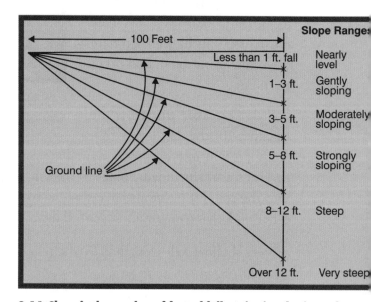

8-14. Slope is the number of feet of fall or rise in a horizontal distance of 100 feet.

- Surface runoff—**Surface runoff** is the water from rain, snow, or other precipitation that does not soak into the ground and runs off. It collects in streams, ponds, and wetlands areas. Runoff depends on the soil texture and slope of the land. Cultural and conservation practices can be used to reduce surface runoff. For example, terraces reduce the rate of runoff and help conserve the soil. Chopping stalks after harvest to cover the land reduces runoff.

As runoff is reduced, infiltration (soaking in) increases. This helps restore ground water levels and provide moisture for future crop growth. Four categories of surface runoff are typically used: very slow—deep sandy soils with low areas and little slope present, slow—clayey soils with little slope, moderate—non-clay soils with 1 to 3 percent slope, and rapid—usually on land with more than 3 percent slope with surface water that flows rapidly causing some erosion. Soil texture and cultural practices have considerable impact on surface runoff.

8-15. Land near the Great Salt Lake in Utah is too salty for most uses.

8-16. A flexible, perforated, drainage tube has been installed under this field to promote internal and surface drainage.

LAND CAPABILITY CLASSIFICATION

Land capability classification is assigning a class number to land. Eight land capability classes are used. These are based on factors in land capability. The numbers are I to

8-17. How would you class this land? (It is Class I unless the slope would be more than 1 percent and then it would be Class II.) (Courtesy, David Pettry, Mississippi Agricultural and Forestry Experiment Station)

8-18. Papaya is well suited for growth on this Class II land in Maui, Hawaii.

VIII, with I being the best in terms of arability and VIII being the worst. The classes are further divided into cultivable and noncultivable groups. Classes I to IV can be cultivated, though some conservation practices may be needed. Classes V to VIII should not be cultivated. These classes tend to have high slope or to be low and wet.

CLASS I: VERY GOOD LAND

Class I land has few limitations. It is nearly level and has deep soil, good internal drainage, and good surface drainage. Class I land can be cropped year after year without the need for special practices to control erosion and runoff. The soil has good tilth—loose with no large clods, not too wet or dry, and easy to work. The land is suitable for most crops.

Class I land is the ideal to which other classes of land are compared. The six capability factors are compared with the ideal. As listed earlier, these factors are: surface texture, internal drainage, soil depth, erosion, slope, and surface runoff.

CLASS II: GOOD LAND

Class II is good land, but not quite as good as Class I. Class II land has deep soil with a few limitations. The soil requires moderate attention to conservation practices. Contour plowing and other easy-to-use practices are often appropriate. The producer has fewer cropping alternatives with Class II land.

CLASS III: MODERATELY GOOD LAND

The land in Class III has more limitations than Class II. Crops must be more carefully selected. Class III land is often on gently sloping hills. Increased attention must be given to conservation practices, such as terraces and strip cropping. Class III land can be used for many purposes if given the appropriate soil and water management. Class III land can be productive with proper management.

8-19. A small water and erosion control pond has been constructed on this Class III land in Mississippi.

CLASS IV: FAIRLY GOOD LAND

Class IV land is the lowest class that should be cultivated. It has very severe limitations that restrict the choice of crops and require special conservation management practices. Class IV land is on hills with more slope than Class III. The land is frequently subject to erosion, especially gullies. Careful use of terraces, strip cropping, and contour irrigation and plowing can result in good crop productivity from Class IV land.

8-20. Poor soil conservation measures are being used on this Class IV land. The rows and strips on this strawberry farm should be on a contour rather than run up and down the hills. Notice the deep ruts forming between the rows, indicating erosion.

CLASS V: UNSUITED FOR CULTIVATION

Class V land can be used for pasture crops and cattle grazing, hay crops, and tree farming. The land is often used for wildlife and recreation areas. The soil typically has good tilth and fertility, but is restricted in use by rock outcrops or frequent overflow from nearby streams.

8-21. This Class V land is best used for grazing or forestry because of the rock outcrops.

The land is often gently sloping or nearly level. Some Class V land may border or be in wetlands areas. Though contrasting in appearance, the presence of rocks on gently sloping land or overflow near streams are two major characteristics of Class V land.

CLASS VI: *NOT SUITED FOR ROW CROPS*

Class VI land has too much slope for growing row crops, though the slope may not be steep. The soil may have fair productivity if it has not been damaged by erosion. Gullies often

CONNECTION

JUDGING A PIT

The subsurface features of land are important. Soil depth, restrictions, and internal drainage must be considered. Pits are often dug and used in judging land.

Most pits are several feet deep (1-1.5 m). They are cut to make it easy to see the soil profile. Roots of plants and the subsoil environment should be visible. The presence of a hardpan, rock, shale, and other obstructions can be detected. In some places, water seepage may indicate a shallow water level.

Youth groups often have land judging contests. Assistance with the contests is available from the local office of the Natural Resource Conservation Service (NRCS) of the U.S. Department of Agriculture and the local soil and water conservation district. (Courtesy, Jeanine May, Mississippi)

quickly form if not carefully managed. Class VI land can be used for trees, pasture, wildlife habitat, watershed areas, and recreation.

CLASS VII: HIGHLY UNSUITED FOR CULTIVATION

Class VII land has severe limitations. This class should not be cultivated. Best uses are permanent pasture, forestry, and wildlife. Residential and resort development may be possible. The land should be carefully managed to prevent erosion and conserve water. Slope is often well over 12 percent. The soil is very shallow. Large rock surfaces and boulders may be found. Little soil may be present. Sometimes, Class VII land is in dry areas and may only have precipitation in the winter. The limited watershed may provide runoff for irrigation reservoirs, cattle ponds, or recreational use.

8-22. Large gullies have washed away any productive capacity of this Class VI land. It should have never been plowed! (Courtesy, U.S. Department of Agriculture)

CLASS VIII: UNSUITED FOR PLANT PRODUCTION

The land in Class VIII cannot be used for row crops or other crops where the land is tilled. It is often lowland covered with water most or all of the time. The soil may be wet and high in sand or clay. Some aquatic crops

8-23. This rocky mountain in the Arizona desert is unsuited for any type of cultivation.

can be grown on it, such as cranberries, wild rice, and crawfish. A few forestry crops can be grown, such as cypress or selected bottomland hardwood species. Class VIII land is often used for waterfowl habitat.

8-24. This land in the Okefenokee Swamp in South Georgia stays under water most of the time. A few cypress trees can be grown. It is primarily used for recreation. (Courtesy, Stephen J. Lee, Chapel Hill, North Carolina)

REVIEWING

MAIN IDEAS

Land is all of the features of an area of the earth's surface. It includes both natural and artificial features. These features are typically viewed for economic value. Cropland is the land used for growing crops. The use of land for crops is influenced by soil, climate, topography, water supply, subsurface conditions, and pollution.

"Best land use" is the term used to describe the use of land that provides the most benefits to society. A rapidly growing human population is resulting in increased attention to sites for factories, homes, shopping centers, and other purposes.

Land capability is the suitability of land for agricultural purposes. Arable land is land that can be safely used for row crops. Sometimes, arable land may need irrigation, drainage, forming, and erosion control to make it productive. The major capability factors are surface texture, internal drainage, soil depth, erosion, slope, and surface runoff.

Eight land capability classes are used, with Class I being the best and Class VIII being the least suited to cultivation.

QUESTIONS

Answer the following questions using complete sentences and correct spelling.

1. What is land? How is it related to soil?

2. What is cropland?

3. What are the six major characteristics of land in terms of cropland?

4. What is "best land use?"

5. What improvements are often made in land to make it arable?

6. What is soil tilth? Why is it an important feature of cropland?

7. What are the six major factors in land capability? Briefly explain each.

8. What classes of land can be cultivated? What limitations restrict the cultivation of the other classes?

EVALUATING

Match the term with the correct definition. Write the letter by the term in the blank that is provided.

a. slope
b. soil depth
c. land

d. climate
e. topography
f. arable land

g. land forming
h. soil tilth

1. _____ thickness of the soil layers that are important to crops

2. _____ physical condition of the soil that makes it easy or difficult to work

3. _____ percent of rise and fall in the elevation of the land

4. _____ average of the weather conditions

5. _____ characteristics of an area of the earth's surface

6. _____ outline of the surface of the earth

7. _____ land that can be used for row crops

8. _____ smoothing or reshaping the surface of land

EXPLORING

1. Organize a land judging team. Obtain the instructions for your state from an agriculture teacher or the land-grant university. The local office of the Cooperative Extension Service may have the details. Practice judging land near your school and at nearby schools. The local office of the U.S. Department of Agriculture may help. In some cases, the local soil and water conservation district can help.

2. Assess the overall capability classification of the land in your area. Get an aerial photograph from the local U.S. Department of Agriculture office. Determine the slope, surface texture, depth, and other features of fields identified on the map. Occasionally, the office may have a soil scientist or conservationist who can help with this project. Prepare a bulletin board that shows what you have found.

CHAPTER 9

Soil Fertility and Management

OBJECTIVES

This chapter defines soil fertility. It has the following objectives:

1 Describe the meaning and importance of soil fertility.

2 Explain soil pH as related to fertility.

3 Describe nutrient diagnostic procedures.

4 Explain the use and application of soil amendments.

5 Describe the use of precision farming practices.

TERMS

adsorption
anions
band application
broadcast
cation
cation exchange capacity
   (CEC)

fertilizer analysis
ion
nutrient cycle
organic fertilizer
side-dressing
soil fertility
soil pH

soil test
top-dressing

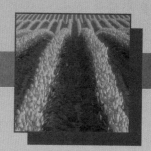

9-1. Corn yields are highly dependent on soil fertility.

T IS recommended that you eat many kinds of foods to achieve a balanced diet. Sometimes, you might find yourself eating nutritious foods, sometimes, you might not. When you do not eat a good diet, you may become deficient in certain vitamins and minerals. To be sure you are getting all the vitamins and minerals your body needs, you might take a daily vitamin.

Plants depend on the soil to provide the essential elements, or nutrients, for healthy growth. Some soils are rich in nutrients, and the plants growing in them thrive. Other soils are nutrient poor and do not support healthy plant growth. With proper management and use of fertilizers, nutrient levels in these soils can be improved.

WHAT IS SOIL FERTILITY?

Soil fertility is the ability of a soil to provide nutrients for plant growth. Soil fertility involves the storage of nutrients. It also refers to how available those nutrients are for plants.

An ideal soil will provide a plant with all the nutrients for healthy growth. A plant will suffer if it cannot obtain all the essential nutrients from a soil. Even if only one necessary nutrient is missing, the plant can develop deficiency symptoms. For example, nitrogen promotes the growth of green, leafy tissue. If the available nitrogen is low, the plant may take on a yellow appearance. This will happen even when all the other essential nutrients are available.

THE NUTRIENT CYCLE

In a natural setting, plants obtain their nutrients from the soil. The plant roots absorb water and nutrients. The plants grow, die, and decompose. As fungi and bacteria decompose dead plants, nutrients are released. The released nutrients become available for living plants. This recycling of essential elements used by plants is called the *nutrient cycle*.

The nutrient cycle is particularly important in many tropical rainforests. Rainforest soils tend to be thin, old, and infertile. Many essential plant nutrients are tied up in the plants.

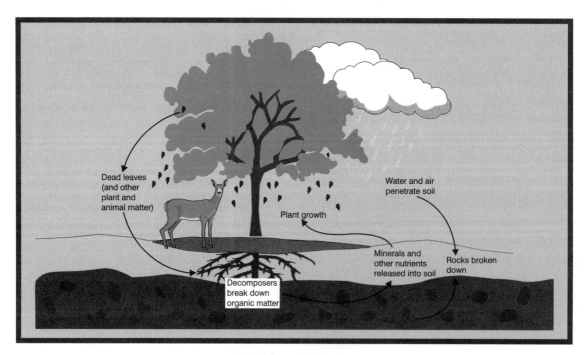

9-2. The nutrient cycle.

People obtain a plot of rainforest land and burn the rainforest to release the limited nutrients for their crops. For a few years, yields are good. In a short period, however, they must move to another section of the rainforest as their soil becomes depleted of nutrients. The nutrient cycle is broken. The abandoned land is easily eroded, and regeneration of natural vegetation is slow. This "slash and burn" destruction of the rainforests has caused great concern worldwide.

ELEMENTS ARE HELD IN THE SOIL

A great quality of soil is that it serves as a reservoir for plant nutrients. As plant roots grow through the soil, they can tap into the nutrients. Essential elements for plant growth are held in the soil in four ways.

The most available nutrients to plants are those held in solution. These nutrients are dissolved in the soil water. The plants can absorb these nutrients quickly. Of course, in a well-drained, sandy soil these nutrients will leach out of the soil as the water percolates downward.

The mineral matter that makes up soil is a source of nutrients. The weathering of these particles releases elements used by plants. Newer soils tend to have a high level of fertility. Older soils have fewer nutrients to release and are less fertile. Weathering and leaching over a long period depletes a soil of many nutrients.

Organic matter in the soil supplies nutrients. As the organic matter decomposes, nutrients are released. Organic matter supplies many of the negatively charged ions used by plants. **Ions** are electrically charged atoms. They may carry a negative charge or a positive charge. The term for negatively charged ions is **anion**.

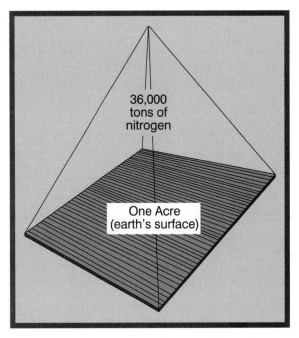

9-3. The air above each acre of the earth's surface contains 36,000 tons of atmospheric nitrogen! Nitrogen-fixing plants can convert the nitrogen into forms useful in plant growth. (Atmospheric nitrogen must be combined with other elements to form fertilizer.)

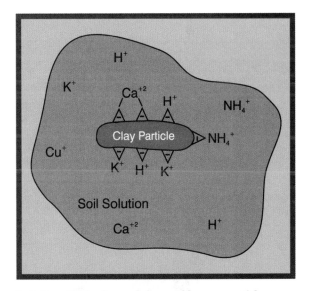

9-4. Negatively charged clay and humus particles attract and hold positively charged cations.

Many plant nutrients are held to negatively charged clay and humus particles. The positively charged ions, called **cations**, are attracted to the negatively charged soil particles. Although they are held tightly to the soil particles, these nutrients are considered available for plant use.

Cation Exchange Capacity

Soil particles determine the availability of nutrients in a soil. Clay and humus particles in the soil are negatively charged. These particles attract positively charged cations. This same principle of attraction applies to magnets. The negative pole of a magnet is attracted to the positive pole. The negative pole also repels another negative pole. Cations, such as potassium (K^+), calcium (Ca^{+2}), magnesium (Mg^+), copper (Cu^+), iron (Fe^{+2} or Fe^{+3}), manganese (Mn^{+2}), and zinc (Zn^{+2}), are attracted to these negatively charged sites on soil particles. This bonding of cations to the soil particle is called **adsorption**.

The cations can leave the soil particle and be replaced by a cation held in soil solution. For instance, a potassium atom may leave the soil particle and be replaced by a copper atom dissolved in the soil water. This replacement of one cation for another is called cation exchange. The total measure of exchangeable cations that a soil can hold is the **cation exchange capacity (CEC)**. The type of clay, the amount of clay, and the amount of humus determine the number of cations held within a soil. The fertility of a soil is directly related to the number of cations a soil can attract and hold. The greater the cation exchange capacity, the greater the fertility of the soil.

CAREER PROFILE

SOIL TESTING TECHNICIAN

Soil testing technicians carefully analyze soil samples using sophisticated laboratory equipment. The findings are used to determine the nutrients needed for a crop. Accuracy is extremely important. Errors in testing can cause false reports. Using false reports results in the use of the wrong fertilizer.

Soil testing technicians need training in chemical laboratory procedures. They usually have at least baccalaureate degrees in soils, chemistry, or related area. Advancement is often based on having a masters, or doctoral degree. The technicians must also get training in new equipment and procedures that are developed.

This photograph shows several samples being tested at a soil testing laboratory. (Courtesy, Mississippi State University)

SOIL pH

Soil pH is the measure of acidity or alkalinity of the soil. A fourteen-point scale is used to measure pH. A neutral pH is 7.0. Any reading between zero and 7.0 is acid. A pH of 1.0 indicates a very acid solution, and 6.0 is considered slightly acid. A solution between 7.0 and 14.0 is said to be alkaline or basic. A reading of 8.0 is slightly basic and a reading of 14.0 indicates a strong base. The pH is determined by the concentration of hydrogen (H^+) ions and hydroxyl ions (OH^-) in the soil solution. A sample of pure water has an equal number of H^+ and OH^- ions and is therefore neutral.

Soil pH plays a large role in the availability of nutrients in the soil. Most essential elements for plant growth are available to plants when the soil pH is

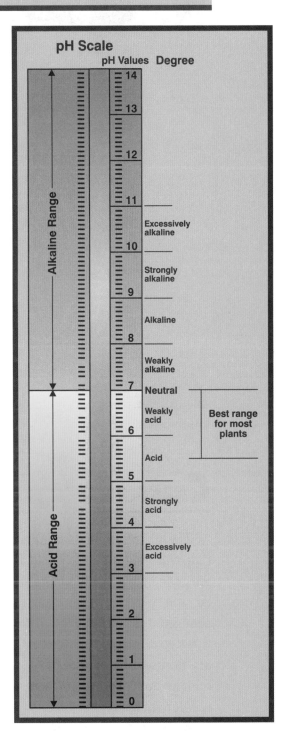

9-5. The pH scale measures acidity and alkalinity of a soil.

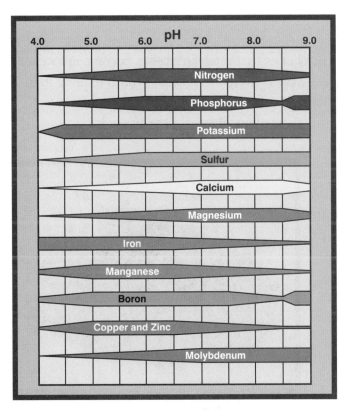

9-6. The pH range in which selected nutrients are most available for plants.

Table 9-1. Soil pH Preferences for Selected Plants

Plants in the Neutral Range	Preferred pH Range	Plants in the Neutral Range	Preferred pH Range
Broccoli	6.0–7.0	Tomatoes	6.0–7.0
Cabbage	6.0–7.0	Wheat	6.0–7.0
Carrot tops	6.0–7.0	Bluegrass	6.0–8.0
Celery	6.0–7.0	Cucumber	6.0–8.0
Corn	6.0–7.0	Parsnip	6.0–8.0
Lettuce	6.0–7.0	Pea	6.0–8.0
Oats	6.0–7.0	Radish	6.0–8.0
Onions	6.0–7.0	Squash	6.0–8.0
Rye	6.0–7.0	Turnip	6.0–8.0
Soybeans	6.0–7.0		

Plants in the Slightly Acid Range	Preferred pH Range	Plants in the Acid Range	Preferred pH Range
Carrot	5.5–6.5	Bunchberry	4.0–5.0
Lima Beans	5.5–6.5	Hartford fern	4.0–5.0
Potatoes	4.8–6.5	Mountain ash	4.0–5.0
Strawberries	5.0–6.0	Venus Flytrap	4.0–5.0
		White cedar	4.0–5.0

between 5.5 and 7.0. When the pH is above or below that range, nutrients become unavailable to some plants. For example, the pin oak prefers a slightly acid soil. If it is planted in a soil that is alkaline, it displays nutrient deficiencies. The nutrients may exist in large quantity in the soil, but the high pH makes the nutrients unavailable for the pin oak.

CHANGING SOIL pH

The pH often limits the types of plants that will grow well in a soil. It is best to select and grow plants adapted to a particular soil pH. However, soil pH can be changed if necessary. First, the pH of a soil is tested to determine how much the soil pH needs to be adjusted. To do this, soil samples are collected and sent to a laboratory for analysis.

Acid soils can be adjusted by adding alkaline materials to the soil. Lime is com-

9-7. A common form of agricultural lime is of pea-size or smaller particles.

monly used to raise the pH of a soil. Two forms of lime frequently used are ground limestone ($CaCO_3$) and dolomitic limestone ($CaCO_3$ and $MgCO_3$).

Soils with an alkaline pH can be altered by adding sulfur. Aluminum sulfate ($Al_2[SO_4]_3$) and iron sulfate ($Fe_2[SO_4]_3$) are two products often used to lower pH. The amount of lime or sulfur material to apply to a soil depends on the type of soil and the extent to which the pH needs to be changed. Generally, the material is spread on top of the soil and then mixed into the soil.

NUTRIENT DIAGNOSTIC PROCEDURES

Before fertilizers or other soil amendments are applied, the need for nutrients should be carefully determined. If not, money will be wasted in buying and applying the nutrients and plant production may be harmed by using the wrong nutrients.

Two approaches are widely used in determining the nutrients that are needed: soil analysis and plant tissue testing. Other approaches include observing a field of crops for signs of deficiencies and assessing yields as a crop is harvested. Applying nutrients after deficiency signs show is often too late—the crop yield has already been reduced. Making an assessment at harvest time is useful only for the next crop year.

SOIL ANALYSIS

Soil analysis involves making chemical tests to determine the nutrients present as well as deficient in the soil. These analyses are often known as soil tests. A **soil test** by a profes-

CONNECTION

COLLECTING A SOIL SAMPLE

A soil sample must be representative of the field or turf that is being tested. The sample must also be kept free of impurities that would give inaccurate results. Special containers are provided for sending samples to laboratories, such as the University of Georgia.

This photograph shows a sample being taken from a lawn area. The sample is collected about 4 inches below the surface. Small amounts of soil from several locations are taken to assure that the sample is representative of the lawn. A clean hand shovel is used. The shovel is cleaned between each sample taken.

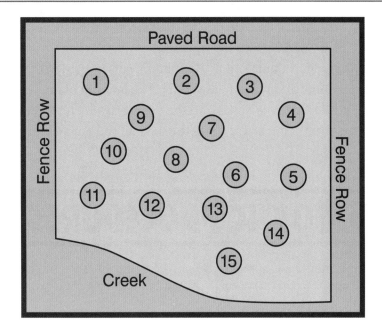

9-8. A large field of similar soil type and typography should have soil from several sites in the field used to make up the sample.

sional laboratory is used to indicate the level of soil fertility. How much and what type of fertilizer to apply depends on the soil test results, the type of soil, and the crop being grown. It is important that the laboratory be told the intended crops to be grown.

The first step in soil testing is to collect a soil sample. With many crops, the samples are taken in the winter or before land preparation begins. One or two collection sites may be sufficient for small areas. Large areas of different soil types and typography are divided into smaller areas based on variations in soil conditions. One sample should generally represent no more than 20 to 40 acres.

The depth of a sample should be representative of the area where most of the roots will grow. Turf samples are 3 to 6 inches deep. Row crop samples are taken 6 to 12 inches deep. Use a soil auger, tube, or spade.

Special care is needed in sampling a larger area with similar soil type and typography. With these areas, small amounts of soil are taken from a number of places in a field. The collected soil is mixed together in a clean bucket. The mixing is to provide one sample representative of an entire field. The soil to be tested is often placed in a special container for delivery to the laboratory. The sample may be sent to a qualified soil analysis laboratory. Some people use soil test kits and make determinations in the field. Handheld computers and global positioning systems may be used to identify precise locations in a large field where the samples are taken.

Be sure to keep careful records on each sample. The identity of the sample should be written on the container. This is needed to assure that you can match the report from the soil lab with the location of the area sampled.

9-9. Soil analysis uses precise methods of determining nutrient content and deficiencies in a soil sample. (Courtesy, University of Georgia)

The soil laboratory will analyze the soil sample for nutrient content. Based on the crop to be grown, the laboratory will prepare a recommendation on the amounts of nutrients to add to the soil that will provide an optimum level of fertility.

TISSUE TESTING

Plant and plant tissue analysis involve using plants or plant parts. Laboratory analysis is made on the plant material to determine deficiencies, if any, of nutrients. The plant materials must be sent to a laboratory that is equipped to conduct such analysis. The general procedure is to "ash" (heat and drive out all moisture) plant material. The ash is placed in solutions that are chemically analyzed.

Timing is important in plant tissue testing. By the time plant tissue is analyzed and the results are obtained, deficiencies may have caused substantial reductions in production. Adding fertilizers may result in only a partial benefit in terms of yield.

FERTILIZERS

Agricultural crops use nutrients held in the soil. As the crops are harvested and removed from the land, nutrients are removed with the plant tissues. Cultivation opens the door for erosion and the loss of fertile topsoil. Also, natural flooding that might renew soils is often prevented. To maintain high yields, nutrients must be added to the soil. The nutrients are

9-10. A truck equipped with a spreader is applying chicken litter to a frozen field. (Courtesy, Potash and Phosphate Institute)

added as fertilizers. Fertilizers are organic or inorganic materials applied to soils or water, which provide nutrients that increase plant growth, yield, and nutritional quality.

ORGANIC FERTILIZER

For thousands of years, agriculture producers have fertilized their crops with organic fertilizers. **Organic fertilizers** originate as plant or animal tissue. Long ago in Egypt, animal wastes or manure were spread on the fields. This practice continues throughout the world today. Many Americans are also familiar with the report of how the Native American Indians helped the Pilgrims with their first crops. The Indians showed the Pilgrims how a dead fish planted with corn would fertilize the plant.

Organic fertilizers have certain characteristics. Organic fertilizers release nutrients slowly as the material decays. Nutrients released slowly are not likely to cause root damage from a high concentration of nutrient salts. Another benefit is the organic fertilizer contributes to the organic matter content in the soil. Some disadvantages of organic fertilizers are: the amounts of macronutrients in the fertilizer are relatively low, the material is bulky, and the exact amount of fertilizer applied is difficult to measure. Some commercially produced

Table 9-2. Nutrient Content of Selected Organic Fertilizers

Organic Material	Nitrogen Average Percent N	Phosphorus Average Percent P_2O_5	Potassium Average Percent K_2O
Bat guano	10.0	4.5	2.0
Bone meal	4.0	23.0	0.0
Cattle manure (fresh)	0.55	0.15	0.45
Cotton seed meal	6.0	2.5	1.5
Dried blood	13.0	1.5	0.8
Sewage sludge	3.0	2.5	0.4
Soybean meal	7.0	1.2	1.5
Wood ash	0.0	2.0	6.0

organic fertilizers include bone meal, cotton seed meal, dried blood, dried manure, sewage sludge, and soybean meal.

INORGANIC FERTILIZERS

The introduction of inorganic fertilizers in the middle 1900s greatly improved crop yields. Inorganic fertilizers are those fertilizers from a nonliving source. Large fertilizer manufacturing plants use natural minerals as well as processed materials in producing fertilizer.

Inorganic fertilizer is manufactured in dry, liquid, or gaseous forms. The dry fertilizer is made as granules or pellets. Nutrients are dissolved in water before application to make liquid fertilizer. Anhydrous ammonia (82 percent N) is a gaseous form used with some crops.

Inorganic fertilizers differ from organic fertilizers. Inorganic fertilizers have a relatively high level of plant nutrients. The nutrients are immediately available for plant use following application. It is also easy to measure the exact amounts of specific nutrients that need to be applied. Disadvantages are chemical fertilizers are more costly and errors in application can damage crops and the environment.

9-11. Ammonium nitrate particles are known as prills and are somewhat smaller than BBs.

FERTILIZER ANALYSIS

The nutrient content of a fertilizer is important if the producer is going to try to apply the recommended amounts. Different fertilizers have different nutrients. The percentage of nutrients in a fertilizer is called **fertilizer analysis**. The percentage of the three macronutrients is always listed on the fertilizer label in the same order. They appear as nitrogen, phosphorus, and potassium. The other nutrients the fertilizer might contain are also listed in terms of percent content.

For example, a 100 hundred-pound bag of 16-17-18 fertilizer contains 16 percent nitrogen or 16 pounds, 17 percent phosphorus in the form of P_2O_5 or 17 pounds, and 18 percent potassium in the form of K_2O or 18 pounds. Thus, 51 percent of the 100 hundred pounds is nutrients. The other 49 percent are inert materials that carry the nutrients. Micronutrients

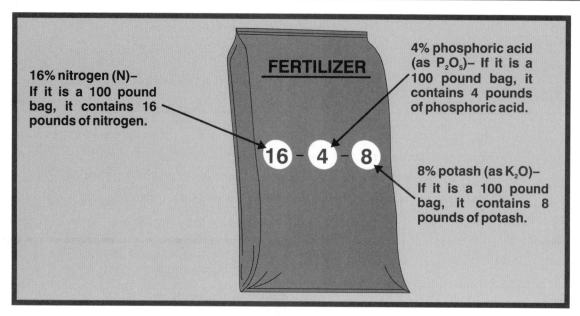

9-12. Analysis of a commercial fertilizer is printed on the label.

may or may not be included in a fertilizer. A fertilizer with micronutrients might contain .07 percent copper. One hundred pounds of this fertilizer would contain .07 pounds of copper.

To determine the actual phosphorus being applied, multiply the percent P_2O_5 by .44. To determine the actual potassium being applied, multiply the percent K_2O by .83.

9-13. This machinery is being used to apply fertilizer to a field. (Courtesy, Top Soil Testing Service Company)

*F*ERTILIZER *A*PPLICATION

Fertilizers can be applied in many ways. With each method, the goal is to get the fertilizers placed where the roots of the plants can absorb the nutrients. Fertilizers may be applied before the crops are planted or after the plants are established.

Fertilizers applied before the crop is planted can be worked into the soil where the roots will be growing. The fertilizer is **broadcast**, or spread evenly, over the ground. Then, the soil is worked. In the

Table 9-3. Nutrient Content of Selected Inorganic Fertilizers

Inorganic Material	Nitrogen Average Percent N	Phosphorus Average Percent P_2O_5	Potassium Average Percent K_2O
Ammonium nitrate	33.5	0	0
Amonium sulfate	21	0	0
Anhydrous ammonia	82	0	0
Sodium nitrate	16	0	0
Urea	45	0	0
Superphosphate	0	20	0.2
Ammonium phosphate	11	48	0
Muriate of potash	0	0	60
Potassium sulfate	0	0	50
Potassium nitrate	13	0	44

home garden, this might be done with a spade or a rotary tiller. In the field, the soil might be worked after application.

Fertilizers are sometimes applied at planting time. The fertilizer is placed in the soil in narrow bands 2 to 6 inches from the seed. This is called *band application*.

After plants have begun to grow, fertilizers are applied on the soil surface. Working the soil at this time would damage the plant roots. These nutrients must become soluble in the soil water and move down into the root zone. The method of application in which bands of fertilizers are applied along side existing crops is *side-dressing*. If a fertilizer is broadcast over an existing crop, it is *top-dressing*.

Procedures with ornamental and greenhouse crops vary from those used to apply fertilizers in large fields. Slow-release forms may be used to assure long-term availability of nutrients. Application methods may be used that target individual plants. (More details are given in later chapters.)

SITE-SPECIFIC AGRICULTURE

Space age technology used in site-specific crop and soil management is rapidly changing agricultural practices. Site-specific agriculture or precision farming gives the producer greater control with crop management. In terms of fertilizer, producers have always known that there were good areas and bad areas in a field. Typically, fertilizers have been applied in

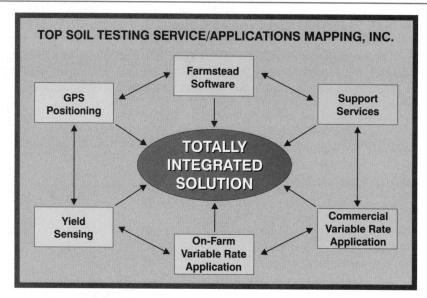

9-14. Site-specific agriculture is a complex system involving GIS and GPS. (Courtesy, Top Soil Testing Service Company)

9-15. This tractor has been equipped with a global positioning system. (Courtesy, Top Soil Testing Service Company)

equal amounts throughout a field, even if one part of the field has high nutrient levels and another part has low nutrient levels.

Site-specific agriculture is now enabling producers to apply exact amounts of fertilizer to different areas within the field. Site-specific agriculture is made possible by global positioning systems (GPS). Twenty-four satellites in orbit of the earth are used in site-specific agriculture. The satellites beam radio waves to GPS receivers mounted on farm machinery. The system pinpoints the machinery to within 30 feet.

In addition to the satellites, site-specific agriculture depends upon geographic information systems called GIS. Geographic information systems are designed to store, display, and analyze digital map data. GIS uses computer equipment and powerful computer software. Agricultural machinery is equipped with devices that can vary the rate of applications of fertilizers, lime, etc.

In site-specific agriculture, a field is divided into small square plots (usually 2.5 acres) forming a grid. Soil samples are taken from each grid throughout the

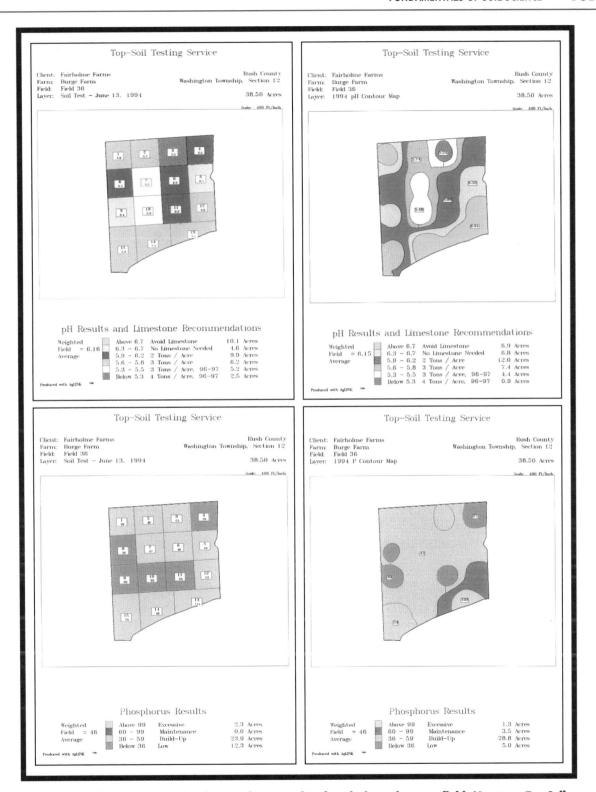

9-16. A series of computer-produced maps shows results of analysis on the same field. (Courtesy, Top Soil Testing Service Company)

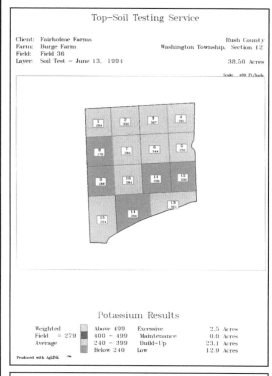

Top–Soil Testing Service

Client: Fairholme Farms   Rush County
Farm: Burge Farm   Washington Township, Section 12
Field: Field 36
Layer: Soil Test – June 13, 1994   38.50 Acres

Scale: 400 Ft./Inch

Potassium Results

Weighted	Above 499	Excessive	2.5 Acres
Field = 279	400 – 499	Maintenance	0.0 Acres
Average	240 – 399	Build–Up	23.1 Acres
	Below 240	Low	12.9 Acres

Produced with AgLINK ™

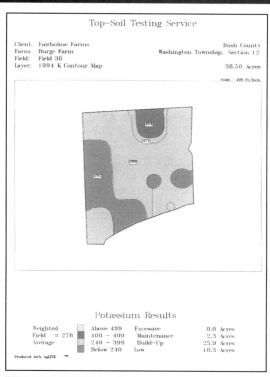

Top–Soil Testing Service

Client: Fairholme Farms   Rush County
Farm: Burge Farm   Washington Township, Section 12
Field: Field 36
Layer: 1994 K Contour Map   38.50 Acres

Scale: 400 Ft./Inch

Potassium Results

Weighted	Above 499	Excessive	0.0 Acres
Field = 278	400 – 499	Maintenance	2.3 Acres
Average	240 – 399	Build–Up	25.9 Acres
	Below 240	Low	10.3 Acres

Produced with AgLINK ™

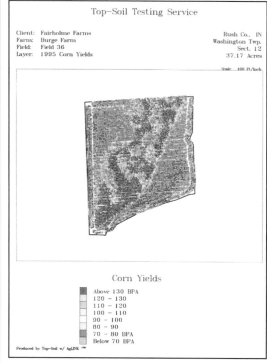

Top–Soil Testing Service

Client: Fairholme Farms   Rush Co., IN
Farm: Burge Farm   Washington Twp.
Field: Field 36   Sect. 12
Layer: 1995 Corn Yields   37.17 Acres

Scale: 400 Ft./Inch

Corn Yields

	Above 130 BPA
	120 – 130
	110 – 120
	100 – 110
	90 – 100
	80 – 90
	70 – 80 BPA
	Below 70 BPA

Produced by Top–Soil w/ AgLINK ™

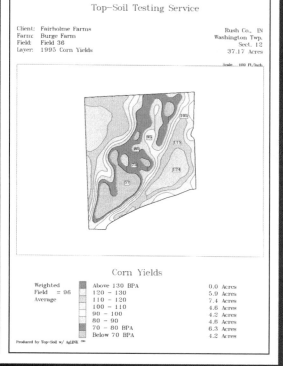

Top–Soil Testing Service

Client: Fairholme Farms   Rush Co., IN
Farm: Burge Farm   Washington Twp.
Field: Field 36   Sect. 12
Layer: 1995 Corn Yields   37.17 Acres

Scale: 400 Ft./Inch

Corn Yields

Weighted	Above 130 BPA	0.0 Acres
Field = 96	120 – 130	5.9 Acres
Average	110 – 120	7.4 Acres
	100 – 110	4.6 Acres
	90 – 100	4.2 Acres
	80 – 90	4.6 Acres
	70 – 80 BPA	6.3 Acres
	Below 70 BPA	4.2 Acres

Produced by Top–Soil w/ AgLINK ™

9-16. (Continued)

field. This sampling is done next to the equipment that has been precisely located in the field by the global satellite system. Three or more satellites are used in locating exact points in the field, a technique known as triangulation.

The soil samples are analyzed in a laboratory for nutrient levels and pH. Computers are used to produce a map of the field based on the fertility level for each grid soil sample. The map may be in a grid format or a contour format. The contour map gives a view of how the fertility and pH levels flow with soil types and the lay of the land. Recommendations for limestone and fertilizers can be made based on the data collected and the type of soil.

Fertilizer application rates are adjusted as the tractor moves across the field, a practice known as variable-rate technology. The computers store information as to how much fertilizer is required for each grid. The global positioning system determines the exact position of the tractor on the field. Then, variable-rate controllers on application equipment adjust the amount of fertilizer being applied based on the position of the machinery and the results of the soil tests.

The global positioning system is also in use during harvest. The practice of monitoring yields as crops are harvested is known as yield sensing. As crops are harvested, yields are recorded. The yield is determined throughout the field. The producer can look for direct relations with the fertility program and yields by studying the fertility and yield maps. On average, soil testing for site-specific agriculture is repeated every three to four years.

There are advantages to site-specific agriculture. Production costs are cut because less fertilizer is used and the environment is protected from the excess use of fertilizers that can cause pollution. Site-specific agriculture is beneficial in that producers gain knowledge about their fields. Also, site-specific agriculture technology allows a producer to monitor yields as the crops are harvested. With so many advantages, it is easy to see why site-specific agriculture is a rapidly expanding practice used to manage crops and soil.

REVIEWING

MAIN IDEAS

Soil fertility is the ability of soil to hold nutrients and the degree to which those nutrients are available to plants. The nutrient cycle explains how nutrients are absorbed and used by plants and returned to the soil through the decomposition of dead plants.

Nutrients are available in the soil in four ways. Nutrients are held in soil solution. Weathering of mineral matter releases nutrients. Organic matter holds negatively charged anions. Clay and humus particles attract and hold positively charged cations.

The ability of a soil to attract and hold cations is referred to as the cation exchange capacity. The greater the capacity to hold cations, the more fertile a soil may be.

Soil pH is the measure of acidity or alkalinity of a soil. The pH scale has 14 points. A soil is acidic if the pH measures 1 to 7, neutral if it is 7.0, and alkaline or basic if it measures 7 to 14. Most nutrients are available for plant use when the pH is in the range of 5.5 to 7.0. Soils can be made more alkaline by adding lime and more acidic by adding sulfur.

Nutrient levels can be raised in soils to improve fertility by adding fertilizers. Organic fertilizers originated as a plant or animal. They are bulky, have relatively low nutrient content, add organic matter to the soil, and release the nutrients slowly as the material decays. Inorganic fertilizers are from nonliving material. Inorganic fertilizers have higher nutrient content than organic fertilizers. It is easier to calculate the amount of inorganic fertilizer to apply.

Before applying a fertilizer, the soil is tested in a professional laboratory to determine what nutrients are needed. Site-specific agriculture uses computers and satellites to apply exact amounts of fertilizers on a field. Fertilizers can be broadcast over the soil and worked in, applied when seeds are planted, used to side-dress existing plants, or top-dressed on soil around existing plants.

QUESTIONS

Answer the following questions using complete sentences and correct spelling.

1. What is soil fertility?

2. How does the nutrient cycle work?

3. Why is the cation exchange capacity important?

4. What is pH?

5. How can soil pH be changed?

6. What are some characteristics of organic fertilizers?

7. What are some characteristics of inorganic fertilizers?

8. What is fertilizer analysis?

9. How are fertilizers applied?

10. What is site-specific agriculture?

EVALUATING

Match the term with the correct definition. Write the letter by the term in the blank that is provided.

a. cation exchange capacity
b. adsorption
c. soil test
d. broadcast

e. fertilizer
f. inorganic fertilizer
g. organic fertilizer
h. ions

i. soil pH
j. fertilizer analysis

1. _____ laboratory analysis of soil to determine nutrient content

2. _____ electrically charged atoms

3. _____ to spread a fertilizer over the surface of the ground

4. _____ percentage of nutrients in a fertilizer

5. _____ fertilizer originating as a plant or animal

6. _____ ability of a soil to hold cations

7. _____ fertilizer obtained from mineral matter

8. _____ bonding of cation to a soil particle

9. _____ organic or inorganic materials added to soil to increase nutrient content

10. _____ measure of acidity or alkalinity in a soil

EXPLORING

1. Take two cuttings of a houseplant, such as a philodendron or a coleus. Place one in a glass of distilled water. Place the other in a glass of tap water. Set these near a window. As the cuttings develop roots and begin to grow, observe any differences in the plants. When water evaporates from the glass of distilled water, add only distilled water. To the other, add a small amount of soluble fertilizer.

2. Practice using a handheld global positioning system. Go into a field or area around the school. Make readings at different locations. Plot your readings on paper.

3. Job shadow a soil technician or other person in collecting soil samples for analysis. Write a report on your experiences.

CHAPTER 10

Irrigation

OBJECTIVES

This chapter introduces the importance of irrigation. Sources of water and methods of application are also included. It has the following objectives:

1 Explain soil and plant relationships in moisture management.

2 Describe the benefits and use of irrigation.

3 List and evaluate sources of water.

4 Explain methods of water application.

5 Describe how to make efficient use of water.

6 Explain the meaning and use of chemigation.

TERMS

acre-foot
available soil moisture
border irrigation
chemigation
emitter
flood irrigation

furrow irrigation
ground water
irrigation scheduling
precipitation
root zone
soil moisture

soil moisture management
soil moisture tension
sprinkler irrigation
surface water
trickle irrigation
wilting point

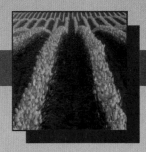

10-1. Sprinklers are being used to water large numbers of newly set chrysanthemums at this horticulture facility.

PLANTS need water to grow. When they do not have enough water, yields drop and they may die. In one form or another, irrigation has been used for thousands of years. In some countries, more water is used for irrigation than any other purpose. About 40 percent of the freshwater in the United States is used for irrigation.

Just how much water do plants need? Scientists say up to 500 pounds (227 kg) of water may be needed for a plant to produce one pound (0.45 kg) of dry matter. An acre of corn requires about 500,000 gallons (nearly 2 million liters) of water. Of course, water requirements vary with the corn plant population—more stalks need more water!

Plants require more water on hot, dry days. Wind also increases the need for water. Transpiration rates go up with wind and hot weather. To keep a plant growing, the moisture must be available in the soil for absorption by roots.

SOIL MOISTURE MANAGEMENT

Water in the soil is absorbed by the roots of plants. The water is needed for the plant to grow and produce a crop. Plant stems, leaves, fruit, and roots may be 90 percent water. Plants use a lot of water in transpiration—the loss of water through the stomata in the plant leaves. Transpiration serves to regulate plant temperature much the way that people perspire when they are warm.

10-2. Crops, such as alfalfa, would not grow in the Arizona desert without irrigation.

10-3. Leaves on this sunflower have wilted due to warm weather and low soil moisture.

Moisture is usually provided by precipitation or irrigation. **Precipitation** is the natural way water is deposited on the earth. It includes rain, snow, and other lesser forms. Sometimes precipitation is not adequate. Irrigation is used to artificially maintain a level of soil moisture for normal plant growth and production. Irrigation is used when the soil moisture level drops below what the plant needs.

Some moisture is lost as runoff, by evaporation, and by deep percolation in the soil. Good soil moisture management can help reduce losses and provide more water for plants. **Soil moisture management** is following practices that make good use of the natural and artificial sources of water for crops. Understanding moisture concepts is needed for good soil moisture management.

SOIL WATER

Soil moisture is the moisture found in soil. Soil particles hold moisture for use by plants. Some soil

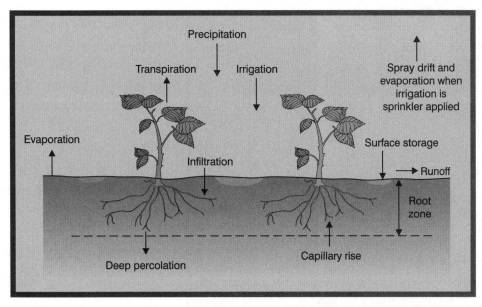

10-4. Sources and losses of soil moisture are shown here. Losses occur when moisture leaves the root zone by evaporation, transpiration, and deep percolation. Sources of water include precipitation, irrigation, and capillary rise.

moisture is available to plants; other soil moisture is not available. Availability depends on how the moisture is held in the soil.

Moisture is held in the soil in three ways: gravitational, capillary, and hygroscopic. Water that moves downward through the soil is known as gravitational water. It may help replenish groundwater supplies. Capillary water is held in the pore spaces between the tiny soil particles. Gravitational and capillary waters are the most important sources of soil water for plant

CAREER PROFILE

IRRIGATION ENGINEER

An irrigation engineer designs irrigation systems. The work may involve researching systems to determine which are best as well as installing new systems. In some cases, they work with existing systems to solve problems and improve the efficiency of the system.

Irrigation engineers typically have baccalaureate degrees in agricultural engineering, agronomy, soil and water management, or a related area. Those in higher-level positions have masters and doctors degrees in related areas. Practical experience working with irrigation, water, and crops is essential.

This shows an irrigation engineer working with small research rice plots.(Courtesy, Agricultural Research Service, USDA)

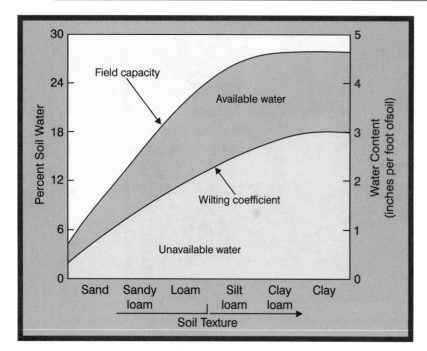

10-5. The amounts of available and unavailable water increase as the clay content of soil increases.

growth. Hygroscopic water is soil moisture that tightly clings to the soil particles. This moisture is not available to plants. If the moisture level gets too low in the soil, plants cannot get the needed moisture. They will wilt and stop growing.

Available soil moisture is the water in the soil that can be used by plants. When moisture levels are high, plants can easily extract the moisture from the soil. As the moisture is used, soil moisture tension increases. ***Soil moisture tension*** is the force by which soil particles hold on to moisture. As moisture in the soil decreases, soil moisture tension increases. The soil con-

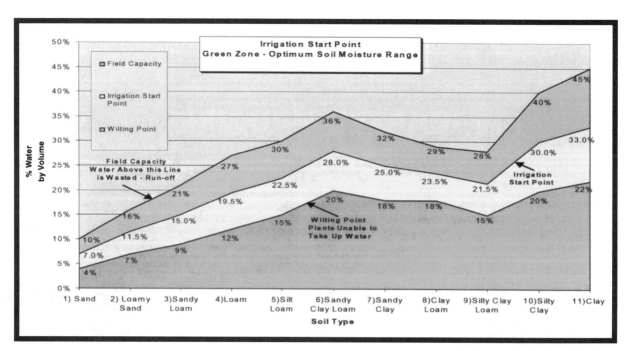

10-6. Soil moisture content as related to soil type and efficient use of irrigation water. (Courtesy, Dynamax, Inc., Houston)

tains less available soil moisture. Plants cannot get the water they need from the soil when the moisture supply is low.

Soil texture influences water availability. Sandy soils hold less water and result in plants wilting quicker than in loam and clay soils. Water management must consider the soil texture. Irrigation water is less efficiently used in sandy soils. Soils high in clay hold water and keep it from percolating out of the root zone.

ROOT ZONE

The **root zone** is the part of the soil where plant roots are found. Some plants have deep root systems and can get water and nutrients from a larger area. Other plants have shallow root systems and depend on the soil in a small area for nutrients. The root zone is the most important location for moisture. Moisture out of the root zone is not available to plants.

The depths of root zones for selected crops are: alfalfa, 4 to 6 feet; corn, 2.5 to 4 feet; cotton, 3 to 4 feet; turf, 1 to 2.5 feet; potatoes, 2 to 3 feet; wheat, 3 to 4 feet; and vegetable crops, 2 to 4 feet.

Soil moisture management is based on the moisture in the root zone. More moisture is extracted by plants from the upper part of the root zone. The upper one-fourth of a root zone provides 40 percent of the water for a plant. The bottom one-fourth of a root zone provides only 10 percent of the water for a plant.

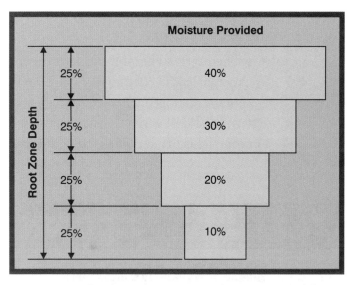

10-7. Sources of moisture in a root zone. (Forty percent of the moisture for plants is from the upper 25 percent of the root zone.)

SOIL MOISTURE BALANCE

Soil moisture management involves trying to maintain high levels of available soil moisture. Water used by plants and lost by evaporation must be restored to the soil by rainfall or irrigation. Maintaining a good soil moisture balance requires knowing the water needs of crops and the levels of moisture in the soil. Crops at different stages of maturity also require more water. For example, a crop that has just come up does not need as much moisture as one that is flowering and developing fruit.

Typical water needs for selected crops have been calculated on a daily basis. Here are a few examples of inches of moisture needed each day during the peak growing season: alfalfa, 0.26 to 0.36; permanent pasture, 0.23 to 0.32; corn, 0.26 to 0.47; cotton, 0.22 to 0.38; and shallow-rooted vegetables, 0.21 to 0.34. Crops growing in soil that cannot provide this amount of moisture will not grow well. Irrigation is needed to supplement the soil moisture.

WHEN TO APPLY WATER

Knowing when to irrigate is important in soil moisture balance. Do not wait until the plant shows signs of stress, such as wilting. **Wilting point** is the amount of water in the soil when a plant wilts. It occurs more quickly in sandy soils. Irrigation should be used before the wilting point. When plants wilt, damage has already been done to production. Always assess moisture in the soil before applying water.

Ways of determining the need for irrigation without stressing plants include:

- Ribbon test—The ribbon test is determining soil moisture content by "feel." A little experience is needed to use it accurately. Small amounts of soil are rolled between the thumb and fingers. The feel and appearance of the soil are used to assess moisture content. Soil that is slightly sticky and holds together does not need irrigation. Soil that crumbles is dry and needs to be irrigated.

CONNECTION

UNIFORM COVERAGE

Irrigation water needs to be uniformly distributed but varied to meet differences in soil and typography. What happens if a sprinkler system is not designed for adequate coverage? The answer: some areas do not receive the water that is needed.

Look at this photo of a residential lawn sodded with fescue. You will see a distinct T-shaped area in the middle that is brown. On each side of the brown area you will see circular-shaped green areas. What caused the pattern? It is the result of round sprinkler patterns that do not meet. This results in a lawn that is not aesthetically pleasing.

Variations in coverage are due to nozzle placement and water pressure. Careful design is essential. All areas should get the needed water.

Appearance of sandy clay loam, loam, and silt loam soils at various soil moisture conditions.

Available Water Capacity 1.5-2.1 inches/foot

Percent Available: Currently available soil moisture as a percent of available water capacity.

In./ft. Depleted: Inches of water needed to refill a foot of soil to field capacity.

0-25 percent available
2.1-1.1 in./ft. depleted

Dry, soil aggregations break away easily, no staining on fingers, clods crumble with applied pressure. (Not pictured)

50-75 percent available
1.1-0.4 in./ft. depleted

Moist, forms a ball, very light staining on fingers, darkened color, pliable, forms a weak ribbon between the thumb and forefinger.

25-50 percent available
1.6-0.8 in./ft. depleted

Slightly moist, forms a weak ball with rough surfaces, no water staining on fingers, few aggregated soil grains break away.

75-100 percent available
0.5-0.0 in./ft. depleted

Wet, forms a ball with well-defined finger marks, light to heavy soil/water coating on fingers, ribbons between thumb and forefinger.

100 percent available
0.0 in./ft. depleted (field capacity)

Wet, forms a soft ball, free water appears briefly on soil surface after squeezing or shaking, medium to heavy soil/water coating on fingers. (Not pictured)

10-8. An example of the ribbon test is shown here for three soil types. Refer to the USDA Program Aid Number 1619 Publication for other soil types and additional information.

- Moisture sensors—A moisture sensor is an electronic instrument that is used to probe the soil and provide a reading on moisture content. It involves using a single probe.

- Sap flow sensors—A sap flow sensor is a device that measures the movement of water (sap) inside the stem of a plant. The process is used on small herbaceous plants as well as the branches and trunks of trees. The device clamps onto the stem of the plant.

10-9. A tensiometer is used to assess moisture content in soil.

- Tensiometers—A tensiometer is a device that assesses moisture content by determining the "pull" of soil particles in the soil. Tensiometers are permanently placed in the soil and can be damaged by freezing weather.

- Moisture meters—A soil moisture meter is a device that assesses moisture based on the flow of low-level electric current between its two probes.

- Information sources—Weather information and published reports on soil moisture can be used as sources of information. Remote sensing information on soil moisture is increasingly being used. Also, scientific labs may use ovens to dry soil and assess the amount of moisture that is driven out.

USE OF IRRIGATION

Irrigation is an important part of soil moisture management in many locations. As the artificial application of water to promote plant growth, irrigation can have several important benefits:

- Provides water when adequate water would not otherwise be available for plant growth.

- Promotes plant growth by applying fertilizer, growth regulators, and other materials with the water.

- Disposes of waste water by land application.

- Protects plants from extreme cold temperatures, such as preventing frost damage in fruit and vegetable crops.

- Reduces dust from field and other surfaces.

A deficiency of water influences plant growth. Productivity is lost. Susceptibility to diseases and other pests is increased. Overall, plants suffering from water deficiency will not achieve their genetic potential.

Table 10-1. How Water Deficiency Affects Plants

The effects of water deficiency include:

- Poor plant growth
- Stunted mature plants
- Lower crop yields and loss of potential profit
- Death of plants, including rowcrops, trees, ornamental plants, and turf
- Stress, which lowers disease and insect resistance
- Loss of aesthetics caused by dead plants, such as ornamentals, turf, and trees

IRRIGATION AND CLIMATE

Climate and weather are big factors in the approaches taken with moisture management. In general, irrigation is used in three kinds of climates in moisture management:

- Dryland cropping—This involves near total reliance on irrigation for the needed water. It is used in the desert areas where there is little precipitation. Some major production areas in the southwestern United States could not produce a crop without irrigation. Large dams and canals are used to collect and distribute irrigation water.

- Seasonal precipitation—Some areas have rainfall during only a few months of the year. In the other months, no rainfall occurs. Irrigation is used in these areas to assure adequate moisture year round. The moisture from precipitation is lost to evaporation or is not enough for the crop. The fertile Salinas and San Joaquin Valleys of California have seasonal precipitation.

- Dry periods—Some locations where natural moisture is adequate may have short dry spells in the summer. Irrigation is needed to assure sufficient soil moisture. Much of the eastern United States uses irrigation on a limited basis when soil moisture supplies are low. Some years, natural moisture may be adequate and no irrigation will be needed.

10-10. Alternating furrow flood irrigation is used on cotton in Arkansas during a summer dry spell.

IRRIGATION SCHEDULING

Irrigation scheduling is providing the right amount of water at the right time. Water should be available when a plant needs it. The greatest need for water is usually during the middle of the growing season for a crop. This means that water supplies are increased before the peak need.

Good scheduling helps make efficient use of scarce water supplies. Good scheduling prevents over irrigating and supplies water to the crop just before the time of highest demand. Waiting until plants wilt to begin irrigation results in some damage to the plants and loss of production.

Most crops are not watered each day. Water needs are calculated and the amount of water is applied to meet the need for a specific period, such as two weeks. For example, if corn is using 0.25 inches (0.64 cm) of water a day, 3.5 inches (8.9 cm) may be applied to provide water for two weeks.

A few specialty crops have daily irrigation. This often depends on the irrigation system used. For example, grapes or ornamental plants continuously receive water through drip irrigation.

10-11. Poinsettias in a greenhouse are continuously provided small amounts of water through a system of "spaghetti" tubes and emitters.

Scheduling the use of water may involve gaining an allocation and time with the local water management district.

QUALITY IRRIGATION WATER

Water quality is important when irrigating crops. Chemical composition is the major factor in the quality of water. All irrigation water contains some chemical compounds that have dissolved from rocks and some minerals that the water has come in contact with. Water with the lowest amount of damaging chemicals should be used.

SOLUBLE SALT PROBLEMS

Most of the minerals in irrigation water are soluble salts. As such, they are charged particles known as ions. Using irrigation water high in salts results in the accumulation of the salts in the soil. Plants remove water from the soil, but not salts. Water evaporates, leaving the salts behind. The accumulation of salts also depends on the amount of irrigation water applied. It is the salt content (salinity) of the soil water, rather than of the irrigation water, which affects crop production. In areas with winter rains, the rainfall will leach some salt from the soil.

The common chemical elements that may form salts in water are:

- Calcium—Water gets calcium from limestone and other rocks with which it is in contact. Plants need some calcium to grow. Soils benefit from the calcium in irrigation water.

- Magnesium—Magnesium is similar to calcium and is generally abundant in the soil. It does not damage crops.

- Sodium—Sodium is a major source of salt in soil from irrigation water. It greatly affects the uptake of water by plants. Sodium concentrations may result in spots in fields where no plants will grow. As the major constituent of table salt, sodium is most damaging to soil. Water high in sodium should not be used. If soil is contaminated with sodium, some improvement can be made by adding gypsum (a form of calcium) to the soil and applying a substantial amount of water. The water leaches (washes) the salt away.

- Chloride—Chloride, also a constituent of table salt, damages plants. It is toxic to some plants. Avoid using irrigation water with high chloride content.

- Sulphate—Though plants need sulfur for growth, its action can increase the overall soil salinity. Irrigation water with low sulphate content is preferred.

Salts affect the availability of water to plant roots. The higher the salt content in the water, the more difficult it is for plants to get water from the soil. Chemical analysis of water should be done to determine its suitability for irrigation.

10-12. An irrigation water treatment unit removes mineral ions at a South Carolina greenhouse facility.

OTHER WATER CONTENTS

Irrigation water may contain non-salt minerals, gases, heavy metals, and other pollutants that can cause problems. Some originate with the source of the water and others may get into the water after it leaves its source. However, these are usually not problems.

Boron is a mineral typically found in irrigation water. In high concentrations, boron is toxic to plants. Fortunately, most irrigation water contains only small amounts of boron.

The two gases likely to be found in irrigation water that may cause problems are carbonate and bicarbonate. High amounts of carbonates can reduce the quality of irrigation water. If carbonate combines with sodium, it can form a substance that is very toxic to plants.

Heavy metals get into irrigation water from improper disposal of wastes. Heavy metals are highly poisonous elements usually with an atomic mass over 100. Lead, mercury, and cadmium are the most common. Throwing a used flashlight battery into irrigation water is one step toward pollution. Heavy metals are very poisonous in small amounts. They can accumulate in the liver, brain, kidneys, and other organs of humans, wildlife, and farm animals. The long-term effect of heavy metals on irrigated land may be a major loss of productive farm land.

Any type of pollution in irrigation water can be potentially damaging. Pesticide residues, detergents, and runoff from city streets are potential hazards.

SOURCES OF IRRIGATION WATER

Water for irrigation is from two main sources: surface water and ground water. Other sources are also available depending on the amount of water needed, quality of the water,

10-13. The water line around the rock at Watson Lake near Chino Valley, Arizona, shows a large water draw-down for irrigation. The water will need to be replenished by melting snow from nearby mountains the following spring.

and the ease of using the water. A homeowner with a small lawn area may use water from the city water system to irrigate the turf. This likely would not be economical nor allowed on a large-scale basis. Recycled wastewater may be used for some irrigation.

SURFACE WATER

Surface water is the accumulated water from rainfall, melting snow, springs, and other sources found in lakes, streams, and reservoirs. It is the major source of irrigation water. Large dams across streams in the western United States create huge reservoirs of water for irrigation. Most of the water is from melting snow high in the mountains some distance from the fields that are irrigated.

10-14. This irrigation canal at the Alpaugh Irrigation District is protected with a locked gate.

Systems of canals guide the water from its origin to the fields where it is to be used. Since most farms are below the elevation of the reservoir, the water flows by gravity. The large canals may be known as aqueducts. The smaller canals may be known as laterals or irrigation ditches. Steps are taken to reduce the loss of water by evaporation or seepage. Most canals are made of concrete and have control devices that allow easy use of the water.

The larger reservoirs are often organized into water districts. Each grower in the district gets an amount of water based on water-use history. Irrigation water is measured in acre-feet. An **acre-foot** is the amount needed to cover one acre of

10-15. An irrigation canal flows through Mesa, Arizona.

land with 1 foot of water. Acre-feet may also be known as units, with one unit being one acre-foot. Acre-feet are shared over larger acreages. For example, one acre-foot can provide 2 inches (5.1 cm) of water for six acres of land.

Known as water rights, the allocation of water goes with the land if it is sold. Sometimes, water rights may be more valuable than the land!

GROUND WATER

Ground water is pumped from aquifers in the earth. An aquifer is an underground stream or pool of water. The water is often held in spaces between sand, rocks, and other underground formations. Ground water is not available for widespread use in some locations. Permits are needed to dig wells and use the water in an aquifer.

Getting ground water involves drilling a well into an aquifer. A pump is used to lift the water to the surface. A system of pipes or ditches transports the water to where it is needed.

10-16. Water well with an electric-powered pump on a California farm.

Pumping water is expensive because electric motors or gas engines must be used to operate the pumps. The amount of water from wells may be limited. As the water is pumped out, the level of the aquifer goes down. The pipes from some pumps may no longer extend into the water.

Careful testing is needed for ground water as well as surface water. Ground water is less likely to be polluted, though it may contain some salts and other substances that can build up in the soil.

APPLICATION METHODS

Several methods are used to apply irrigation water. The method selected should economically provide the irrigation water. Overall, the methods vary based on how water is applied to the land or medium.

Three general methods are used: flood, sprinkler, and trickle. Each of these has several variations. In addition, below-ground systems are sometimes used.

10-17. Examples of common water application methods.

Flood Irrigation

Flood irrigation is covering the surface of a field with water and allowing it to soak into the soil. The surface must be smooth and have a slight slope from the point where the water is applied. Fields that are not level will result in water not being uniformly distributed over the area.

If a field has a raised border or dike surrounding it, this approach to flood irrigation is known as basin irrigation. The water is introduced into the basin through an inlet and allowed to spread uniformly across the basin. It is most effective in fields with uniform soils that have been land formed to gain a desired typography. Alfalfa and rice are examples of crops where basin irrigation is used.

Large fields are often divided into smaller fields to assure good water distribution. Known as **border irrigation**, small earthen ridges are made through the field dividing it into strips. The ridges direct the water to the area in the strip. Border irrigation is used with pastures and hay crops where the land has a small slope.

Furrow irrigation is a modification of flood irrigation. It uses the rows on which crops are planted to divert the water the entire length of the row. Water is applied between the rows and flows through the field soaking into the soil. Sometimes, alternating rows are flooded.

Flood irrigation is often more economical to use. Expensive equipment needed for sprinkler irrigation is not used. Water may be siphoned from an irrigation ditch into the field. The area irrigated is soaked and the irrigation is stopped until it is needed again.

Sprinkler Irrigation

Sprinkler irrigation is applying water through the air in droplet form. Nozzles form the droplets similar to rain. Sprinkler irrigation works well on land that is not level. The water must have sufficient pressure to form the drops and project them out over the field. Pumps are commonly used to create the pressure. Some reservoir systems will provide enough pressure.

Systems

Several kinds of sprinkler systems are used: portable sprinklers, side roll wheel move systems, traveling guns, center-pivot systems, and linear move systems. Large fields may use center-pivot irrigation equipment. This is an automated system on wheels that pivots around the field. Speed of movement can be varied based on water needs. The wheels on the system are powered by electricity or by oil or water systems.

10-18. Installing a sprinkler system for a residential lawn involves trenching, placing pipe, using a backflow prevention device (top right), and positioning underground pop-up nozzles.

Sprinkler systems in smaller fields may be on wheels or stationary. Turf areas may have emitters that are in the ground. These effectively provide water and do not block the movement of mowers and other equipment. Various kinds of nozzles are used to assure good coverage of the area to be irrigated.

10-19. A traveling gun sprinkler.

10-20. A rain gauge can be used to determine the amount of water applied by sprinkler irrigation.

10-21. A drag sock can be used on center pivot and linear move systems to eliminate the need for sprinkling water into the air. This reduces the amount lost to evaporation.

A simple method of determining the amount of water applied with sprinkler irrigation is to use a rain gauge. The gauge is set in the path of the sprinkler. It collects the drops just as it would collect rainfall. Reading the rain gauge tells the amount of water that has been applied.

Efficiency

Methods have been developed to conserve water and energy. LEPA is Low Energy Precision Application. It is not a separate system but makes use of existing center-pivot and linear move systems.

Drop tubes and low-pressure emitters are positioned 8 to 18 inches above the soil surface. Closeness to the soil reduces evaporation in the air. Research has found that 95 to 98 percent of the moisture enters the soil with LEPA.

More efficient use of water is being obtained in some areas by fitting linear move and center-pivot systems with "drag socks." Water is released through the drag sock rather than sprayed into the air. The drag socks are attached to the water application devices. Water is slowly released. Runoff is virtually eliminated.

Low VOLUME IRRIGATION

Low volume irrigation is the application of a lower volume of water over an extended time. Emitters (devices for releasing water) precisely place the water near plants. Evaporation and runoff are reduced. The methods may be used on the surface or below the surface of the soil.

Trickle Irrigation

Trickle irrigation is the application of water to the root zones of plants. A system of tubes is used to deliver the water at very low pressure. Also known as drip irrigation, trickle irrigation makes efficient use of limited water. It is more expensive to install because of the elaborate system of tubes. Smaller plants may have one tube. Larger plants may be irrigated by several tubes.

Trickle irrigation is commonly used with ornamental and high-value crops. Emitters are at the end of individual tubes or spaced along the tube. An **emitter** is a small opening that releases irrigation water—often a drop at a time. Ornamental gardens may have complex systems of tubes for trickle irrigation.

Sub-surface Drip Irrigation

Sub-surface drip irrigation is similar to regular drip irrigation except that the emitters are below ground level. Water is applied in low volumes directly to the root zones. Lines are placed 12 to 18 inches below the soil surface or below normal tillage range. Plows can damage sub-surface tubes. Emitters are made into the tubing.

Micro-Spray Irrigation

Micro-spray irrigation involves spraying very low amounts of water. The emitters produce a mist and work effectively with tree crops produced on sandy soils. The system can also be used to protect tree and vine crops from frost or freeze damage.

10-22. Several emitters have been placed in one container for this tree that is to be transplanted.

10-23. Sub-surface irrigation hose (also known as tape or flexible tubing) is manufactured in rolls. Emitters are spaced along the hose.

10-24. A mini-spray sprinkler head.

MAKING EFFICIENT USE OF WATER

10-25. Control structures should be in good condition on canals and ditches so water is not wasted.

Irrigation water is valuable. It should be used properly and not wasted. Using water more efficiently results in greater production.

Here are several approaches in making more efficient use of irrigation water:

10-26. Grapes are irrigated with an emitter on a hose (tape). With this tape, micro-irrigation emitters are attached only as needed. (Courtesy, Toro Agricultural Irrigation)

- Use sprinkler irrigation during the cooler part of the day and when the wind is not blowing. More water is lost to evaporation in heat and when the wind is blowing. For example, 9 percent of the water is lost if the temperature is 80°F (27°C) and the wind is blowing at 5 miles per hour. The loss increases to 20 percent at the same temperature if the wind is blowing 15 miles per hour. At 100°F (57°C) and a wind of 15 miles per hour, 26 percent of the sprinkler irrigation water is lost.

- Monitor moisture in the root zone. Irrigate enough to have good root zone moisture. Stop adding water when the water has penetrated into the root zone. Adding more water will result in loss by percolation of the water outside the root zone of the plants.

- Keep the irrigation system in good condition. Leaks in pipes, canals, and other water structures should be prevented. Be sure all connections fit properly to avoid leaks.

- Apply water uniformly. All areas of a field need to receive the appropriate water. High places may not be irrigated adequately. Low places may have too much water. Land forming may be needed to efficiently use flood irrigation. Water needs vary within a field and adjustments should be made accordingly in the rate of application.

- Apply only the amount of water that can be used. Avoid over irrigation where excess water runs from the field into nearby creeks. This water is wasted; in addition, it may carry nutrients, sediment, or pesticides from the field as it runs off.

CHEMIGATION

Chemigation is the application of agricultural chemicals mixed with irrigation water. The irrigation water is used to carry fertilizer (known as fertigation), fungicides, herbicides, and insecticides to crops. The chemical is injected into the irrigation water system. It is carefully metered within the system to assure uniform distribution with the irrigation water.

Using proper methods, chemicals can be applied to the foliage of plants in sprinkler irrigation systems. Flood irrigation can be used to apply fertilizer and other chemicals needed in the soil, such as nematocides. Trickle irrigation can also be used to apply fertilizer.

10-27. A center-pivot system using chemigation on cotton in the high-plains of Texas. (The chemical tank is in the foreground.)

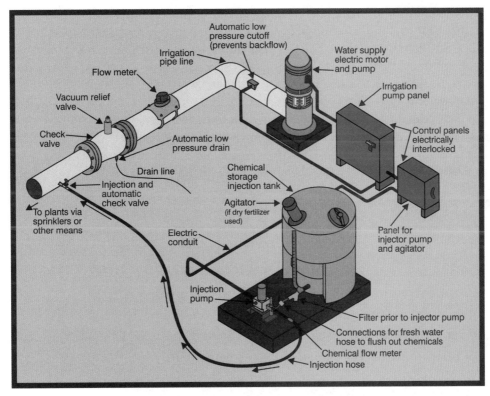

10-28. Diagram of a possible layout for a chemigation system.

The chemical must be soluble in water or one that will stay in suspension. In some cases, chemicals can foul the irrigation system. Algae and bacteria may grow in pipes and emitters used to apply nitrogen fertilizer and clog emitters.

Chemigation requires considerable attention to safety, equipment calibration, and other aspects of application. A sample layout is shown in Figure 10-28. The chemical injection pump must deliver the chemicals into the water within plus or minus 5 percent accuracy. Injection rates are varied with the type of crop and chemical being applied.

The system should have a backflow prevention device to stop contamination of the water source. In case of system failure, water containing chemicals can move backward into the ground or lake if a backflow device is not used.

REVIEWING

MAIN IDEAS

Irrigation is the artificial application of water for plant growth. It increases soil moisture available to plants. Adequate moisture is particularly important in the root zone. Moisture outside the root zone is of no benefit to plants.

Water needs of crops must be met or they will not grow and produce. How irrigation is used varies with the climate. In some areas of the western United States, dryland cropping relies on irrigation as the source of water. In other areas, seasonal precipitation is supplemented with irrigation. Elsewhere, irrigation may be used during short droughts.

Irrigation scheduling is providing the right amount of water at the right time. Water should be in the root zone when plants need it. This means that irrigation is based on anticipated plant needs. Only quality water should be used. Water high in salt or other damaging chemicals should be avoided.

Irrigation water is typically from two major sources: surface water and ground water. Complex systems of dams, canals, and ditches have been built to carry irrigation water.

Water is applied in three major ways: flood, sprinkler, and trickle. Each of these has several variations depending upon the plant, characteristics of the terrain, and the size of the area being irrigated. These systems can also be used to apply other chemicals, such as fertilizer and insecticide, by chemigation.

QUESTIONS

Answer the following questions using complete sentences and correct spelling.

1. What is soil moisture management? Why is it important in plant production?
2. Distinguish between soil moisture and available soil moisture.
3. Why is soil moisture important in the root zone?
4. What are the three kinds of climates in which irrigation is used? Distinguish between them.
5. What is irrigation scheduling? Why is it important?
6. Why is quality irrigation water important?
7. What are the sources of irrigation water?

8. What general methods are used to apply irrigation water? Briefly describe each.
9. What can be done to make efficient use of irrigation water?
10. What is chemigation?

EVALUATING

Match the term with the correct definition. Write the letter by the term in the blank that is provided.

a. root zone
b. soluble salt
c. surface water
d. acre-foot

e. trickle irrigation
f. available soil moisture
g. soil moisture tension
h. irrigation scheduling

i. border irrigation
j. chemigation

1. _____ force by which soil particles hold on to moisture
2. _____ using irrigation water to apply chemicals to crops
3. _____ part of the soil where roots are found
4. _____ providing the right amount of water at the right time
5. _____ ion forms of minerals found in irrigation water
6. _____ accumulated water from precipitation, springs, and other sources found in lakes, streams, and reservoirs
7. _____ amount of water needed to cover one acre of land one foot deep in water
8. _____ dividing larger fields into smaller fields with earthen ridges that direct the flow of irrigation water
9. _____ the slow application of water to the root zones of plants
10. _____ water in the soil that can be used by plants

EXPLORING

1. Take a field trip to a farm, golf course, or other place where irrigation is used. Interview the manager about irrigation scheduling, soil moisture management, and sources of water. Determine the methods used to apply the water. Prepare a report on your observations.

2. Test a local water source to determine suitability for irrigation. Use a water testing kit or meter. In some cases, a sample will need to be collected and sent to a testing laboratory.

3. Visit a farm equipment show. Collect brochures about irrigation equipment. Interview a representative of an irrigation equipment company about the kinds of equipment, cost, and other details. Prepare a written report on your visit.

4. Use the Internet to contact the International Irrigation Center (www.ee.usu.edu /iichome0.html). This source at Utah State University has information on soil water balance, assured rainfall, and irrigation requirements.

Hydroponics

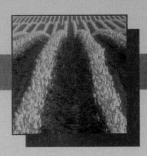

11-1. Lettuce is being grown without soil in this hydroponics facility.

HAVE you ever tried to get a potato or an avocado to sprout in a glass of water? Maybe you have taken a stem cutting from a houseplant or a leaf from an African violet and gotten it to grow in water. If you have done one of these, you have been practicing a simple form of hydroponic culture.

Practical application of hydroponics is on the increase. Some companies have constructed many acres of hydroponic systems operating within greenhouse structures. Hydroponically grown vegetables are produced and supplied to supermarkets. Also, extensive research is taking place to use hydroponics in the space program. It is likely that hydroponic systems will be used on space stations in the future to grow fresh foods for the astronauts.

HYDROPONICS AND ITS ADVANTAGES

11-2. Hydroponics means growing plants without soil.

What is hydroponics and why does hydroponics stir the imagination? The term hydroponics was coined in the 1930s in California by W. F. Gerike. It is a combination of two Greek words. "Hydro" means water and "ponics" means labor. Together, they mean water labor.

Simply defined, **hydroponics** is growing plants with their roots in a medium other than soil. Sometimes, hydroponics is referred to as **soilless culture** because soil is not used. Nutrients essential for plant growth are dissolved in water, and the solution is delivered directly to the roots.

Although hydroponics sounds like something new, it really is not. The first recorded experiments in hydroponics were conducted over three hundred years ago in 1699. Developments in large-scale hydroponic systems were successful in the late 1920s and early 1930s. In recent years, there has been a widespread expansion of hydroponics systems by the home hobbyist and agriculture industry. This growth is due to a better understanding of plant growth and the nutrient needs of plants, as well as technological advances.

A VARIATION

A combination of fish culture and plant culture is sometimes carried out in the same system. This is known as **aquaponics**. The term is derived from aquaculture (fish culture) and hydroponics. The plants and aquatic species share the water and often benefit from the presence of each other. For example, the plants may remove nutrients from the water to provide a better water environment for the fish.

ADVANTAGES AND DISADVANTAGES OF HYDROPONICS

Hydroponic production has some advantages over the production of plants in soil. It should be noted, however, that equally good crops can be produced in soil.

Advantages of hydroponic production include:

- Some insects harmful to crops live in the soil and damage plants below the soil level. Because hydroponics does not involve the use of soil, these insect pests are not a problem.

- Soils contain dormant weed seeds. Given the right conditions, they germinate and compete with crops for water, nutrients, and light. Hydroponic systems do not have weed seeds that might germinate.

- Because optimal nutrient needs can be provided to every plant in a system, competition between plants is reduced. This allows the plants to be grown closer together than they would in a field, leading to higher yields per acre.

- Plants require different nutrient needs as they grow and develop. Obviously, a seedling does not require the same amount of nutrient a mature plant requires. In hydroponic systems, the amounts of certain nutrients supplied to the plants can be adjusted relatively quickly.

CAREER PROFILE

HYDROPONIC PLANT PRODUCER

A hydroponic plant producer grows plants for setting into a field. Plant producers select the varieties to grow, plant seed, maintain proper growing environments, remove plants for setting, and otherwise operate a facility. This photograph shows burley tobacco plants being started on this Kentucky farm in a system that "floats" on a shallow water bed.

Plant producers need practical experience with hydroponic plant production. Education should include plant science classes, agriculture classes, and related areas. A high school education is needed. College study is very beneficial.

Jobs are found where plants are produced in hydroponic systems. In some cases, the work is quite seasonal and requires that a person have other jobs in the off-season. (Courtesy, Ann S. Clark, Kentucky)

- The pH level of the solution can affect the uptake of nutrients. In a hydroponic system, the pH of the solution can be monitored and easily adjusted to a range best for the type of plant being grown.

- Hydroponics can result in high-quality yields in parts of the world where there is nonproductive land or poor growing conditions.

Hydroponic systems are not without disadvantages. Disadvantages include:

- Commercial systems have a high initial cost.

- Some diseases, such as *Fusarium* and *Verticillium*, can spread rapidly throughout a system.

- The pollination of flowers to set fruit can be difficult in greenhouses where many hydroponics systems are housed.

CROPS GROWN IN HYDROPONIC SYSTEMS

Theoretically, any plant can be grown hydroponically. Of course, it is hard to imagine or see any sense in growing an oak tree or an ornamental landscape shrub in a hydroponic system. The plants usually grown in hydroponic systems include house plants, annual flowering plants, herbs, and vegetables. These do well in hydroponic systems. The most common plants grown in commercial operations are vegetables, some cut flowers (roses and carnations), and herbs.

There has been much success growing vegetables hydroponically. Tomatoes, peppers, cucumbers, lettuce, spinach, squash, eggplants, and melons are produced in commercial hydroponic operations. Other plants grown commercially are snow peas, snap beans, basil, parsley, onions, broccoli, cabbage, Brussels sprouts, beets, asparagus, corn, kale, carrots, celery, potatoes, and sweet potatoes.

11-3. A greenhouse with commercial hydroponic lettuce production.

1-4. Herbs, such as sweet basil, are grown commercially using hydroponics systems.

11-5. This hydroponically grown lettuce has been harvested and packaged for sale in a supermarket.

Vegetables are a natural choice for application in hydroponic operations. Great quantities of vegetables are consumed daily so there is a market for the produce. Vegetables are also best when they are fresh. Hydroponic operations near urban areas produce vegetables that can be quickly transported to the markets. These are also shipped at less cost than vegetables grown in another part of the United States or in another country.

FACILITIES MEET PLANT NEEDS

Most commercial hydroponic systems are inside greenhouses. Greenhouses allow high light transmission. It is also easier to control the temperature and humidity. The biggest challenge in producing vegetables in a greenhouse is pollinating the flowers to get a good set of fruit. Pollination is often achieved by tapping flowers with a stick or using a device to vibrate the flowers.

Hydroponically grown plants have the same basic requirements as plants grown in soil. Therefore, all hydroponic systems must supply the things normally supplied by

11-6. Weeds and insects that live in soil are not problems with hydroponic systems.

soil: support, water, nutrients, and air. The major differences are in the way plants receive support and the way in which nutrients are made available.

TEMPERATURE

Since most hydroponic systems are in greenhouses or confined structures, temperatures can be set. Most plants grow best when temperatures are kept within a certain range. Temperatures outside the range slow plant growth and reduce yields. Warm season vegetables, such as tomatoes, cucumbers, and peppers, do best when the temperatures are maintained between 60 and 75 to 80°F (23 to 28°C). Cool season vegetables, which include lettuce, spinach, cabbage, and broccoli, do best when temperatures are between 50 and 70°F (10 to 21°C).

11-7. Temperatures are controlled in the greenhouse to encourage optimum growth.

LIGHT

All vegetables and most flowering plants need large amounts of light. Hydroponically grown vegetables require 8 to 10 hours of direct sunlight each day for healthy growth. In commercial operations, high-powered lamps, such as sodium vapor lamps or metal halide systems, are sometimes used to provide necessary light intensity. These can supplement sunlight, or, in some cases, provide enough light to produce crops.

Another factor to consider is the spacing of the plants. Adequate spacing of plants to allow maximum lighting for each plant is important. For example, pruned and staked tomatoes need

11-8. Greenhouses allow a high level of light to reach the plants.

about 4 square feet and cucumbers about 8 square feet, while lettuce plants need to be about 9 inches (23 cm) apart.

Vegetables and flowers grown in the winter months do not produce as well as those grown in the summer. This is because the days are shorter. Most vegetables produce best when they are started in January and harvested in June or started in July and harvested in December.

WATER

Providing plants with enough water is not a problem with water culture systems. It can be a problem with aggregate culture systems if the aggregates and the roots are allowed to dry. Drying damages roots, which, in severe cases, may cause the death of plants.

The water quality is also important to provide a balance of nutrients needed to produce healthy growth. The pH of the water should be tested, and if necessary, adjusted for the crop. Softened water may contain harmful amounts of sodium and, therefore, should be avoided.

OXYGEN

Perhaps the most critical factor is that of providing the root system with enough oxygen. Plants and plant roots in particular require oxygen for respiration. Respiration is the chemical process in which a plant can convert stored energy to carry out plant functions. The function of the roots is to uptake water and nutrients. Dissolved oxygen in water is quickly used by the roots. To maintain high levels of oxygen, it is a common practice to bubble air through the water. Aeroponic systems are open to air and nutrient film technique (NFT) systems have constantly moving water so it is not usually necessary to provide additional aeration.

NUTRIENTS

Hydroponically grown plants have the same nutrient needs as those grown in soil. The difference is the lack of soil to serve as a reservoir for nutrients. The essential nutri-

11-9. Prepared nutrient solutions are mixed with water to provide plants with the essential nutrients.

ents must be provided in the water solution. Careful calculations are made to provide the optimal amount of each macronutrient and micronutrient in solution. The real advantage of hydroponics is the ability to test the nutrients in solution and make immediate adjustments as needed.

An understanding of chemistry is valuable when working with nutrient solutions. Nutrients in solution are measured in ***parts per million (ppm)***. This means that if a solution had 200 ppm of nitrogen, 200 out of 1 million molecules would be nitrogen. Each of the essential nutrients can be calculated to determine the ppm of that nutrient in a solution.

SUPPORT

The importance soil plays in supporting plants is often overlooked. Soil provides a firm anchor for plants to grow upright. In hydroponic systems, artificial support must be provided. This can be accomplished with string, stakes, mesh material, trellises, etc. One advantage to aggregate systems is they provide some support for the plants.

11-10. Strings, wrapped around the plant stems, provide support to these eggplant.

COMMON HYDROPONIC SYSTEMS

Hydroponics is used to describe many different types of systems. Since most are unique designs, they vary in size, appearance, and in the way they operate. All the systems can be placed in one of two main categories. One group, called ***aggregate culture***, involves the

1-11. Seeds are typically sown in horticubes, as shown, or in rockwool cubes.

11-12. After the seedlings have developed leaves (later than the picture indicates) the cubes are moved to the hydroponic system.

use of aggregate or substrate materials that help support the plants. The other group, known as **water culture**, does not use aggregate.

AGGREGATE CULTURE

Aggregate culture involves materials in which plants take root. The materials used are known as **substrates**. Some substrate materials include sand, perlite, vermiculite, marbles, gravel, peat moss, and rockwool. Substrate materials also provide support for the plants. **Rockwool** is one of the most commonly used materials. It is a spongy, fibrous material spun

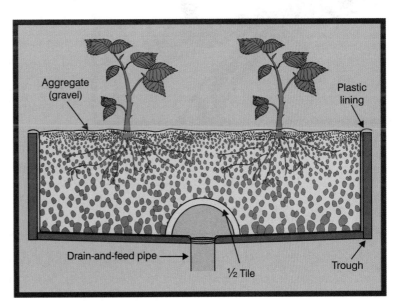

11-13. A cross-section of an aggregate culture system.

Aggregate (gravel)

Plastic lining

Drain-and-feed pipe

½ Tile

Trough

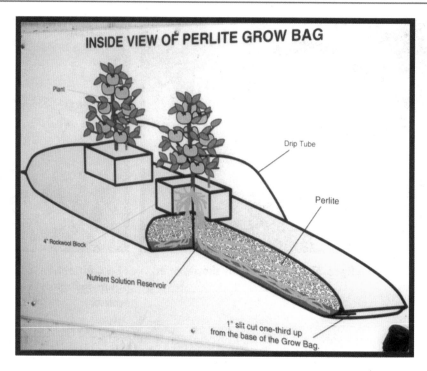

11-14. Some aggregate systems involve growing the plant inside the bag in which the aggregate is packaged.

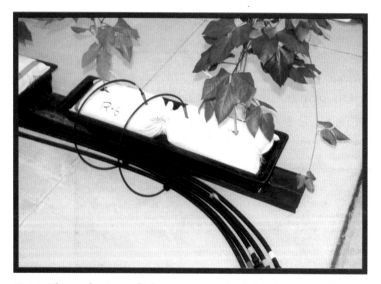

11-15. These plants are being grown in plastic bags containing an aggregate.

from molten volcanic rock. These materials are considered inert and do not provide nutrients for the plants.

Nutrient solutions provide the plants with essential nutrients. Drip, trickle, flush, or sub-irrigation are methods in which nutrient solution is delivered to the aggregate. One method calls for flooding the aggregate for ten minutes. The aggregate is allowed to drain for 30 minutes before flooding the aggregate again. Drainage is swifter in aggregate systems when large aggregate is used. Many commercial systems do not recycle the nutrient solution.

WATER CULTURE

Water culture is sometimes called nutriculture. In this type of system, there is no substrate used. Although plants may be started in rockwool cubes, most of the roots are growing in a nutri-

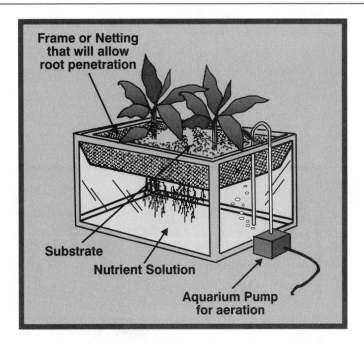

Frame or Netting
that will allow
root penetration

Substrate

Nutrient Solution

Aquarium Pump
for aeration

11-16. A small water-culture system in an aquarium as may be used in a classroom.

ent solution. Most water culture systems have a continuous flow or mist of nutrient solution that is recycled. These are said to be circulating systems. Some systems do not have circulating nutrient solution and plant roots just hang in the nutrient solution. Non-circulating systems are considered low-tech.

CONNECTION

BEGIN LEARNING ON A SMALL SCALE

Hydroponics requires unique skills for success. Begin learning these skills on a small scale. Kits are available that provide all of the materials needed to construct and operate a small hydroponics unit.

This shows a tray-sized, windowsill hydroponics unit being assembled. The tubing and nutrient water circulation system is in place. Aggregate is being added to provide a place for plant roots to anchor the plants. Only a few plants can be grown in the system, but it provides a valuable learning tool.

Water culture systems are the simplest systems to set up on a small scale. Disadvantages are large amounts of water are needed for large-scale operations, the water must be aerated, plants need support, and light reaching the nutrient solution promotes algae growth.

The water culture system most commonly used in commercial operations is called ***nutrient film technique (NFT)***. In NFT systems, a continuous flow of nutrient solution runs through a series of tubes, troughs, or channels washing over the plant roots. The system

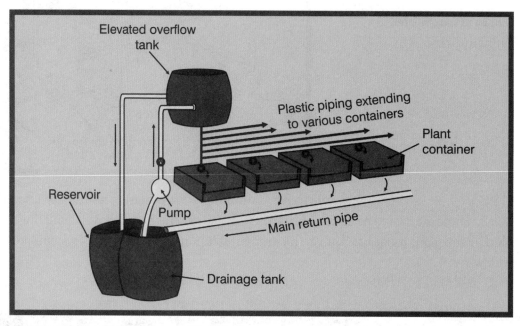

11-17. Nutrient film technique (NFT) is the system most often used in commercial operations.

11-18. These lettuce plants are produced commercially using the nutrient film technique.

relies on a pump to raise the nutrient solution from a holding tank to the channels within growing trays or pipes. The force of gravity moves the solution through the system and back to the holding tank. The system is constantly recycling the nutrient solution. Plants are often started in root cubes or rockwool cubes before being transplanted with the cube to this system.

Another water culture system that is very interesting is aeroponics. **Aeroponic** systems are designed to have plant roots suspended in the air within a closed container. Inside the container, spray nozzles are used to mist the roots continuously or at regular intervals with a nutrient solution. The nutrient solution forms a film on the roots. The inside of the container is kept at nearly 100 percent humidity so the roots do not dry.

11-19. The plant roots are enclosed in a chamber in this aeroponics system. Their roots are misted with a nutrient solution.

11-20. This photo taken at Epcot Center in Disneyworld shows squash grown using aeroponics. For the benefit of viewers, the plants move in and out of the system so the roots can be seen.

The common denominator of both aggregate culture and water culture systems is the method in which plants receive nutrients and water. Nutrients and water are supplied by the solution, not the aggregates (if any are used).

11-21. Harvest-size cucumbers grown hydroponically by an Illinois producer.

Costs versus benefits

The question producers must answer before investing in hydroponics is whether the costs will outweigh the benefits. New techniques in hydroponics are constantly being developed, producing higher-quality crops.

New systems are being designed with new materials. There is improved control over plant growth and development. There is also less labor involved. When combined, these result in higher yields. However, the cost of setting up hydroponic systems on a large scale is expensive. There is also an increased requirement for higher paid people with the technical skills necessary to operate and maintain hydroponic systems.

REVIEWING

MAIN IDEAS

Hydroponics is the growing of plants in a medium other than soil. Application of hydroponics is found in producing vegetable crops. Tomatoes, cucumbers, lettuce, spinach, herbs, and many other vegetables are successfully produced in commercial operations. Advantages of hydroponic systems include the easy adjustment of nutrient levels and pH of the solution, the absence of soil-borne insects and weed pests, plants can be grown closer together, and food can be produced where conditions prevent field production.

Soil is not used. Water, nutrients, air, and support normally received from soil must be provided in other ways. Water is seldom a problem, but problems can arise if aggregate systems are allowed to dry. Nutrient levels best for the growth of particular crops can be easily adjusted in solutions. In water culture systems, the solution is aerated to provide adequate oxygen for the roots. Support also needs to be provided to plants in the form of strings or stakes.

Hydroponic systems can be placed in one of two groups. Aggregate systems use substrate materials like sand, perlite or gravel in which plant roots grow. Nutrient solutions are delivered to the aggregate by dripping, trickling, or flooding the aggregate at regular intervals. Water culture systems do not use substrate materials. The nutrient solutions in water culture may be circulating or non-circulating. Some non-circulating systems involve the immersing of plant roots in a nutrient solution. Circulating systems, like aeroponic systems, involve misting the plant roots with a nutrient solution. Nutrient film technique (NFT) is the most commonly used system in the industry. It, too, is a circulating system involving a continuous flow of nutrient solution around the plant roots.

QUESTIONS

Answer the following questions using complete sentences and correct spelling.

1. What is hydroponics?

2. What are some advantages and disadvantages of hydroponic systems?

3. How do the two categories of hydroponic systems differ?

4. What must hydroponic systems provide plants that are normally provided by soil?

5. What are some benefits and costs of large-scale hydroponic systems producers must consider?

EVALUATING

Match the term with the correct definition. Write the letter by the term in the blank that is provided.

a. substrate   d. hydroponics
b. aeroponics   e. parts per million
c. rockwool   f. aggregate culture

1. _____ measure of nutrients in solution

2. _____ growing plants in a medium other than soil

3. _____ spongy, fibrous material spun from molten volcanic rock

4. _____ materials in which plants take root, such as sand, perlite, and rockwool

5. _____ roots suspended in air are misted with nutrient solution

6. _____ hydroponic system in which plants take root in substrates

EXPLORING

1. Build two identical small hydroponic systems in which experiments can be done. Some experiment ideas include:

    a. Testing how pH affects plant growth. Water pH can be adjusted by adding a little vinegar to lower pH, or baking soda to raise pH.

    b. To one system, add nutrients, and to the other, provide water only.

    c. Test the effects of well water versus city water versus pond water versus distilled water on plant growth.

    d. Depending on the type of system you build, you could try aerating the solution in one system and not the other.

    e. Add only nitrogen, phosphorous, or potassium to one system and a complete fertilizer with micronutrients to the other.

    f. Devise your own experiment.

2. Go to the local supermarket. See if any of the fresh produce was grown in hydroponic systems. Often, packaged vegetables will state that they have been grown hydroponically.

3. Access the World Wide Web to learn more about hydroponics. Some sites you may want to try include:

    a. Hydro/Aquaponic/Aquaculture— http://www.intercom.net/biz/aquaedu/hatech/pages/airs1.html

    b. Growing Edge Magazine — http://www.teleport.com/~tomalex

    c. The Growroom — http://a1.com/growroom

    d. Stanford University — http://hawg.stanford.edu/~rfrench/hydro.html

    e. To subscribe to the hydro mailing list, which is a newsgroup, simply send an E-mail to the following address that says: SUBSCRIBE HYDRO: majordomo@hawg.stanford.edu

Part Four

Plant Pests

PLANT pests are their enemies. Pests attack and destroy plants. They may damage the ability of a plant to grow and produce good crops. In some cases, pests kill plants. Fortunately, methods are available to minimize or eliminate the losses pests cause. Growers need to know the best approaches and how to use them. Methods are used that manage pest populations to keep losses below a level where profit is reduced. Promoting plant growth involves insect, weed, and disease management using integrated approaches. The goal is to assure a good environment for plant growth.

Integrated Pest Management

This chapter introduces the practice of integrated pest management. It has the following objectives:

1 Explain plant health and pests.

2 Describe tactics of integrated pest management.

3 Identify the benefits of IPM to agriculture and the environment.

4 Explain the meaning and importance of agroecosystems.

5 Describe the role of genetically modified crops in IPM.

6 Identify important safety practices with pesticides.

┤ TERMS ├

agroecosystem	injury threshold	plant pest
carcinogen	key pest	restricted-use pesticide
general-use pesticide	pesticide	scouting
host plant resistance	plant health	toxicity

12-1. A mesquite plant is being examined for signs of health problems.

PLANT pests cause huge losses and reducing those losses is challenging to growers. But, suppose a pest could work for you. What would you let the pest do? Would you be willing to tolerate a few pests in a crop?

New ways of providing for the health of plants are being used. These protect the environment from excessive hazardous chemicals and rely on natural processes. The new methods are based on long-term study of the environment of plants.

Changing how we feel about pest control is not easy. As we look toward a sustainable future, some traditions must change. One big area of change is pest management. Growers are changing how they view and control pests.

PLANT HEALTH AND PESTS

12-2. A fungus disease is impairing growth of a tree.

Plants grow best in an environment that promotes good health. A healthy plant is not diseased. Healthy plants produce high-quality crops. Diseased crops have little value and provide small yields. The wise producer carefully monitors plant health and selects appropriate methods of pest control.

Plant health is the condition of a plant as related to disease. Plants that are free of disease are growing rapidly and making efficient use of nutrients. Diseased plants fail to grow as they should. Most plant health problems are caused by pests that can be easily identified.

PLANT PESTS

A **plant pest** is anything that causes injury or loss to a plant. Most plant pests are living organisms. They attack plants and damage the growth of the plant. In some cases, they may kill the plant or make it worthless.

Plant pests can be placed in three major groups: insects and nematodes, diseases, and weeds. Each of these is presented in detail in other chapters of the book.

In addition, environmental problems and rodents and other animals may damage plant health. Plants also fail to grow properly if they are not provided the nutrients they need. Plants that do not have adequate nutrients may have nutritional diseases.

HOW PESTS CAUSE LOSSES

Plant pests cause losses in different ways. All usually result in decreased production. Here are a few examples:

12-3. A tiny borer in the stem of a pumpkin plant has destroyed the plant's vascular system. The damage caused the plant to slowly die. (The photo has been enlarged.)

- Damage plant parts—Some pests attack plants. They may eat holes in

12-5. The sunflower is a weed in this wheat field. It has competed for space with the wheat that is now mature. Sunflowers use the same nutrients as the wheat. If enough sunflowers were present, the harvest process could be inefficient.

12-4. An ear of corn has been damaged by a worm.

leaves, buds, roots, fruit, and other plant structures. This damage makes the plant less productive. Leaves with holes cannot carry out photosynthesis as efficiently. Heavily damaged leaves may die and not be productive at all. Insects and nematodes are particularly known for causing physical damage to plants.

- Compete for space and nutrients—Weeds grow in the fields where crop plants grow. They use space and nutrients needed by the crop plant. Weeds use water, fertilizer, and light that the crop plants need to grow.

- Reduce quality of harvested crop—Pests may contaminate the products of plants. Harvested food crops may contain insects. Fiber crops may be dirty with weed leaf fragments and other trash. The presence of any impurities lowers the quality of harvested crops. The prices paid to growers are discounted.

- Increased production cost—Pests increase the cost of production. They reduce yields and lower the quality of harvested products. Pests also make cultural practices more difficult, such as trying to harvest when many weeds are present in a field. Control methods are also expensive cultural practices.

PEST MANAGEMENT TACTICS

Various approaches can be used in limiting damage to plants by pests. The goal is to keep losses to a minimum and productivity as high as possible. Growers are increasingly using tactics to manage pests.

12-6. Scouting cotton helps provide accurate information for integrated pest management. (Courtesy, Marco Nicovich, Mississippi State University)

12-7. A handheld computer is being used with a global positioning system to record insect damage and location. (Courtesy, Communication Services, North Carolina State University)

Integrated pest management (IPM) is using methods to reduce pest populations that produce favorable consequences. The methods selected ensure favorable economic, ecological, and sociological outcomes. Integrated pest management involves using methods that cause minimal damage to the environment and provide satisfactory economic return.

IPM includes two major procedures:

- Scouting—**Scouting** (monitoring) is carefully observing crops for the presence of pests. Damage may or may not be evident. Measures are selected on the basis of information gained from scouting. If damage appears likely to cause economic loss greater than the cost of a treatment, the appropriate measures are used. Good scouting is an essential component of IPM.

- Treatment—Treatment measures are selected based on monitoring information. The goal is to get the population below the economic threshold level. A combination of measures may be used, including biological, cultural, and chemical. Most IPM approaches make greater use of biological and cultural practices. Chemical treatments are minimized. The measures selected are economically and socially acceptable.

CHEMICAL TREATMENTS

Chemical pest treatment expanded rapidly for three decades following 1945. By the mid-1970s, scientists observed that some pests were no longer being effectively managed by chemicals. New chemicals were developed and used in greater amounts. Pests continued to develop resistance. Alternatives to chemicals were needed.

Early resistance by insects to pesticides was observed with the housefly. Regular spray programs initially killed houseflies. After a couple of years, the flies were not killed. New chemicals were developed and used, with the same results. Other insect species that

12-8. Even with the greatest care, some pesticide materials will drift and likely damage the environment. (Courtesy, U.S. Department of Agriculture)

CAREER PROFILE

PLANT PATHOLOGIST

Plant pathologists study plant functions, diseases, and other processes. They often carry out research in fields as well as in greenhouses. The research often addresses problems of pests. The plant pathologist shown here is examining wheat plants for a fungus disease.

Plant pathologists have baccalaureate and advanced degrees in plant pathology or a closely related area. They must have practical experience in plant production as well as laboratory skills. Most plant pathologists work with agricultural research stations or large agricultural corporations. (Courtesy, Agricultural Research Service, USDA)

developed resistance included the citrus red mite in California and the cotton bollworm in Texas.

Pests have shown that they can adapt to a wide range of attempts to manage them. Resistance to a pesticide results in it no longer being effective.

ENVIRONMENTAL CONTAMINATION

The widespread use of large amounts of pesticides was causing damage to the environment. Soil, water, and air were being contaminated. Pesticide residues were found in food, feed, and organisms at all levels in the food chain. Damage to human health was suspected.

Scientists realized that new approaches were needed. They began studying the natural environment of pests. They investigated the biological characteristics and the natural predators of insects. These approaches have resulted in less environmental contamination.

USING IPM

Since IPM involves using a wide range of pest strategies, making decisions requires good information. The **key pests** must be identified. These are the pests that regularly cause losses in crops. The presence of an insect or a weed does not necessarily make it a key pest.

12-9. Parasitic wasps can be used on insect pests. This wasp is laying an egg inside a caterpillar. The egg will hatch and feed on the caterpillar, destroying it. (Courtesy, Agricultural Research Service, USDA)

Biological characteristics of the crop and of the key pest are important in using IPM. Knowing the cultural requirements needed for a crop is the first step. Knowing the biological characteristics of the key pests is the second step in using IPM. Practices can be selected to promote crop growth and disrupt reproduction and growth of the pest.

The chapters on methods of pest management and production of major crops give more information on IPM. The biological characteristics of pests and cultural requirements of crops are integrated into a system of production.

Table 12-1. Summary of Methods Used in IPM

Methods	Practices
cultural	use resistant varieties; rotate crops; chop stalks and dispose of refuse after harvest; tillage approaches; times for planting and harvesting; pruning and thinning with some crops; fertilizing based on crop needs; sanitation; water and runoff control; using trap crops
mechanical	trapping and collecting; mowing, chopping, crushing, and grinding plant residues, pests, and other forms; hand pulling and picking
physical	using high and low temperatures; irradiation, particularly with seed and food grains; light traps
biological	using natural predators, such as beneficial insects; using parasites, such as bacteria; using genetically engineered crops; releasing sterile or incompatible pests
chemical	poisons; growth regulators; attractants and repellants; sterilants
regulations	quarantines; government-sponsored eradication and supression programs

BENEFITS OF IPM

IPM is said to have many benefits to both agriculture and the environment. These benefits help sustain the ability of the earth to meet the needs of an increasing human population.

BENEFITS TO AGRICULTURE

The benefits to agriculture vary with the crop and the extent to which pests interfere with economical production. Careful planning is required to make effective use of IPM. Good planning also helps growers in overall efficiency.

The major benefits of IPM include:

- Reduced pesticide costs—Fewer pesticides are used with IPM. Not only is the cost of pesticide reduced, but less equipment is needed.

- Reduced application costs—Time and cost of labor for pesticide application are reduced.

- Less pesticide resistance—Insects, weeds, and other pests will be less likely to develop pesticide resistance. This means that a pesticide is more effective when its use is required.

12-10. Chopping the stalks on this field has destroyed hiding places for insects through the winter.

- Promote sustainable agriculture—Growers are more likely to be able to continue production on the same land on a long-term basis. The soil is not degraded by pesticide residues.

- Reduced crop damage—Beneficial insects are allowed to help control harmful insects. This would not happen if pesticides were used. Pesticides kill beneficial insects!

- Stronger social and political support—Citizens are not "turned off" by negative notions of the use of pesticides. They are more likely to support research programs and other activities that benefit growers.

BENEFITS TO THE ENVIRONMENT

Benefits of IPM to agriculture are also environmental benefits. The environment is made more sustainable and friendly to people.

Three benefits of IPM to the environment are:

12-11. Young brachnid wasps are developing on a tomato hornworm. The wasps gain their food from the worm and will eventually destroy it.

- Reduced contamination—The environment suffers less degradation through the use of IPM. Pesticide residues do not build up in soil, water, and other natural resources. Wildlife are protected from health problems associated with excessive pesticide use.

- Fewer residues on food—Food products will have less pesticide residue with IPM. This reduces the chance of people contracting diseases associated with pesticides.

- Improved human health—IPM supposedly results in food products that promote good health. Cancer-causing residues are present in smaller amounts or are not on food at all.

AGROECOSYSTEMS

Fields of crops, orchards, pastures, and other places where cultured plants and animals grow form an **agroecosystem**. This is a shortened form of "agricultural ecosystems." Important relationships exist between living organisms and between the organisms and their nonliving environment. It is the ecological aspect of pest management.

The agroecosystem can be used to benefit crops and reduce pests. When growers understand the needs of crops, they can provide for a good environment. Likewise, understanding the nature of pests helps put conditions into place that are not favorable to their growth.

CHARACTERISTICS OF AGROECOSYSTEMS

The agroecosystems have fewer species than most natural ecosystems. One species usually predominates—the cultured crop plant. In any field or orchard, however, numerous insects, birds, microorganisms, and other species live with the crops. Some of these organisms are harmful to the crop plants; others are beneficial.

Humans intensively manipulate the agroecosystem. Plowing, seeding, mowing, fertilizing, and other cultural activities alter the system. Using pesticides definitely has an impact on the agroecosystem. Some intensive practices are essential if the needs of people for food, clothing, and shelter are to be met.

Agroecosystems have less species diversity than natural ecosystems. This creates a condition that insects, weeds, and pests find beneficial. For example, the fertilizer applied to a crop also gives additional nutrients for weeds.

Crop density has a positive effect on an agroecosystem. Within plant population guidelines, fields have more plants per area than do natural ecosystems. An increase in density offsets attacks by insects and other pests. Fields are complex biological systems.

12-12. A cherry orchard creates an agroecosystem with one species (cherry) dominant. Many species of insects, birds, and small organisms are found in the orchard.

AGROECOSYSTEM PLANNING

Pest problems can be reduced by using selected crop varieties and other factors in an agroecosystem. Some problems can be anticipated and avoided with good planning. Consider the climate and other environmental factors. Select crops that are less susceptible to damage from pests.

Some plants are resistant to damage by species of insects and other pests. These plant varieties have developed **host plant resistance**. Selective breeding is often used to obtain varieties that are resistant to a pest.

Cropping practices can be used that reduce pest problems. Using plant populations that shade the ground helps prevent weed growth. Planting trap crops near fields allows insect pests to be controlled before they reach the crop.

Some tolerance of pests may be needed. Loss to pests often cannot be completely prevented regardless of the control effort. The economic benefits of a practice must be considered. Minor damage to a crop may not be sufficient to classify an insect or weed as a pest.

THRESHOLDS

A threshold is the point at which an event or change occurs. IPM involves several thresholds, with injury and economic levels being key points.

Injury threshold is based on the damage caused by a potential pest to a crop to be classified as a pest. Some insects, for example, can live in crop plants and cause no damage or very little damage to the crop. In addition to the species of potential pest, the number of pests is another factor in injury threshold. One insect or one weed in a field does not create crop injury and loss. Many insects or weeds of the same species would likely create injury and loss. Using some pesticides kills beneficial organisms and, in the long run, increases loss to insects unless repeated pesticide applications are used.

Economic threshold is based on the returns to be gained from using a pesticide. Controlling a few pests would not usually be sufficient justification for using a pesticide. A treatment should be used when the returns from the treatment will be greater than the cost of the treatment.

12-13. Infestation of this rose bush with caterpillars has gone beyond the threshold level. Severe damage has been done.

CONNECTION

APPLYING FUNGICIDE WITH A HELICOPTER

Even when the best pest management practices are used, aerial application of pesticides is sometimes needed. A helicopter is a fast, efficient way to reach large groves of citrus. Quick action is needed to control the fungus disease that is affecting the grove. If not, the crop can be lost. (Courtesy, U.S. Department of Agriculture)

GENETICALLY MODIFIED CROPS

Plants can be modified to resist disease, insect attack, and other problems. The process involves altering the genetic material of a plant so it has a quality that is useful in assuring plant health and productivity. In effect, the resulting plant is transgenic–a new genetic factor has been transferred into it.

Commonly, a transgenic plant is known as a genetically modified organism or GMO, for short. The process involves recombinant DNA, which is taking DNA from one organism and moving it to another. The gene that creates a desired trait in one organism can be identified, removed, and placed in another organism. This results in the transgenic organism having the trait. Several examples are now used in crops:

- Cotton—A gene from *Bacillus thuringiensis* bacteria has been used to create transgenic cotton that repels damage by certain species of worms. The toxin genes of the bacteria have been identified and used in genetically engineering the cotton.

- Corn—Corn has been modified to resist insect damage, have nutrients specifically needed by certain animals, and tolerate a common herbicide known as RoundUp®.

- Soybeans—Soybeans have been modified to tolerate RoundUp® and in other ways.

- Tomatoes—Tomatoes have been modified to retain firmness for a longer time after harvest. This quality is known as extending the shelf-life of the tomatoes.

- Potatoes—Similar to cotton, potatoes have been modified to resists certain insect pests, particularly the pesky Colorado potato beetle.

By engineering a plant to resist a pest, the need to use a pesticide for controlling the pest is eliminated. This protects the environment from pesticide residues and contamination.

Some people, however, feel that GMOs pose dangers to the environment. They feel that the engineered plants may breed with wild plants. The process may reduce the diversity in the genetic material in both wild and domestic plants.

SAFELY USING PESTICIDES

With IPM, pesticides are frequently used. How they are used is carefully planned. Safety procedures are closely followed.

A **pesticide** is a poison. When properly used, they efficiently control pests. When used improperly, pesticides damage the environment and injure people and other organisms. Water, soil, and other natural resources are unfit if contaminated. Properly using pesticides and disposing of empty containers are essential.

USE CLASSIFICATION OF PESTICIDES

Pesticide materials vary in the safety hazards posed. Some are much more toxic (poisonous) than others. **Toxicity** refers to the degree of poison in a material. It is influenced by the amount of active ingredient in a material as well as the chemical nature of the poison. Pesticides are classified into two use categories:

12-14. Wear appropriate protective clothing and equipment when mixing and applying pesticides. (Courtesy, U.S. Department of Agriculture)

- General-use pesticides—A **general-use pesticide** can be more widely used by following instructions on the label. These pesticides pose less danger to the environment. Special training in application may not be required. Information on the label is usually adequate to safely use the pesticide.

- Restricted-use pesticides—A **restricted-use pesticide** has higher toxicity than general-use pesticides. Risk is greater to

Table 12-2. Pesticides Used

Pesticide Group	Pest
insecticide	insects
miticide	mites
acaracide	ticks and spiders
molluscicide	snails and slugs
fungicide	fungi
rodenticide	rodents
nematocide	nematodes
bactericide	bacteria
herbicide	weeds

humans as well as the environment. People who apply restricted-use pesticides must have training in proper use of the material. All regulations must be carefully followed.

POSSIBLE HAZARDS TO PEOPLE

Materials used in making pesticides may injure human health. Precautions should be followed in using them. Some may cause death. Others may cause various disorders. A pesticide suspected of causing cancer is known as a **carcinogen**. Pesticides should be carefully transported, mixed, and applied. This reduces the danger from exposure.

Pesticides enter the human body through the skin and by mouth and inhalation. Ninety percent of all exposure occurs through the skin. Once on the skin, rates of absorption vary with different parts of the body. The forearm has the lowest absorption rate. Absorption in the scrotal area of males is over 11 times faster. Absorption continues to occur as long as pesticide is on the skin. Figure 12-8 compares selected parts of the body with the forearm.

Protection should be used to reduce exposure to pesticides. Any contact with the human body should be avoided. This includes contact with the skin, in the air, and in water and food. Gloves, boots, hats, overalls, goggles, face shields, and respirators should be used.

After use, clothing and protection equipment should be properly washed before its next use. People should take showers to rinse any materials off their bodies before they go into their homes. Be careful not to contaminate the home with dirty clothes, boots, and other contaminated safety wear.

FOLLOWING SAFETY PROCEDURES

Hazards in using pesticides can be reduced by following a few safety rules. First, the pesticide must be properly identified because pesticides are not alike. Second, pesticides must only be used as approved. Read the label!

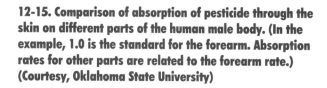

Dermal Exposure

Scalp 3.7

Forehead 4.2

Ear canal 5.4

Abdomen 2.1

Forearm 1.0

Scrotal area 11.8

Palm 1.3

Absorption rates compared to forearm which is 1.0

Ball of foot 1.6

12-15. Comparison of absorption of pesticide through the skin on different parts of the human male body. (In the example, 1.0 is the standard for the forearm. Absorption rates for other parts are related to the forearm rate.) (Courtesy, Oklahoma State University)

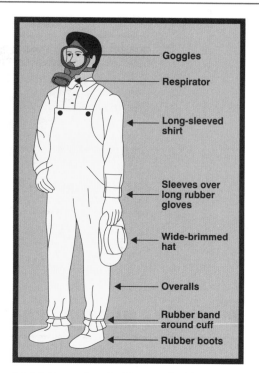

Goggles

Respirator

Long-sleeved shirt

Sleeves over long rubber gloves

Wide-brimmed hat

Overalls

Rubber band around cuff

Rubber boots

12-16. Protective clothing needed for applying pesticides.

12-17. Pesticide container labels usually have detailed information about the use of the product.

Here are several important rules for using pesticides:

- Use only approved pesticides in approved ways—Obey all regulations. Pesticides approved for use to control a pest in one crop might not be approved for controlling the same pest in another crop.

- Read and follow instructions—Labels provide important information about the pesticide and its safe use. Following the instructions on the label reduces the risk to some extent.

- Use only when needed—Using a pesticide when it is not needed wastes money and causes undue exposure. Improper use creates additional damage to the environment.

- Use low toxicity pesticides—Pesticides should be used to keep the amount of poison released into the environment to a minimum and yet control the pest. Using more than is needed is a waste and poses additional hazards.

- Consider the weather—Apply pesticides when they will have the most effect. Do not apply just before a rain or on a windy day.

- Properly use equipment—Application equipment should be in good order and properly calibrated. Equipment should be adjusted to apply no more than needed. The material should be directed to the plant or other place where the pest is located.

- Properly dispose of empty containers—Empty containers retain a small amount of pesticide. They should never be thrown into creeks or along roadsides. Some manufacturers will take returned containers and recycle them. Containers that are not recycled should be rinsed at least three times. Always buy the amount of pesticide needed in the fewest possible containers. This reduces the number of empty containers as well as prevents having left over pesticide.

- Avoid contaminating the environment—Applying a pesticide is a form of pollution. The purpose is to create a condition under which the pest cannot survive. Water, air, soil, and other natural resources can be damaged by pesticides.

- Protect from exposure—People who apply pesticides should take precaution to prevent exposure. Dress properly and wear respirators, gloves, boots, and other safety devices.

- Post warning signs—Signs indicating that a poison has been applied should be posted in fields, greenhouses, or other places where it has been used.

- Know emergency procedures—Pesticide applicators should know the procedures to follow in case of an accident. Post emergency telephone numbers.

12-18. A warning sign indicating that a pesticide has been used in the field.

12-19. Pesticide containers being collected for proper disposal.

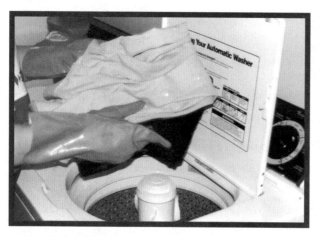

12-20. Clothing worn to apply pesticides being carefully handled and washed.

12-21. Always thoroughly clean application equipment after use. (Courtesy, U.S. Department of Agriculture)

REVIEWING

MAIN IDEAS

Plant pests cause large losses each year. New approaches of control are being used. Integrated pest management (IPM) is using pest control methods that have favorable long-term outcomes. Methods used to control weeds, insects, and other pests are carefully selected. The goal is to protect the environment and have satisfactory economic return. IPM also assures a sustainable agriculture and environment.

IPM makes much use of ecosystem concepts. The altered environment in a crop field is an agroecosystem. One or a few species predominate. Many other species are found including birds, insects, and plants.

IPM is best used when key pests are identified. The methods used to control the key pests begin with cultural practices and include mechanical, physical, biological, chemical, and regulatory requirements.

Benefits of IPM are important in agriculture and the environment. In both cases, IPM contributes to a sustainable quality of life.

Pesticides are classified as general-use and restricted-use. Restricted use pesticides pose greater hazards to the environment than general use pesticides. With all pesticides, proper attention to safety is essential. Avoiding exposure to pesticides is top priority.

QUESTIONS

Answer the following questions using complete sentences and correct spelling.

1. What is integrated pest management (IPM)? Why has it become important?

2. How have chemical pest treatments contributed to the use of IPM?

3. What is an agroecosystem? How does it differ from a natural ecosystem?

4. Why are injury threshold and economic threshold important in IPM?

5. Select an agronomic, horticultural, or forestry crop in your area. Use Table 12-1 and identify possible methods and practices that could be used to control pests in your selected crop.

6. What are the benefits of IPM to agriculture? The environment?

7. Distinguish between a general-use and restricted-use pesticide.

8. What safety procedures should be followed in applying pesticides?

EVALUATING

Match the term with the correct definition. Write the letter by the term in the blank that is provided.

a. agroecosystem   c. host plant resistance   e. key pests
b. carcinogen   d. injury threshold   f. restricted-use pesticides

1. _____ pesticide suspected to cause cancer
2. _____ amount of damage caused by a pest to crop
3. _____ plants that have resistance to pests
4. _____ pests that regularly cause crop losses
5. _____ relationships between living and nonliving factors in a field, orchard, or ornamental area
6. _____ pesticides with higher toxicity

EXPLORING

1. Make a field trip to a farm, orchard, greenhouse, or other crop production facility that uses IPM. Interview the manager and determine the nature of the IPM that is carried out. Ask the manager to give a personal assessment of the success of IPM. Prepare a written report on your findings.

2. Study a crop field to determine the components of the agroecosystem. Note the plants, insects, birds, and other living organisms. Also, note the nonliving characteristics of the area. Give an oral report in class on your observations.

3. Prepare a report on the use of IPM in a field, vegetable, or ornamental crop that grows in your area. Obtain bulletins and other information from the local office of the Cooperative Extension Service. Share your findings in an oral report in class.

4. Investigate the use of genetically modified crops in your area. Determine which crops have been modified and the desired trait achieved by the modification. Write a report on your findings.

Insects and Nematodes

13-1. Near-perfect broccoli heads are grown when pests are managed. (Courtesy, Agricultural Research Service, USDA)

INSECTS and nematodes cause major plant losses. Annually, the value of lost production and money spent to manage pests in the United States amounts to billions of dollars! These losses have a major impact on humans. The food lost to these pests could help prevent hunger in the world.

Producers of plants must be smarter than the insects and nematodes that are causing damage. If not, the producer will fail and the pests will win. Pests must have food to survive. They compete for the needed nutrients. They use soil fertility and other natural resources in the foods they use.

Fortunately, methods are available to limit losses to insects and nematodes. Selecting and properly using the best methods are essential. Knowing what to do is important to success.

INSECT BIOLOGY

Knowing insect biology helps explain why insects damage plants or perform beneficial services. Basic insect biology helps in selecting control methods.

An insect's exoskeleton is made of chitin, which serves much like a suit of armor. Muscles and organs are attached to the inside wall of the strong chitin. The chitin gives shape to the body and protects the organs. Compared to bone, chitin is much stronger on a weight basis.

The number of segments in the exoskeleton varies but is about 20 in most insects. Some segments are easy to see; others are fused tightly together and difficult to see. The segments form the three major body sections: head, thorax, and abdomen. The head has eyes, antennae, and mouthparts. The thorax has wings and three pairs of legs attached to it. The abdomen contains organs of food digestion, reproduction, and excretion.

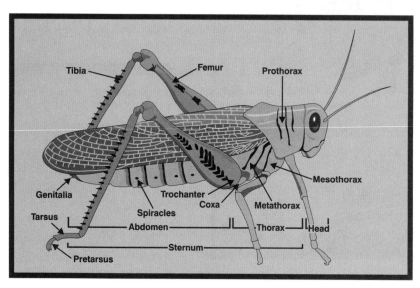

13-2. Major parts of an insect.

MOUTHPARTS

The mouthparts of insects tend to be one of two types: chewing and sucking.

A **chewing insect** bites off, chews, and swallows plant parts. Holes in leaves, buds, flowers, or other parts indicate the damage was by a chewing insect. Examples include armyworms, cutworms, cotton bollworms, Colorado potato beetles, Mexican bean beetles, and peach tree borers.

A **sucking insect** pierces the outer layer of a plant and sucks sap from it. The insect makes a tiny hole and uses the plant juice as its food. Examples of sucking insects include aphids (also known as plant lice), leafhoppers, chinch bugs, thrips, squash bugs, mealy bugs, and scales. (These insects are sometimes called piercing-sucking insects.)

Some variations are found. A few insects both chew and suck. Some moths have long tubes for sucking. Insect mouths vary with the stage of development or maturity. Some insects are chewing at one stage of their life cycle and sucking at another stage. An example is the hornworm (*Protoparce sexta*). As a larva, it feeds on tomato or similar plants by chewing. The adult moth feeds by sucking.

REPRODUCTION

Insects reproduce sexually. Females produce eggs and males produce sperms. Following mating, hatching and development vary.

Insects go through stages of development known as **metamorphosis**. The changes are distinct as they go from egg to adult. Two kinds of metamorphosis are common: incomplete and complete.

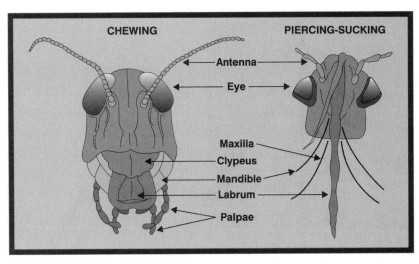

13-3. Mouthparts of chewing and piercing-sucking insects.

Incomplete metamorphosis has three stages of development: egg, nymph, and adult. Eggs hatch into nymphs, which are immature forms that resemble the adult. Nymphs usually molt (lose and regrow their exoskeleton) several times before reaching adult stage. Grasshoppers, for example, have an incomplete metamorphosis.

CAREER PROFILE

ENTOMOLOGIST

Entomologists study insects, the problems they cause, and how they can be controlled. The work may involve breeding insects in a laboratory or engaged in practical work in the field. They often work with producers to solve insect damage problems and prevent losses.

Entomologists have baccalaureate degrees in entomology or a closely related area. Many have masters' and doctors' degrees in entomology. Practical experience with crops and insect pests is essential in relating to the needs of growers.

This shows an entomologist working with an aerial applicator investigating the influence of wind on drift of the spray. This is important in preventing unwanted contamination of pastures, streams, and lakes. (Courtesy, Mississippi State University)

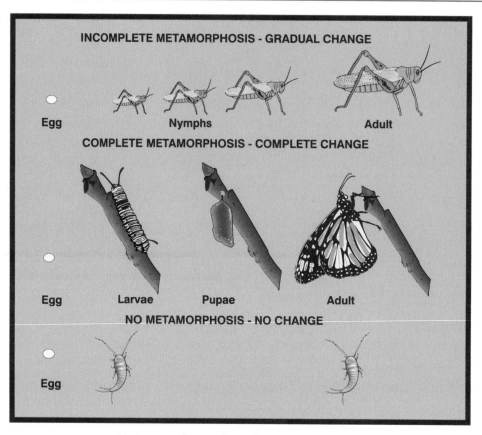

13-4. Incomplete and complete metamorphosis.

Complete metamorphosis has four stages of development: egg, larvae, pupa, and adult. Larvae are segmented, wormlike forms that often inflict considerable damage on plants. After an active larva stage, a pupa is formed. The pupa is in a resting stage for a while before it becomes an adult. Most pupae are surrounded by a cocoon or protective case. The hornworm is the larva stage of a moth species. Caterpillars, grubs, and maggots are examples of larvae stages of insects.

Successfully managing many insect pests involves understanding the type of metamorphosis. Measures are selected based on a combination of mouthparts and metamorphosis.

INSECT CLASSIFICATION

Insects and nematodes are similar in that they are small animals. They are also similar in that some of them damage plants. The similarity quickly ends.

Insects and nematodes are in different phyla, or basic divisions, in the Animalia Kingdom. Insects are in the phylum Arthropoda, while nematodes are in the phylum Nematoda.

An **insect** is a small boneless animal whose body is divided into three sections. Most insects are very small and can easily hide. They are in many shapes and colors. How insects live in the environment varies. Some cause considerable damage; others are beneficial. Scientists have identified more than 800,000 different species of insects or about 80 percent of all animals on the earth!

Insects are classified in many ways. Five ways are included here: scientific, benefit, mouthpart, reproduction, and feeding location. Understanding the classifications is an important part of successfully managing insect populations.

13-5. Worms (immature insect forms) have damaged these developing tomatoes resulting in a total loss.

SCIENTIFIC CLASSIFICATION OF INSECTS

The Arthropoda phylum includes animals with exoskeletons and segmented bodies. The phylum is further divided into classes, with insects being in the Insecta class. The class has several orders, with each of these having fami-

Table 13-1. Examples of Insect Classifications

Common Names	Order	Mouth Parts
grasshoppers, crickets	Orthoptera	chewing
"true bugs"	Hemiptera	sucking
aphids, leafhoppers, and whiteflies	Homoptera	sucking
beetles, boll weevils, and ladybug beetles	Coleoptera	chewing
butterflies, moths	Lepidoptera	sucking/sucking tube
flies, mosquitoes, and houseflies	Diptera	sucking
ants, bees, wasps	Hymenoptera	sucking (bees); chewing (ants and wasps)
cockroaches	Blattodea	chewing
mantids	Mantodea	chewing
dragonflies	Odonata	chewing
thrips	Thysanoptera	sucking

13-6. Insect damage to leaves has almost destroyed this corn. (The damage is too great for a crop to be produced.)

13-7. The bee—a beneficial insect—is pollinating a wild azalea flower.

lies, genus, and species. The genus and species form the scientific names of insects. These divisions within the Insecta class are based on similarities and differences among the animals. Table 13-1 presents a summary of the common orders of insects.

Mites and spiders are often associated with insects, but they are not in the Insecta class. Mites, spiders, ticks, tarantulas, and scorpions are in the Arthropoda phylum, but they are in a different class: Arachnid. Arachnids have four pairs of legs and other body structures different from insects.

BENEFIT CLASSIFICATION OF INSECTS

A **beneficial insect** is one that is of value for the role it fills in the environment. These insects perform activities that help humans in providing for their needs. The benefits may include pollinating plants, preying on other insects, and producing products, such as honey, that people use. Table 13-2 lists examples of beneficial insects.

13-8. The tiny phorid fly imported from Brazil attacks fire ants and injects an egg into their heads. The developing larva eats the brain and other contents of the head destroying the ant. The fly was released in the United States in 2000 as a biological control of fire ants and is only about one-sixteenth of an inch long. (Courtesy, Agricultural Research Service, USDA)

Table 13-2. Examples of Beneficial Insects

Common Name	Benefit
bees, butterflies, moths, and flies	pollinate plants
honeybees	produce honey and beeswax
ants	aerate soil
ladybug beetles, mantids, and lacewings	prey on harmful insects
silkworm moth	cocoons provide silk fiber
honey ants, flying ants, and and grasshoppers	human food
scarab beetles	help decompose carrion (dead flesh), dung, and vegetation

A **harmful insect** is one that causes damage to plants, animals, or property. They injure and destroy what they attack. Some insects inflict pain on humans and other animals, with the sting of a wasp being an example. In other cases, they feed on crops, ornamental plants, or the houses we live in. Insects may transmit disease and contaminate food. Only about 1 percent of all insect species are labeled as harmful. Table 13-3 lists examples of insects that harm plants.

Table 13-3. Examples of Insects Harmful to Plants

Common Name	Damage Caused
grasshoppers, aphids, some beetles, and caterpillars	feed on plant foliage
ear worms and bud worms	feed on fruit and buds of plants
some weevils	feed on grain and other fruit, such as cotton bolls
ants, including fire ants	undermine plant root systems, damage turf and ornamental plants

MOUTHPARTS CLASSIFICATION

Insects can be classified according to mouthparts. The kind of mouthpart determines how an insect feeds. Control measures must be selected based on the way they feed. Insects are classified on the basis of feeding, such as chewing or sucking.

REPRODUCTION CLASSIFICATION

Insects are classified by the kind of metamorphosis they have. The stage of development of an insect is important in assessing the damage they cause plants. These stages also influence the methods used to manage insect pests and control the damage they cause.

FEEDING LOCATION CLASSIFICATION

Insects feed in different locations on plants. Where an insect feeds may or may not be something that can be seen immediately. Feeding location also determines the control methods that can be effective.

External feeding insects chew or suck from the exterior of the plant. They feed on the leaves, stems, buds, or fruit. The damage is often easy to see. Holes in plant parts are good signs. In some cases, leaves can be eaten from a plant in a matter of a few hours if a large number of insects are present. Armyworms are an example.

13-9. Cotton bollworms are external feeders. (Courtesy, Agricultural Research Service, USDA)

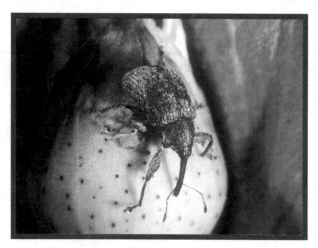

13-10. The cotton boll weevil lays eggs in a cotton boll or square (flower bud). When the eggs hatch, the larvae feed on the inside of the boll, destroying it. (Courtesy, Agricultural Research Service, USDA)

Internal feeding insects are of the chewing type that make an opening in the plant and go inside. The insects feed internally on the plant tissues. The damage may not be evident for several days or longer. Plants may wilt, die, or break over because of a weakened stem. The pine borer damages

13-11. Damage from borers resulted in the death of these pine trees in the Chincoteague National Wildlife Refuge in Virginia. (The borers went under the bark and around the tree, disrupting the flow of sap. The tree died because water could not get to the leaves and food could not get to the roots.)

CONNECTION

COMMUNICATION PROVIDES IMPORTANT DETAILS

Today's crop producers need good information to make "smart" decisions. Radio, television, magazines, and other media are important sources of helpful details. In many cases, the media events are prepared by a trained agricultural communicator working with a scientist and a producer. This keeps information practical as well as technically accurate.

This shows a television production underway in a field. The information will provide practical, up-to-minute decision-making details. Many states have statewide television networks that share in program delivery in agriculture. (Courtesy, Mississippi State University)

pine trees by feeding internally between the bark and the wood. After a long enough time, the tree dies because the borer has girdled the tree so water and food cannot move through its trunk. Other examples include the European corn borer, cotton boll weevil, and corn earworm.

Subterranean insects are species in the soil that attack the roots of plants. In some cases, they may attack root-type structures, such as potatoes or peanuts. Both chewing and sucking insects may be involved. Damage is not readily apparent. Examples of subterranean insects are corn rootworms, wireworms, and white grubs.

NEMATODE CLASSIFICATION AND BIOLOGY

As members of the Nematoda phylum, nematodes are in the same phylum as pinworms and hookworms. Those that cause damage to plants are typically very tiny species that live in the soil, though some are on leaves, stems, and buds. Species that attack aboveground plant parts are known as foliar nematodes. Because of their shape, they are sometimes known as roundworms, threadworms, or eel worms.

Many plant pest nematodes are so small that they can only be seen with a microscope. Nematodes are often no larger than 0.2 to 10.0 mm, with most of those that attack plants being smaller than 2.0 mm. Scientists have identified 132 different species of nematodes that attack plants. Nematodes cause far less damage to plants than insects. However, an infestation of nematodes in a field can destroy a crop.

Nematodes damage plants by piercing and sucking juice or tunneling inside the roots. They secrete a substance that injures roots. The injury allows bacteria and fungi to enter, which can cause disease. **Root knot nematodes** are most common. Enlarged areas appear (knots) on the roots of plants where the nematodes have fed.

Nematodes have a complete reproductive cycle. Some species bear young, while females of other species lay eggs. Most reproduce sexually, though some reproduce asexually. Plant debris, roots, and soil protect eggs and larvae through cold winter months. After hatching, the nymphs go through

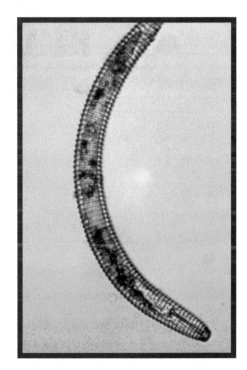

13-12. Greatly enlarged photograph of a nematode. (Courtesy, Frank Killebrew, Mississippi State University)

13-13. Golden nematode cysts (knots) on roots of plants. (Courtesy, Agricultural Research Service, USDA)

four molts before reaching adult. Adults have a sharp spear that punctures plant cells. Their activity can transfer disease from one plant to another.

The typical early symptoms of nematode damage are lower plant vigor and reduced resistance to disease. Damaged plants may die or have a low crop yield. Foliar nematodes cause malformed stems and leaves and dead spots on leaves. Some cause leaf galls. A leaf gall is an abnormal growth resembling a bubble or cyst in the leaf or other structure of a plant.

METHODS OF INSECT AND NEMATODE MANAGEMENT

Preventing damage by insects and nematodes requires good information. Four questions should be answered. These help in selecting and using a method of pest management. The questions are:

- What is the pest species to be managed? Proper identification of the pest is essential. Measures used are based on the species and the way it feeds.

- What methods of management are used? Different methods can be successful. Select a method that is appropriate.

- What is the level of damage? Damage by pests should be at a level that merits action. Minor damage may not justify using pesticides.

- How are management methods properly used? Some methods of pest management can be dangerous to people, other living organisms, and the environment. Proper use is essential.

SCOUTING AND THRESHOLD

The suspected presence of insects does not provide enough information. Management measures are expensive and have other effects, such as killing beneficial insects. Making a good decision is based on accurate information. Two methods are used in determining if and when to take action against insect pests: scouting and threshold.

Scouting

Scouting is the process of visually inspecting for the presence of insect pests and damage. This requires going into fields or greenhouses. Looking at plants closely for evidence of damage or eggs of pests is one step. Opening leaf folds and areas around buds allows closer inspection. Another process is using nets and traps to collect samples of insects.

Finding a number of insects may mean that the population is high. Seeing evidence of damage is also a sign. Some damage may not justify costly measures. Growers have learned to tolerate some damage.

13-14. Scouting canola involves assessing insect populations and damage. (Courtesy, U.S. Department of Agriculture)

Threshold

Threshold is the density of the pest population that will justify using pest management measures. Two kinds of thresholds are important: economic and aesthetic. An **economic threshold** is the balance of cost with returns. Minor damage or a low population density does not usually justify spending money on management. There are exceptions. Early in the season, the presence of adults may indicate that future problems are likely. An aesthetic threshold deals with the appearance of plants, such as turf. Some measures are used based on damage being objectionable to people.

Some crops are also more sensitive to damage. Consumers will not tolerate any damage to some crops, such as cut flowers and sweet corn. Consumers may not be aware of insect damage in other crops, such as cotton or soybeans.

13-15. Insect traps are used to assess pest populations.

13-16. The small damage by chewing insects to the leaves of these purple hull peas is not great enough to justify control.

13-17. All luggage being checked at the Honolulu Airport is inspected to prevent tourists from bringing pests to the mainland.

Low pest population densities merit action in those crops where consumer acceptance of injury is unacceptable.

SELECTING MANAGEMENT METHODS

Many different methods can be used to manage insects and nematodes. Selecting a method is an important decision. Growers want good results as well as to safeguard the environment.

Quarantine

Quarantine is the isolation or exclusion of a pest or pest problem. It includes keeping pests away through inspection programs for fruit, seed, and other plant products. Using only clean seed or plants is a big step. Insects and nematodes can sometimes hide in flowers, leaves, soil, or other materials.

Another part of quarantine is making the growing or possession of a certain crop illegal. This has been widely practiced with cotton. Growing a few stalks of ornamental cotton in a flower bed or garden is prohibited. Boll weevils can hide in these plants and later attack cotton grown for fiber. This would delay attempts to eradicate the pest.

Cultural Management

Cultural management is preventing insect and nematode problems by the practices used in growing the plants. These are based on knowing the life cycles of insects and nematodes as well as other habits of the pest. Good knowledge of the plant being grown is also needed.

Several methods of cultural insect management are:

- Crop rotation—Planting similar crops in succession on the same land builds up pest populations. **Crop rotation** is planting different crops on land in alternating years or seasons. This reduces infestations of pests. In addition, crop rotation reduces disease problems and makes better use of nutrients in the soil. Rotating corn and soybeans—a monocot and a dicot legume—is one example.

- Residue management—Leaves, stalks, roots, and other plant materials provide a place for pests to hide. Chopping these materials makes them less protective of insects. Practices of burning straw and other residue are not widely used because of air pollution problems. Removing all of the residue from a field is not a good practice either because the soil is left unprotected from erosion.

13-18. White flies on this gardenia need to be controlled with a pesticide.

- Trap crops—**Trap cropping** is planting a small plot of a crop near a field several days before the crop itself will be planted. Insects are attracted to this plot where they can be destroyed. This gets rid of the adult population before it reproduces larger numbers that will attack the main crop.

- Using resistant varieties—New varieties of some crops have been developed that repel pests. In some cases, these plants produce insect repellants. Genetic engineering techniques have been used to produce resistant varieties of potatoes, cotton, and other crops.

- Sanitation—Sanitation is keeping areas around fields or other places where plants are grown clean. Cutting weeds and bushes that serve as hiding places helps. Using healthy plants and seeds that are free of insects and nematodes helps prevent infestations. Sanitation also includes properly disposing of wastes at processing plants. Insects may hide in bean hulls, leaf and stalk fragments, and other discarded parts.

Biological Management

Biological pest management is using living organisms or processes to control insects and other pests. Some of this naturally occurs when a beneficial insect eats a pest. (Ladybugs and some wasps are well known for this role.) Other biological pest control requires the careful work of scientists. Here are a few examples:

13-19. A parasitic wasp is flying toward a worm to attack and destroy. (Courtesy, Agricultural Research Service, USDA)

• Releasing beneficial insects—Releasing ladybugs, mantids, and other pest-eating insects is effective in some situations. Chemicals to kill insects cannot be used because the beneficial insect will also be killed.

• Disrupting reproduction—Disrupting the reproductive process of pests helps control some pests. Knowing the processes of the species is essential. With some species, a female mates only once. If the male is sterile, no fertile eggs are produced. Males of these species have been sterilized in a laboratory and released. Mating results in no offspring.

• Using bacteria and fungi—Some forms of bacteria and fungi attack worms and other pests. Releasing these in the field may be helpful in pest control. A good example is a bacterium known as *Bacillus thuringinensis*. It is used to help control some worm (larvae) forms of insects.

Chemical Use

Approved chemicals can be used to manage insects and nematodes. Many pesticides are highly poisonous. This means that they should be used carefully and only as approved.

Pesticides used especially with insects are known as insecticides. Those used to control nematodes are **nematocides**.

Insecticides

Insecticides are classified by how they get into an insect's body. The grower must also know the feeding characteristics of the insect, such as chewing and sucking. Three general groups of insecticides, based on how they enter the body, are:

13-20. Insecticide is sprayed on peach trees. (Courtesy, U.S. Department of Agriculture)

- Contact—A **contact insecticide** is absorbed through the skin or exterior of an insect. The insecticide must come into contact with the insect. This means that the material must be applied directly on the insect. Those that are hiding or out of the application area are not contacted by the poison. In some cases, the insect may move into the treated area and come into contact with the insecticide. Contact poisons are used with sucking insects.

- Stomach—A **stomach insecticide** is effective when eaten. The material must be on the leaves, seed, fruit, or other parts of the plant that are eaten. Stomach poisons work best on chewing insects, such as leafworms, armyworms, and bagworms.

- Respiratory—Respiratory insecticides work by entering the respiratory system of insects. These are known as fumigants. A **fumigant** is an insecticide that is in gaseous form. They are used in enclosed places, such as greenhouses, grain bins, and in the soil covered with plastic.

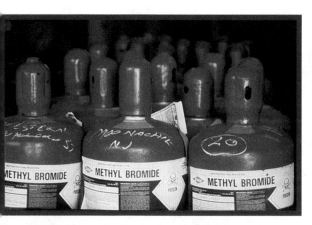

13-21. Methyl bromide has been a widely used fumigant. Its use has been phased out except in special cases because it contributes to depletion of the ozone layer in the atmosphere. (Courtesy, U.S. Department of Agriculture)

13-22. A fumigant is being applied to the soil under the plastic cover.

These three groups may not act independently. A contact poison may also kill an insect if it is eaten. Likewise, a stomach poison may kill an insect if it gets into the respiratory system or is eaten.

Some insecticides are systemic. A **systemic insecticide** is one applied to plants and absorbed into the plant system. If an insect chews or sucks on the plant, poison enters its body and the insect is killed.

Insecticides are manufactured in different forms and with varying amounts of active ingredient. **Formulation** is the way the pesticide product is prepared. It may be dissolved or suspended in a liquid or in a dust or powder form. Insecticides may be as dusts, granules, mixtures with fertilizer, aerosols, wettable powders, or other forms.

Active ingredient is the percent of poison material in an insecticide. The active ingredient is the material that kills insects. The amount of active ingredient may vary widely.

Table 13-4. Common Formulations of Insecticides

Type of Formulation	Description
aerosols	tiny particles suspended in liquefied gas sprayed in the air; used primarily in enclosed areas, such as greenhouses and grain bins
dusts	fine dry particles of materials containing insecticides; may be dusted on plants
emulsifiable concentrates	liquid (often oil-in-water) materials containing insecticide; usually mixed with water and a surfactant and sprayed onto plants
fumigants	gaseous or vapor form; insecticide is held in liquid until released from pressurized container
granules (granular)	similar to dusts, but much larger particles; often used on the ground around plants
solutions	liquids containing insecticides; often concentrated, water-based for dilution and spraying on plants
wettable powders	materials similar to dusts mixed with water for application as a liquid; may be highly toxic

Table 13-5. Examples of Insecticides

Common Name	Trade Name	Toxicity Category*	Crop/Use
methyl parathion	Declare, Pennicap-M, others	I	Corn, cotton, and other field crops to kill aphids, army worms, and others
carbaryl	Sevin	II**	Many vegetables, ornamental plants, and field crops to keep down certain beetles, borers, and other insects.
permethrin	Ambush, Pounce, others	II	Many vegetables, some fruits, landscape plants, and field crops to manage cutworms, corn earworms, certain beetles, thrips, and others.
derris	Rotenone	III**	Fruit and vine crops, flowers, shrubs, and vegetables to keep down a wide range of insects.
piperonyl butoxide	Pbo, brand labels of house and garden insecticides	IV	Houseplants and vegetables to manage whiteflies, thrips, beetles, stinkbugs, and others

Note: Only a few examples are included; no recommendations are implied; refer to labels for information on use; many restrictions exist; changes occur frequently; use insecticides only in approved ways.

*"Toxicity category" has four categories: I=highly toxic, II=moderately toxic, III=slightly toxic, and IV=low toxicity.

**Could have higher or lower toxicity depending on formulation.

Most bottles, cans, and bags of insecticide are filled largely with water or an inert material known as a carrier.

Labels on insecticides provide information about the active ingredient. It is the active ingredient that kills insects. Some products on the market have a low amount of the active ingredient. Use this information as a basis for deciding which product to buy. Reading the label is essential.

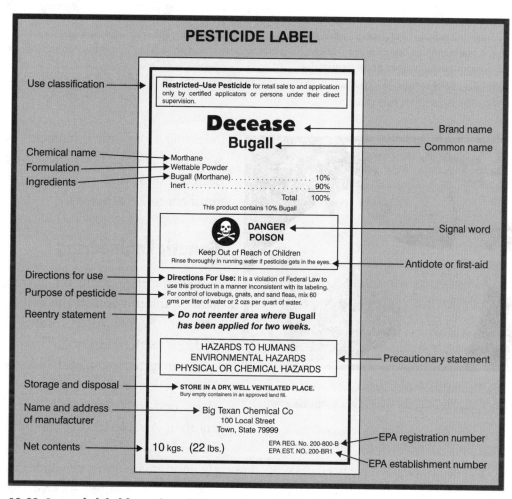

13-23. A sample label for an insecticide container. All labels should present similar information about the approved and safe use of the product.

Nematocides

Most nematodes are in the soil, though a few are on the foliage of plants. Chemicals used to manage soil nematodes must be incorporated into the area where the nematodes live. The soil is usually tilled and the nematocide is injected. Treating small areas often involves

Table 13-6. Examples of Nematocides[1]

Common Name	Trade Name	Class[2]	Crop/Use
aldicarb	Temik	R	nematodes of soybeans, peanuts, cotton, and others
fenamiphos	Nemacur	R	major nematodes in field crops, turf, and vegetables
1,3-Dichloropropylene	Telone	R	nematodes, wire worms, and others in tomatoes, peppers, strawberries melons, grapes, and 112 others

[1]Only a few examples are included; no recommendations are implied; refer to labels for information on use; many restrictions exist; changes occur frequently.

[2]Class refers to classification as follows: R-restricted use product by EPA; G-general use product not restricted by EPA; and S-state or regional restrictions.

13-24. Preparing a soil sample for nematode analysis in a laboratory. After analysis, the lab may assess economic threshold.

releasing a fumigant nematocide under a sheet of plastic. The plastic is tightly sealed around the area being treated until control is achieved.

Genetic Engineering

Genetic engineering is changing the genetic information in the cells of an organism. Sections of DNA are cut out and new sections inserted. This process is known as recombinant DNA. By isolating genes, a gene can be removed from one organism and placed in another. A selected trait can be transferred from one organism to another.

Insect-resistant plants have been developed using genetic engineering. Pesticides are not needed on these plants to control the targeted pest. An example is cotton—a crop that has needed large amounts of pesticide. In 1995, the Environmental Protection Agency approved Bollgard cotton (also known as Bt cotton). This cotton was developed by the Monsanto Company.

Using recombinant DNA, a gene was moved from the *Bacillus thuringinensis* bacterium to cotton. The resulting plants grown from seed have season-long immunity to pests in the Lepidoptera order. The pests include bollworms, budworms, and pink budworms. Among others, the seed is marketed as NuCOTN 33b and NuCOTN 35b. The leading producer of the seed is Delta & Pine Land Co. of Scott, Mississippi.

A similar procedure has been used with soybeans to produce a nematode-resistant variety. The soybean is resistant to common species of cyst nematodes.

Plants altered by genetic engineering are known as transgenic. *Transgenic* means that the natural genetic material of the plant has been changed. Scientists feel that many other pests will be controlled with transgenic methods in the near future.

13-25. Demonstration plots are used to compare insect resistance of transgenic corn varieties to nontransgenic varieties.

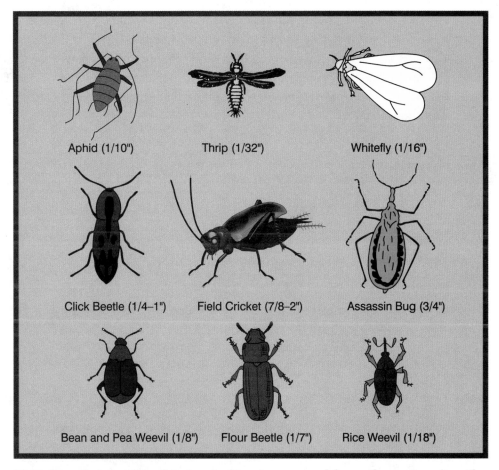

13-26. Examples of common insect pests. (Drawings are much larger than mature sizes. The measurements by the name of the insect provide an average mature length in inches.)

REVIEWING

MAIN IDEAS

Plants can be damaged by insect and nematode pests. Controlling these pests improves the quality and amount of production. Decisions about control should be made based on good information.

Insects are small, boneless animals. Some are beneficial; others are harmful. Knowing the biology of insects helps in selecting methods of control. Insects have bodies with three segments: head, thorax, and abdomen. A hard exoskeleton made of chitin protects the organs and serves as the place for attaching muscles inside the body.

Insects feed in two primary ways: chewing and sucking. Some feed on the exterior of plants while others feed internally and below the soil level. Damage by insects is often done at different stages of their development. Insects have either complete or incomplete metamorphosis. The larva stage often causes the most damage to plants.

Nematodes are tiny worms that live in the soil or on the foliage of plants. Most damage is by cyst nematodes that live in the soil and form knots or cysts on plant roots.

Various methods are used to control insects and nematodes. Scouting helps determine if the population is causing damage that merits control. Threshold is the population level that will justify control. Methods of control include quarantine, cultural control, biological control, chemical control, and genetic engineering.

QUESTIONS

Answer the following questions using complete sentences and correct spelling.

1. What is an insect? Distinguish between useful and harmful insects.

2. What are the major biological characteristics of insects?

3. How are insects classified on the basis of feeding? Why is this important?

4. Distinguish between complete and incomplete metamorphosis. Explain why these differences are important in insect pest control.

5. What are the major biological characteristics of nematodes?

6. What questions should be answered in selecting and using pest control?

7. What are the methods of managing insects and nematodes? Briefly explain each.

8. Genetic engineering is a new method of insect management. Give one example of how it has been used to control insects.

9. How are insecticides classified by how they enter the body?

10. What is active ingredient? Why is it important in buying insecticides?

EVALUATING

Match the term with the correct definition. Write the letter by the term in the blank that is provided.

a. insect
b. beneficial insect
c. chewing insect
d. sucking insect

e. metamorphosis
f. internal feeding insect
g. quarantine
h. economic threshold

i. formulation
j. genetic engineering

1. _____ way a chemical pesticide product is prepared
2. _____ bites off, chews, and swallows plant parts
3. _____ developmental stages of insects
4. _____ pierces the outer layer of a plant and sucks juice from the inside
5. _____ creates an opening and goes inside a plant
6. _____ changing the genetic information in the cells of an organism
7. _____ isolation or exclusion of a pest or pest problem
8. _____ an insect that is of value for the role it fills
9. _____ small, boneless animal with six legs and three body segments
10. _____ density of pest population that will justify control measures

EXPLORING

1. Scout a field or greenhouse for insect pests. Look for damage to leaves, buds, stems, fruit, and other plant structures. Note the presence of insects. Collect specimens of the insect forms and properly identify the pest. Determine if the economic threshold has been met to justify the application of a pesticide. Get the assistance of an entomologist in the local area. Write a report on your findings. Give an oral report in class.

2. Determine the pesticide certification training that is available in your local area. Contact the local office of the Cooperative Extension Service or the state Department of Agriculture for assistance.

3. Select a common insecticide or nematocide. Prepare a report on the product. Include its common and scientific name as well as other details. Indicate the pests for which it is approved. List the percent of active ingredient and the formulation. Note safety practices to be followed in using it.

4. Explore insect identification on the World Wide Web Insects Home Page:

http://web.css.orst.edu/Topics/FQT/Insects/#Identification

CHAPTER 14

Diseases

OBJECTIVES

This chapter covers the fundamentals of plant disease. It has the following objectives:

1. Explain diseases as related to plants.
2. Describe the types and causes of plant diseases.
3. Explain how common plant diseases are identified.
4. Identify common plant diseases.
5. Describe ways plant diseases are spread.
6. Describe ways plant diseases are managed.

TERMS

abiotic disease
aflatoxin
bactericide
biotic disease
blight
canker
damping-off
gall
leaf spot
mosaic disease
pathogen
plant disease
protectant
rot
sign
smut
soil solarization
stunt
systemic
vector
wilt

14-1. Examining mature peanut plants at harvest time for signs of disease. (Courtesy, U.S. Department of Agriculture)

PREVENTING a plant disease is far easier than trying to cure one. Plants are subject to disease much as other organisms. Sometimes, plant diseases are more difficult to identify than insect damage. Diseases are often not as visible as insect damage until an advanced stage has been reached.

Plant diseases have had major impacts on world history. Nearly one-third of the population of Ireland died of starvation in 1845 and 1846 because of the potato famine caused by blight. A similar famine in Germany in 1917 resulted in severe food shortages that led to a quick end to World War I. Ceylon was replaced as the coffee-producing center of the world because of coffee rust. In the United States, chestnut trees have been lost to blight and Dutch elm disease has destroyed most American elm trees.

Plant diseases can create serious problems. Fortunately, ways of managing them are available.

PLANT DISEASE

A ***plant disease*** is any abnormal condition of a living plant. An entire plant or only a small part may be affected. The normal activity of the cells, tissues, or organs is disrupted. The plant is less productive or may not produce at all. Growth and maturity are reduced.

14-2. Tomato damaged by blight—a fungus *(Sclerotaeun rolfsii)*. (Courtesy, Jackie Mullen, Auburn University)

Disease may also create an abnormal appearance and make products from plants unusable. Functions in the plant are reduced. Plant disease may result in the death of the plant.

Plant diseases cost producers large amounts of money each year. The cost is related to control measures and the value of lost production. Lost production may create higher prices to consumers because fewer products are on the market. In the United States, production is abundant. Stored reserves are adequate to sustain the nation for a year or so with most staple foods. A crop failure one year due to widespread disease could result in depletion of food reserves.

As a nation with a large geographical area, regional disease problems do not often threaten the entire nation. For example, losses of citrus due to cold weather in Florida are often overcome by production in Texas, California, and other locations.

14-3. Red spots on this photenia (an ornamental shrub) are signs of a fungus disease known as *Entomosporium.*

TYPES AND CAUSES OF DISEASE

Preventing and treating diseases begins with knowing the types and causes of diseases.

*T*YPES OF DISEASE

Plant diseases are of two major types: biotic and abiotic. Both types result in serious losses among food, fiber, and ornamental plants.

Biotic Diseases

A **biotic disease** is caused by a pathogen. A **pathogen** is a living, disease-producing agent, such as a bacterium, fungus, or virus. Pathogens are typically microscopic in size and are transmissible from one plant to another. These tiny organisms live in or on plants. Laboratory analysis is often needed to confirm biotic disease.

The three causes of biotic disease are briefly described here. (Biotic diseases are also known as parasitic disease. This is because they involve the disease organism gaining nourishment from the plant host.)

14-4. An apple destroyed by rot and bird damage.

- Bacteria—Bacteria are one-celled organisms in the Monera Kingdom. They are so small that 10,000 would need to be placed side by side to make a distance of one inch (2.54 cm).

  Most bacteria do not cause disease. However, several different species of bacteria do cause biotic disease. Only 170 species of bacteria have been identified as plant disease pathogens. They enter plants through cuts, breaks, or insect damage in the epidermis or bark. A few bacteria may enter through flowers or natural openings

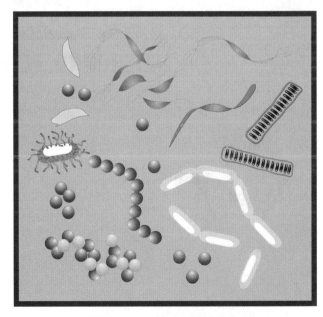

14-5. Drawings of magnified bacteria.

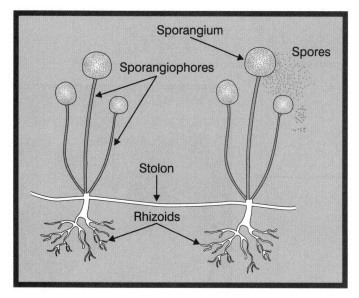

14-6. Greatly enlarged drawing of one kind of fungus—mold. Notice the stolons and spores used in reproduction.

in stems and leaves. Another entry source is the site where fruit has been removed from the plant.

Bacterial diseases include gall, blight, rot, wilt, and retarded growth.

• Fungi—Fungi cause more plant diseases than any other pathogens. Fungi are small, single or multi-celled organisms that attack plants. They may grow on or in plants. Mold on bread is a common example of fungi, though the kind on bread does not usually cause disease in living plants.

Some fungi cause particular problems in grain before and after harvesting. The fungus *Aspergillus flavus* produces a toxin in corn known as **aflatoxin**. If not prevented, the corn cannot be used to manufacture food and feed products. Good production practices, harvesting early, and drying corn to reduce moisture content help prevent the formation of aflatoxin. Reliable methods of detoxification have not been developed.

CONNECTION

MICROPROJECTILE UNITS SHOOT DNA

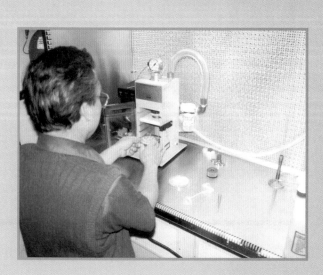

Genetic engineering is being used to develop plants that resist disease. This involves altering the genetic material of the plant. A small section of the desired DNA (plasmid) is inserted into the plant's existing DNA. Because the structures being modified are so small, special devices are needed to do the work.

Microprojectile units, also known as gene or particle guns, "shoot" the desired DNA into the DNA of the plant to be altered. Tiny particles of gold serve as carriers. Propulsion is much like firing a gun or air rifle. Because of the delicacy of the work, the success rate is often not high, though the process is expensive.

This photograph shows a microprojectile unit being used in a laboratory at the University of Illinois.

Fungi grow best in warm temperatures and high humidity. Wind, water, insects, seed, and plants themselves transfer disease. Fungi reproduce by forming spores that germinate like seed and grow strands into plants. Once inside, the strands feed on plant tissues. The tissues gradually die. Fungi that attack living organisms are parasitic. Other species live on decaying organic matter and are known as saprophytic fungi.

Fungi are best known for causing such plant diseases as rust, root rot, leaf spot, canker, and mildew.

- Viruses—Viruses are the smallest causes of biotic disease. Some scientists do not consider viruses living organisms because they do not have organized nuclei in their cells. On the other hand, viruses reproduce and act as living organisms when inside tissue. Powerful microscopes are needed to see a virus.

  Viruses are transferred by insects, nematodes, soil, seed, vegetative propagation, and tools and equipment used with diseased plants. Examples of viral diseases include tomato ring spot, cucumber mosaic, and citrus tristeza. Viruses often create mosaic patterns in leaves.

Table 14-1. Examples of Biotic Diseases

Cause	Disease	Plants Affected
bacteria	wilt (bacterial)	corn, alfalfa, tomato, potato
	gall	crown gall in trees and many grops
	local infections	fire blight of pear, bacterial blight of snapbean, angular leaf spot of cotton
fungi	anthracnose	cotton, cucumber, cantaloupe, bean, watermelon
	downy mildew	grain crops, grape, onion, spinach, lettuce, cucumber
	powdery mildew	grain, cucumber, rose, zinnia, chrysanthemum
	leaf spot	apple scab, rose blockspot, many others
	wilt (fusarium and verticillium)	cotton, tomato, sweet potato, watermelon
viruses	mosaic	tomato, potato, sugarcane
	stunt	corn
	streak	sugarcane

Abiotic Diseases

An **abiotic disease** is not caused by a pathogen. These diseases are caused by elements in a plant's environment that are damaging to the plant. Causes vary widely. Since abiotic diseases are associated with the environment, they are known as environmental diseases.

Causes of abiotic disease include the following:

- Nutrient deficiencies—Plants that do not have enough nutrients are diseased. Nutrient deficiencies often show as discolorations in plant leaves or a definite lack of plant growth. Soil testing and tissue analysis may be needed to define the nutrients needed to correct the problem. Not having enough water will cause plants to wilt and die.

14-7. Crop plants with nutritional diseases, as follows: top left—corn ear with boron deficiency; top right—wheat with manganese deficiency; bottom left—potato leaf with potassium deficiency; and bottom right—soybean with zinc deficiency. (Courtesy, Potash and Phosphate Institute)

14-8. Corn damaged by wind.

Usually, this nutritional disease can be corrected, except in large fields without irrigation or adequate rainfall. (Refer to the chapters on soils for additional information.)

- Weather—Plants are damaged by weather extremes. Hail, wind, rain, and other conditions associated with the weather can damage plants. Late or early frosts can quickly end the growth of tender plants.

- Climate changes—Climate is the average of weather conditions over a long time. Major global climate changes are predicted by some scientists, known as global warming. As this occurs, plants adapted to a certain climate will become diseased unless they can adjust. This will also allow crops that would be diseased in one location to be grown elsewhere without disease. As the earth gradually warms, corn can be grown farther north in North America.

- Pollution—Toxic substances in the environment damage plants. Smoke and exhaust from engines and factories cause injury. Chemicals in runoff can kill grass and other plants. In recent years, many sources of pollution have been controlled. A common source of pollution is pesticide drift resulting from air movement. This is the movement of a pesticide used on an approved crop to a field with a sensitive crop. This results when pesticides used to control weeds along roadways drift into fields or lawns. In the North, salt applied to melt snow on roads also damages nearby soil.

14-9. Plant damaged by pollution.

- Mechanical damage—Mechanical damage occurs when parts of plants are damaged by machinery, animals, or in other ways. Plowing too deeply or too close to the roots of plants cuts and injures plants. Animals, such as cattle or deer, in a field eat leaves and trample plants.

- Mutations—A mutation is a naturally occurring genetic change in organisms. Mutations are sudden and unpredictable changes from the parents' characteristics. A plant may appear that has a characteristic drastically different from its parents. These sometimes create problems; other times, they create special new kinds of plants, such as those with beautiful flowers.

IDENTIFYING PLANT DISEASES

Growers must be able to detect when a plant is diseased. Food processors, lumber and paper manufacturers, and ornamental horticulturists must know the symptoms of plant disease. Consumers also want to know about plant diseases so they can reject products that show disease signs.

SYMPTOMS OF DISEASE

Diseases show themselves in plants as symptoms or signs. A symptom is how a plant responds to disease. A **sign** is the presence of pathogen structures, such as rust spores on a leaf. A plant with a mild disease, such as a nutrient deficiency, may have no symptoms or signs that can be seen.

Anyone working with plants needs to know the normal conditions of plant growth. This comes through experience and education. Since each species varies somewhat, knowing the normal growth characteristics of individual species is important.

Here are several symptoms and signs of disease in plants:

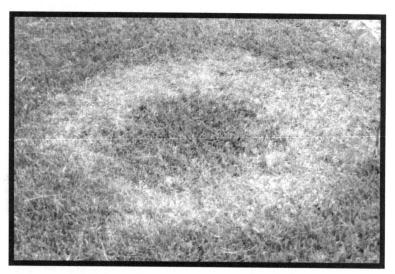

14-10. The round, doughnut-shaped damage in this St. Augustine lawn is most likely due to a fungus disease known as brown spot.

- Leaf color—All healthy plants have leaves and stems within a fairly narrow color range. When the color changes, disease should be suspected. When leaves turn yellow or become mosaic (mottled colors), a plant is most likely diseased.

- Decay—Rotting plant parts, particularly fruit, shows disease. Fruit may rot before maturity, with blossom-end rot of tomatoes being an example. Most of these conditions are due to nutrient defi-

14-11. Cabbage damaged by rot and insects.

ciencies or injuries to plants that allow pathogens to enter. Rotting tree trunks and limbs is also a symptom of disease.

- Wilting—Wilting occurs when a plant does not have enough moisture for the cells to remain turgid. As cell walls become soft or collapse, a plant wilts. The first symptom is for leaves and stems to appear droopy or folded. Unless a plant is provided needed water, it will stop growing, will not flower or bear fruit, and may die. Disease can cause wilting by disrupting normal water movement processes.

14-12. Deformity in leaves is a symptom of disease, such as in this potato leaf. (Courtesy, Potash and Phosphate Institute)

- Deformity—Leaves, stems, fruit, and other plant parts may be deformed. The growth varies from normal. Examples include fruit that is not shaped properly or a plant that has twisted leaves. Deformity may be due to several different causes, including nutrient deficiencies, insect stings, and chemical pollution.

- Reproductive failure—Plants show symptoms of reproductive failure when buds, flowers, and fruit fail to develop properly or fall off before maturity. This can be a serious problem to crop producers. A good example is the pecan. In recent years, pecan trees have been affected by a scab disease. The trees may flower profusely, but no pecan nuts develop.

- Dead plants—Diseases often kill plants. During the normal growing season, plants die of disease. A smart grower keeps a careful watch on plants and would see a problem before the death of a plant.

14-13. Pecan trees with scab disease may flower profusely, but produce no pecans. This pecan tree has an abundance of flowers—known as catkins.

USING LABORATORY ANALYSIS

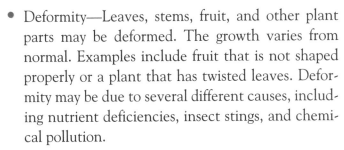

Some plant diseases are very difficult to accurately identify by symptoms and signs. Laboratory analysis of plant tissue may be needed. Samples of diseased tissue are examined using microscopes and other means. Sometimes, organisms may be cultured in a laboratory for additional study.

Tissue testing is carefully examining samples of plant tissue to diagnose disease. It is often used to accurately detect the presence of disease. Testing also helps identify exact causes of environmental disease from pollution. Land-grant universities in most states have plant disease clinics. The local Cooperative Extension Service office can usually provide information on how to send samples for testing.

COMMON PLANT DISEASES

Plants are subject to different diseases. Since plants grow in a wide range of cultural conditions, diseases vary with these conditions. Plants often have diseases that are specific to the species. Other diseases tend to affect more than one species of plant. Diseases of specific crops are included in chapters on those crops.

Several general kinds of plant disease are:

- Gall—**Gall** is an abnormal growth on plants. It is caused by the presence of another organism in the plant. The growth may appear as large rounded swellings on the stems. Galls contain abnormal cells created by the presence of bacteria. Gall is more frequently found on forest trees, fruit trees, grapes, roses, and some species of berries. Insects may injure plants resulting in galls. Galls may also be infected with fungi. The wart-like growths caused by fungi are known as gall rust.

14-14. Gall on a pine tree.

- Blight—**Blight** is the withering and, perhaps, the death of a plant. It typically causes plant colors to change and reduces the green leaf area that carries on photosynthesis. This lowers the productivity of a plant. The plant will die in extreme cases. Blight is usually caused by bacteria, though fungi and viruses are also associated with it. Fire blight in apples and pears is an example.

- Canker—A **canker** is a disease on the bark and in the tissue that is beneath it. Thought to be caused by fungi and viruses, cankers cause spots on the stems of woody plants, such as roses and fruit trees. The canker may begin with a purplish color and change to white as it develops. Affected branches and leaves wilt and fall over.

- Leaf spot—**Leaf spot** is a disease of plants with yellowing around dead areas on leaves, stems, and fruit. It is usually due to bacteria or fungi. The fungal form of the disease may develop in humid locations following wet weather. Bacterial forms result when bacteria enter plants through the stoma and cause a breakdown in the cells. Cotton and tomatoes are two examples of plants affected by bacterial leaf spot.

- Rot—**Rot** is the decay of plant tissue. Bacteria typically cause rot, but fungi may also be a cause. Vegetables and fruit are often attacked by a form known as soft rot. This form typically develops in fleshy tissues, such as tomatoes and potato tubers. The bacteria produce an enzyme that dissolves cell walls. A slimy mass with a bad odor may develop. Rot often attacks fruits and vegetables that have been improperly stored.

- Wilt—**Wilt** is a disease that affects the vascular system of plants. Most wilt is caused by bacteria. Inside the plant, the bacteria reproduce rapidly and block the flow of sap in the xylem. The affected parts of the plant will wilt and may die, depending on severity. Corn, alfalfa, cucumbers, and sugarcane are examples of crops that get wilt disease.

14-15. Leaf spot in the leaf of a holly tree.

14-16. An orange with rot near the blossom end has no value.

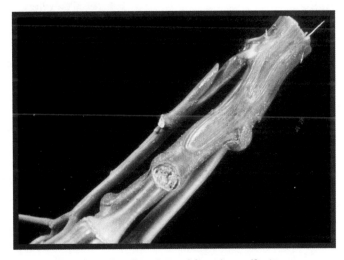

14-17. Okra stem showing signs of fusarium wilt. (Courtesy, Jackie Mullen, Auburn University)

- Smut—**Smut** is a disease caused by fungi. The fungi grow dusty spores that cover grain and other crops. This results in a dark color and offensive odor. Wheat, oats, and barley are primarily affected. Corn and grain sorghum may also have smut. Growers receive a lower price for the grain that has smut.

- Stunt—**Stunt** is a viral disease that causes a plant not to grow. Plants with stunt are not productive. It may also be known as plant dwarfing. Stunt begins in the early stages of growth. The leaves turn yellow and have symptoms similar to those of low nitrogen nutrient supply, low temperature, and too little or too much water. Oats, wheat, and barley are primarily affected. Young plants may be killed quickly by stunt.

- Mosaics—**Mosaic diseases** are caused by viruses that attack cereal grain crops. The chlorophyll in the leaves is destroyed and chlorosis (yellow spots) or other colors appear. The colors are in patches, streaks, or spots. Growth and production are reduced.

- Damping-off (seedling disease)—**Damping-off** is a disease that attacks young seedlings shortly after germination. Small plants that are transplanted are particularly susceptible. Soft plant tissue is destroyed by fungi near the soil line. The stem turns brown, shrinks, and the plant falls over. Tomatoes, sugar beets, cotton, and many other plants are susceptible to damping-off. The disease does not affect older plants.

CAREER PROFILE

PLANT TISSUE TECHNICIAN

Plant tissue technicians examine tissues for disease. The specimens are sent to the laboratory where they are cultured in isolation to prevent contamination from other sources. The work may be done in the aseptic environment of a flow hood.

Plant tissue technicians usually have baccalaureate degrees in plant pathology, botany, or a closely related area. Many have masters or doctoral degrees. Advanced graduate degrees are required for career promotion. Experience in performing laboratory work is essential.

This photograph shows a plant tissue analysis being carried out in the plant pathology tissue laboratory at Auburn University. (Courtesy, Jackie Mullen, Auburn University)

HOW DISEASES ARE SPREAD

Biotic diseases (those caused by pathogens) are spread from one place to another by nature and by human activities. Some natural means can be controlled. Most of the spread of biotic diseases by human activities can be prevented.

INSECTS AND ANIMALS

Plant diseases can be spread from one plant to another by insects and other animals. A **vector** is an organism that transmits disease-causing bacteria, fungi, and viruses. Common vectors, besides insects, are birds, rodents, deer, and humans.

An insect, for example, can chew on a diseased plant and then a healthy plant. Some pathogens may stick on the mouthparts and come off when chewing the healthy plant. The vector organism may attack a healthy plant and it is then diseased. In effect, any organism that can carry organisms from one site to another is a vector.

Humans carry pathogens on their shoes or clothing. Leaves, twigs, or other plant materials stick to the clothing when walking in a diseased field. These may come off while in a healthy field.

14-18. Insects may be disease vectors. A butterfly can transport a disease from one plant to another.

SOIL

Soil provides an environment for the growth of some pathogens. It also serves as a shield to prevent the spread of pathogens. Moving soil from one site to another is a good way to move any pathogens. The pathogens would then infect plants that are not diseased.

Seedling plants can transport pathogens. Soil used with bedding plants is often sterilized to destroy pathogens. This helps prevent the spread of disease.

14-19. Seedling plants, such as these tomatoes, can introduce disease if they are not disease free.

Infected soil from one large bedding plant grower could spread disease over a large area. Inspection is required in many states for producers of bedding plants that are to be sold.

Harvested crops may have soil particles that contain pathogens. Potatoes, sugar beets, onions, and other underground crops generally have some soil on them as moved from the field.

WATER AND AIR

The natural movement of air and water transmits pathogens. Wind may blow spores from one plant or field to another. Diseased leaves may be blown through a forest to infest healthy trees.

Water runoff moves pathogens similarly to the wind. Soil particles, plant parts, and other materials in which pathogens hide may wash from one location to another.

Overall, water and air do not transmit as many plant diseases as do other ways of spreading.

MACHINERY

Plows, planters, harvesters, and other machinery may carry pathogens from one field or farm to another. Even small hedge trimmers or other tools transport plant disease. For example, if a diseased shrub is trimmed with shears, some pathogens will likely get on the shears. The pathogens come off when a healthy plant is trimmed.

All machinery should be disinfected when moved from one farm or major field to another. This may involve washing it with a solution of 10 to 25 percent chlorine bleach water or other disinfectant. Tractor or truck tires, plows, and other parts that touch plants or the soil should be clean.

14-20. Machinery may transport disease from one field to another in soil or plant structures that stick to the equipment.

MANAGING PLANT DISEASES

The same general ways insects are managed apply to plant disease management. The materials used and techniques vary somewhat. Good production practices can help reduce disease problems.

If a disease is prevented, costly treatments to "cure" plants will not be needed. It is far better to prevent than to try to cure a disease.

CULTURAL MANAGEMENT

Cultural methods focus on preventing disease. These methods reduce losses to disease. Managing plant diseases with cultural methods includes at least five procedures.

- Using disease-free and host resistant seedstock—Planting disease-free seed, transplants, and vegetative forms, such as tubers and corms, prevents disease. Seed materials are normally inspected and labeled, indicating any disease material that may be present. Producers who use grafting should be sure that all scions are disease-free. Use seeds treated with chemicals to prevent disease. (Seed treatments may be poisonous. Do not feed treated seed to livestock or use as human food.)

- Using needed soil amendments—Plants need nutrients to grow and be productive. They also have preferences about soil pH and other conditions. Analyzing the soil will be useful in detecting the needed nutrients and the amounts to add to the soil. Precision farming methods help producers balance applications with the needs of the land.

- Planting properly—Plants that get off to a good start are less likely to have a disease. Properly placing the seed or other materials in the soil helps them germinate and grow rapidly. Planting at the proper time of the year prevents damage from the weather. Soil temperature is especially important. Plants that do not come up and grow rapidly are more likely to contract a disease.

14-21. Properly planting good seed allows plants to grow rapidly and resist disease. This young sweet corn shows no signs of disease.

- Rotating crops—Crop rotation prevents the buildup of pathogens in soil. The crops used in rotation should be those that are not subject to the same diseases. For example, corn and soybeans are susceptible to different diseases.

- Preventing injury—Damaged plants are more likely to get a disease. Cuts and bruises on stems and leaves provide openings for pathogens to invade the plant system. Cultivation and other practices should be followed that minimize plant damage. Do not operate plows too deep nor too close to the stems or roots of plants. Be sure the colters and shovels on cultivators are properly adjusted.

- Sanitation—Keeping an area pathogen free and clean helps prevent disease. Destroying diseased plants and plant parts helps prevent transmission to healthy plants.

CHEMICAL MANAGEMENT

Chemical management is the use of chemical materials to prevent and treat diseases. These materials should be used only as approved. Proper safety practices should be followed. (Refer to Chapter 12 for safety information.)

Pesticide Classes

Chemical compounds are selected based on the disease to be controlled. Two common classes of pesticides are used:

14-22. Ground application of fungicide in a citrus grove. (Courtesy, U.S. Department of Agriculture)

- Bactericide—A **bactericide** is a chemical used to control diseases caused by bacteria. These materials may be applied as liquids or dry materials.

- Fungicide—A fungicide is a chemical used to manage fungal diseases. These are used to prevent, as well as treat, diseased plants. More fungicides than bactericides are used to control disease.

Effective treatments for virus diseases are limited. Scientists are continuing to work on compounds to control viruses.

Protectants

Protectants are chemical compounds applied to protect plants against disease. These are applied to the leaves and other surface areas to help prevent a pathogen from invading a plant. Protectants must be used before a plant is diseased. They will not treat a pathogen after it has invaded a plant.

Fumigants are sometimes used to destroy fungi or bacteria. These work best in small areas where the soil can be tightly covered.

Systemics

Systemics are chemical compounds that enter the vascular systems of plants after being sprayed onto the foliage. They move throughout the systems of the plant with the flow of water and food. This means that the chemical compounds (pesticides) reach every part of the plant that could be diseased.

Systemics are used to kill pathogens that are inside a plant. They can help ward-off pathogens before they take over a plant. Over time, fungi and bacteria become resistant to the systemic chemicals. New kinds of pesticides will need to be used.

People are concerned about eating food products from plants on which systemics have been used. Does a tomato from a plant treated with a systemic chemical contain small traces of the compound? The answer is most likely "yes," but not enough to cause problems.

BIOLOGICAL MANAGEMENT

Biological management of plant disease involves several approaches. New plant varieties are being developed that are disease resistant. Genetic engineering methods are being used to find new biological controls.

The environment contains organisms that naturally attack pathogens. Finding a way to use these processes saves chemical compounds as well as protects the environment.

PHYSICAL MANAGEMENT

Physical management is much the same as using disease-preventing cultural practices. Heating soil and other materials destroys pathogens, but will damage other organisms that are present. Avoiding practices that damage plants is important. Using fumigants for entire greenhouses or other plant-growing structures between crops can also be helpful.

Some widely used fumigants are no longer readily available. Alternatives are being investigated. One of the approaches that holds potential is soil solarization. *Soil solarization*

14-23. A soil solarization treatment is being studied by plant pathologists. (Courtesy, Agricultural Research Service, USDA)

is a process of capturing radiant heat from the sun underneath clear polyethelene mulch. Over several weeks, soil temperatures become high enough to kill many damaging soil pests to a depth of about 8 inches. This process reduces the populations of pests; however, it does not destroy all pests. Soil solarization has been used to control nematodes, wilts, rots, and galls. The diseases are those typically caused by fungi in the soil. Soil solarization is also acceptable to organic growers.

REVIEWING

MAIN IDEAS

A plant disease is an abnormal condition in a plant. The health of the plant is threatened. The entire plant or part of it may be diseased. Understanding diseases helps a grower use prevention and control measures.

Diseases are of two types: biotic and abiotic. Biotic diseases are caused by pathogens, which are living disease-producing organisms, such as bacteria, fungi, and viruses. Abiotic diseases are not caused by pathogens. They are caused by the environment in which the plant is growing.

Symptoms of plant disease vary widely. Common signs include abnormal leaf color, decay, wilting, deformity, reproductive failure, and dead plants. Common plant diseases include gall, blight, canker, leaf spot, rot, wilt, smut, stunt, mosaic, and damping-off. Each crop or plant has diseases that tend to be specific to it.

Diseases are spread by insects and animals, soil, water and air, and machinery. Some ways of reducing disease are fairly easily used. Cultural, chemical, biological, and physical control are used.

QUESTIONS

Answer the following questions using complete sentences and correct spelling.

1. What is a plant disease?

2. Why is control of plant disease important?

3. What are the two types of plant disease? Briefly describe each.

4. What is a pathogen? Which pathogens cause plant disease?

5. What are the causes of environmental disease?

6. What are the common symptoms of plant disease?

7. What are the common plant diseases?

8. How are plant diseases spread?

9. What are the ways plant diseases are managed?

EVALUATING

Match the term with the correct definition. Write the letter by the term in the blank that is provided.

a. vector
b. pathogen
c. gall
d. wilt

e. abiotic disease
f. systemic
g. fungicide
h. bactericide

i. smut
j. damping-off

1. _____ disease of young seedlings

2. _____ an organism that transmits pathogens

3. _____ disease-producing agent

4. _____ disease not caused by a pathogen

5. _____ growth of abnormal cells resulting from the presence of a pathogen

6. _____ disease in which the vascular system of a plant is blocked

7. _____ disease in which plants or fruits are covered with dusty spores

8. _____ chemical that controls bacteria

9. _____ chemical that controls fungi

10. _____ chemical that gets inside a plant and is moved to all areas

EXPLORING

1. Make a collection of leaves from diseased plants in the local area. Classify the leaves by the kind of damage. Identify a possible control measure for the disease.

2. Prepare a specimen of diseased plant materials for shipping to a tissue laboratory for analysis. The land-grant university in your state or the state Department of Agriculture can provide information. Interpret the report and determine the corrective action needed to control the disease.

CHAPTER 15

Weeds

OBJECTIVES

This chapter includes basic information on weeds and their safe, economical management. It has the following objectives:

1 Explain weeds and how they cause losses.

2 Describe types of weeds based on life cycle and growth.

3 List ways weeds are spread.

4 Explain methods of weed management.

5 Describe how to select herbicides.

6 Describe how herbicides are used.

TERMS

aerial application
annual weed
biennial weed
calibration
carrier
contact herbicide
directed application
dispersal
ground application
growth regulator herbicide

mechanical weed
   management
mulching
non-selective herbicide
noxious weed
nozzle
perennial weed
postemergence application
preemergence application
preplant application

selective herbicide
soil sterilant
spot application
summer annual weed
surfactant
translocation
weed
winter annual weed

15-1. Careful management has eliminated weeds from the show gardens of DeGoede Bulb Farm in Washington.

WEEDS are pests! In fields of crops or lawns, they are unwanted. Much hard work goes into keeping crops and lawns weed free. Though weeds cost producers billions of dollars a year, not everyone agrees on what to do about them.

In fact, people view the same plant differently. In one place, it is a weed. In another, it is a beautiful, valuable plant. Why the difference of opinion? Weeds can have attractive flowers or foliage. Some weeds even produce useful products. Nevertheless, weeds create pesky problems for people who do not want them.

Managing weeds results in increased food costs for the consumer. Methods of reducing weed populations are carefully selected to minimize damage to plants that are not weeds. But, weeds change—sometimes a plant is a weed and sometimes it is not!

WEEDS AND LOSSES

A **weed** is a plant growing where it is not wanted. A plant can be a weed in some situations and a very desirable plant in another. Some plants tend always to be weeds; others are weeds only part of the time. Another way of explaining this notion is that some plants are hated in some places and cultured in others as very desirable.

Here are two examples of plants that are weeds all and some of the time:

15-2. Good weed management has resulted in a potato field that is virtually weed free.

- Example One: Weed all of the time— The pigweed is a weed all of the time. Pigweed interferes with crops and pastures. It is unsightly on vacant lots and around buildings. Pigweed is never cultured nor does it provide a valuable product.

- Example Two: Weed some of the time—Bermudagrass is a valuable turf and pasture species. Homeowners and cattle producers plant, fertilize, and otherwise culture Bermudagrass. Yet, when it grows in gardens, flower beds, or fields, it is a pesky, hard-to-control weed.

Noxious weeds

Some weeds are greater problems than others. A few weeds cause only minor problems in crops, lawns, or other places. These weeds are typically small plants that do not grow large. Other weeds cause big problems!

A **noxious weed** is a weed that is particularly a problem. These weeds compete more aggressively with crops. They may grow larger and faster than the crops. Of course, the damage a noxious weed does is based on its growth habits as compared with those of the crop. Small weeds in an apple orchard may be of little consequence. These same small weeds in a field of lettuce or strawberries may cause big problems.

Noxious weeds crowd out crops and use nutrients that crop plants need. They use moisture and shade crops so they do not get enough sunlight. Extra expense is required to remove them from crops.

15-3. Examples of common pasture and row-crop weeds widely found in the United States.

```
APM PERENNIAL RYEGRASS
                        LOT # B29-3-36APM
                  PURITY   GERMINATION   ORIGIN
PURE SEED          97.20%      90%        OR.

OTHER CROP SEED     2.36%
INERT MATTER        0.44%
WEED SEED           0.00%
NOXIOUS WEED SEEDS:  NONE

TEST DATE   7-94
                        NET WT. 50 LBS.
                                     PMA 185
             MEDALIST AMERICA
    1490 INDUSTRIAL WAY S.W., ALBANY, OR., 97321

     NOTICE  READ CAREFULLY BEFORE OPENING
1. NOTICE OF REQUIRED ARBITRATION
   Under the laws of some states (including Idaho), arbitration is required as a pre-condition of maintaining certain legal actions, counterclaims or defenses
   against a seller of seed. The buyer must file a complaint along with the filing fee within such time as to permit inspection of the crops, plants or trees and
   notify seller of complaint by certified mail. Information about this requirement, where applicable, may be obtained from a State's chief agricultural official.
2. NOTICE OF EXCLUSION OF WARRANTIES AND LIMITATION OF DAMAGES AND REMEDY
   The labeler warrants that this seed conforms to the label description, as required by federal and state seed laws. WE MAKE NO OTHER WARRANTIES,
   EXPRESS OR IMPLIED, OF MARKETABILITY, FITNESS FOR A PARTICULAR PURPOSE, OR OTHERWISE CONCERNING THE PERFORMANCE OF THIS
   SEED.
   The liability for damages for any cause including, but not limited to, breach of contract or breach of warranty or negligence with respect to the sale of
   seed is limited to a refund of the purchase price of this seed. This remedy is exclusive. IN NO EVENT SHALL THE LABELER BE LIABLE FOR ANY INCI-
   DENTAL OR CONSEQUENTIAL DAMAGES, INCLUDING LOSS OF PROFITS.
```

15-4. Seed label showing no weed or noxious weed seeds.

15-5. Weeds, primarily grass types, are competing with soybean plants for nutrients.

Crop seed labels often list the percentage of noxious weed seed that is present in a bag of seed. State laws define noxious weeds and may limit the presence of noxious weed seed.

LOSSES CAUSED

Weeds cause losses in several ways. Some weeds reduce income to growers, while other weeds cause aesthetic losses for homeowners. Manufacturers make much money selling products to control weeds.

Lower Crop Yields

The greatest problem with weeds on crop farms is they reduce crop yields. Weeds compete for water, nutrients, light, and other plant-growth resources. In effect, weeds steal from the desired plants. Growers in North America lose billions of dollars a year because of lower yields caused by weeds.

A decision that must be made is when is weed control needed? At what point is the weed population in a crop large enough to merit weed control? In some crops and fields, the cropping history is one where weed control is needed. Weed control measures are used before or during planting. In other crops, weeds are controlled only when the number of weeds appears large enough to damage crop yield. Income must be greater after the weed control practices than it would have been had the weeds not been controlled.

Crop Contamination

Weeds contaminate crops when harvested. Weed leaves, stems, seeds, and other parts in harvested crops lower the quality. The weed parts must be cleaned from the crops. Growers receive lower prices for harvested crops containing weeds. Consumers do not want weeds in the foods that they buy. No one wants weed leaves in a green salad or weed seed in baked beans!

The presence of weeds in crops also increases harvest costs. A combine that has to cut and remove weed plants requires more fuel and travels at a slower speed than one in a field without weeds. Extra equipment is often needed in cleaning and removing trash. A good example is cotton. Cotton fibers and weed leaf surfaces tend to stick to each other. Fragments of weed leaves lower the quality of cotton fiber. Cotton gins use expensive cleaning equipment to remove leaf trash.

Poisonous Weeds

Some weeds are poisonous. They not only compete with crop plants and interfere with harvest, they can poison animals. Some weeds are particularly pesky to humans.

Poisonous weeds have toxic substances that may cause injury or death. People allergic to poison ivy should avoid touching or other contact with juices from the plant. Lawns, golf courses, and parks must be kept free of poisonous weeds. These weeds must also be controlled in orchards, Christmas tree farms, and pick-your-own crops.

Some plants are poisonous to livestock, such as larkspur and lupine. Weeds may be poisonous only at certain times of the year. Some weeds in the grass family, such as Sudan grass, Johnson grass, and grain sorghum, may develop prussic acid in the late fall. Cattle that eat these plants may quickly die due to prussic acid poisoning.

CAREER PROFILE

AGRICULTURAL MACHINERY DESIGN TECHNICIAN

Agricultural machinery must be designed to perform a wide range of cultural practices. Design technicians work closely with agricultural engineers, equipment manufacturers, and crop producers in testing new designs. The work is in the field as well as the design laboratory.

Agricultural machinery design technicians need practical experience in agricultural machinery operation. They need to understand how the

machinery is assembled to perform the expected tasks. Most have college degrees in agricultural engineering or a related area. A good understanding of mechanics and applied physics is needed. (Courtesy, Agricultural Research Service, USDA)

Hosts for Diseases and Insects

Weeds can harbor disease, insects, and other pests of crops. Some weeds are closely related to desired plants. The weeds may be subject to the same diseases and serve as a source of disease pathogens for crops.

15-6. Weeds are not attractive growing around this mailbox.

Insects may hide in, feed on, and reproduce in weeds. Insect pests may move from the weeds to the crop and cause damage. Controlling weeds eliminates this problem.

Other pests, such as birds, deer, and rodents, hide and live in weeds. They may go into fields to attack crops only occasionally or serve as vectors of disease.

Growers sometimes plant or set aside strips of land near crops as wildlife habitat or wetland areas. These areas also serve as hiding places for pests. Opinions differ on providing such areas because of the pest habitat that is created.

Keeping areas around a field weed free eliminates weeds as hosts for diseases and insects. Cleaning fence rows and along roads removes weeds.

Degrade Aesthetics

Many weeds are not pretty. Their presence degrades the aesthetic qualities of the lawn, field, park, golf course, or other area. Weeds detract from the appearance of homes, streets, sidewalks, businesses, schools, and other places where they grow. Cutting or controlling in other ways costs money. The cost is well worth the expense to have an attractive lawn.

WHEN WEEDS ARE NOT WEEDS

The corn plant growing in a field of soybeans is a weed. When it is growing in a corn field, it is a desired plant. In fact, a soybean plant in a corn field is a weed. A plant is a weed when it is out of place.

Weeds are sometimes called wild flowers. They have attractive flowers in the spring, summer, and fall. Festivals have been organized to commemorate weeds. One example is the mustard festival in Napa, California. Special community activities are held when the wild mustard has beautiful yellow flowers in the spring. It grows along roads, in fields and vineyards, and other places. The mustard plant is both a pest and a focus of celebration. In Napa, the festival brings in thousands of tourists, but creates hard work for vineyard producers.

15-7. The wisteria, a climbing ornamental, is a beautiful, flowering plant when properly trained. Otherwise, it is a pest that climbs trees and covers the ground. The background here shows that the wisteria has not been managed well and is taking over trees.

15-8. Mustard plants flowering during the Napa mustard festival.

Weeds also contribute to biodiversity. Some people have become concerned about maintaining a diversity of plants and other organisms on the earth. Weeds may provide the genetic material for improved crop plants. They may provide a cure for disease or serve as an alternative for food or clothing. Of course, when they provide benefits and are growing where they are wanted, they are not weeds!

TYPES OF WEEDS

Weeds can be classified in several ways. They may be classified by the crops they damage or whether they are poisonous. Weeds may be classified by how they grow, their life cycles, difficulty of control, and the way they cause damage.

Commonly, weeds are classified by their life cycle and structural growth. Knowing the growth habits of weeds is the first step toward effective control. In addition, some weeds grow in aquatic environments, while others grow on the land.

LIFE CYCLES OF WEEDS

Weeds are plants. Their life cycles are similar to non-weed plants. Three major life cycle classes are used for classifying weeds.

Annual Weeds

An **annual weed** completes its life cycle in one growing season, usually less than a year. Annuals are normally considered easier to control than weeds with other life cycles. Annual weeds are further classified as summer and winter annuals.

A **summer annual weed** is one that primarily grows during the summer. The seeds germinate in the spring and the seedling emerges. The plant will mature in the summer and die in the fall. It produces seed that fall to the soil and repeats the cycle the next spring. Destroying weed seedlings prevents damage to crops and keeps summer annual weeds from producing seed. Summer annuals are problems in corn, soybeans, vegetables, fruit, cotton, and many other crops. Examples of summer annual weeds include morning glory, pigweed, crabgrass, and cocklebur.

15-9. Two common summer annual weeds are cocklebur (left) and crabgrass (right).

A **winter annual weed** is one that primarily grows in the winter. It germinates in the fall and grows through the winter. It produces seed in the late spring and dies as the days get long and warm in late spring or early summer. These weeds are problems in winter wheat, oats, and similar crops. Common winter annual weeds are henbit, chickweed, tansy mustard, and little barley.

Biennial Weeds

A **biennial weed** has a life span of two years. The plant grows rapidly during the first year. Flowers and seeds are produced in the second year. The life cycle is repeated. Biennial weeds are dormant and typically cold tolerant for the winter at the end of the first year. Biennial weeds can be problems in most crops. Examples of biennial weeds include wild carrot, sweet clover, some thistles, and common mullein. As with annuals, disrupting the formation of seed is a good control measure.

Perennial Weeds

A ***perennial weed*** is one that lives for three or more years. The above ground part may die back in the winter and regrow each spring from the underground parts. Perennial weeds reproduce by seed and vegetatively.

15-10. Three common perennial weeds are nutsedge (left), wild onion (center), and curly dock (right).

Simple perennial weeds reproduce by seed. These weeds do not have runners nor other means of vegetatively spreading. If cut into two or more pieces, each piece of some simple perennials may sprout and form a plant. Examples of simple perennial weeds include dandelions and curly dock. Disrupting the production of seed may be useful in limiting the spread of simple perennial weeds.

Creeping perennial weeds reproduce by seed and by creeping roots, tubers, stolons, and rhizomes. These weeds are more difficult to control. They are often viewed as major pests in some crops. Examples include nutsedge, which reproduces with tubers, Johnson grass with rhizomes (below ground stems), and Bermudagrass with aboveground stems. These weeds are among the more difficult to control.

SEED STRUCTURE AND GROWTH OF WEEDS

Weeds that reproduce by seed may be classified by the structure of the seed. (Seed structure of plants was presented Chapter 4.) Weed seeds have similar structure: monocotyledon and dicotyledon. The characteristics of these seeds are present in how the plants grow and develop.

Weed management practices vary with monocots and dicots. Chemical measures are particularly different. For example, a chemical that controls a monocot weed may not control a dicot weed. In other words, chemicals that control weeds with parallel venation leaves may not control broadleaf weeds. This difference is why chemicals used to control broadleaf weeds in corn cannot usually be used to control broadleaf weeds in soybeans.

HOW WEEDS ARE SPREAD

Dispersal is the spread of weeds. Most weeds grow in isolated locations. Dispersal spreads them over wide areas and into places where they have not been a problem before. Weeds can be dispersed over long distances. By knowing how weeds are spread, future spreading can be stopped.

Dispersal occurs in two major ways: natural and artificial. In combination, these ways have resulted in weeds getting into fields and crops long distances from their origin.

NATURAL DISPERSAL

Natural weed dispersal is the movement of weeds by wind, water, and wildlife. Weed seeds often have stickers or feathery features that make natural dispersal easy. Light seeds, such as the dandelion, may be blown about in the wind. Heavier seeds, such as cockleburs, may be washed by water runoff or carried in the hair of animals.

Water from streams that overflow their banks carries weed seeds onto the land. As the water recedes, the seeds are left behind. Keeping streams from overflowing onto crop land is expensive and may require large levees. Regulations on wetlands may prohibit levee construction.

Irrigation water used from creeks, lakes, or canals may have weed seed. Weeds growing along irrigation canals drop seed into the water. As it is used on fields, weed seeds are deposited. The environment in the field is usually ideal for weed growth. Moisture and fertility are high because of irrigation and fertilizer.

Birds, rodents, and other animals move weed seed. Seed eaten by animals may pass through the digestive tract undigested and

15-11. Examples of natural seed dispersal: dandelion seed (left)—light and easily blown by the wind—and cocklebur seed (right)—covered with sharp prongs that cling to fur and clothing.

viable. They may be dropped on the land in feces and germinate when conditions are right. Animals with hair may get weed seed caught in their coat. The seed may come loose in fields and later germinate to grow weeds.

Natural weed dispersal is difficult to control. Preventing the growth of weeds in other places is difficult, but it is a good beginning.

ARTIFICIAL DISPERSAL

Artificial dispersal of weed seed is by people and the activities they carry out in producing and harvesting crops. Both seed and vegetative parts of weeds may be dispersed. How does it happen?

Machinery can transport weeds. Seed and vegetative parts may stick on plows, sprayers, harvesters, and other equipment. As these move from one field to another, the seed and parts are dropped. They grow to become weeds. Cleaning equipment between fields and farms is a good approach in reducing the spread of weeds. Even pickup trucks that get seed on their bumpers or tires in one field can carry them to another!

Weed seed from one field may cling to shoes, trousers, and other clothing. The seed may come loose in another field and fall to the ground. Viable seed will later germinate and produce weeds.

Using crop seeds that are impure is a major artificial way of weed seed dispersal. Unclean crop seed may contain the seed of noxious weeds. The weed seed may be planted along with the crop seed. Always use clean seed from a good source. Read the label on seed for information about the presence of noxious weeds.

Mulch materials may contain seed. A good example is wheat straw. Wheat straw may contain the seed of weeds that grew in the wheat crop. These seeds germinate and infest the mulched area with weeds.

CONNECTION

COMPUTERS AID IN PLANNING

Information from many sources is needed by today's crop producers in controlling weeds and doing other cultural practices. Computers and related technology are important. Increasingly, the information they provide helps make good decisions.

Through a digitizing process, important information can be recorded and analyzed about crops, land, and other features. These approaches help producers in gaining greater efficiency. (Courtesy, Joliet Junior College, Illinois)

METHODS OF WEED MANAGEMENT

As with plant diseases and insects, preventing weeds is far better than trying to get rid of them later. The general methods of weed control are similar to insect and disease control. How the methods are used varies widely.

15-12. A rotary cutter is used to cut weeds in this pasture. (Courtesy, Bush Hog Division of Allied Products, Selma, Alabama)

15-13. Cultivation is used to control weeds in this cotton field. (Courtesy, AGCO Corporation, Duluth, Georgia).

MECHANICAL METHODS

Mechanical weed management is using tools or equipment to eliminate weeds. Sometimes, it involves pulling weeds by hand, but this is impractical on a large scale. Several common methods of mechanical weed control are listed here.

- Mowing—Mowing is using machines to cut weeds. It is effective with tall-growing weeds. Weeds that grow low-to-the-ground are not easily controlled with mowing. Many lawn weeds are not effectively controlled with mowing, such as dandelion and henbit—parts may be cut off, but they grow back. Mowing is effective in preventing seed formation in some weeds if done before the seeds are produced. Mowing is best before flowers are formed because immature seed may germinate. Mowing may not be effective with weeds that have flowers and seed near the ground.

- Cultivation—Cultivation is using tools that cut off and physically lift weeds out of the soil. Roots of weeds exposed to air will usually die. Weeds covered with soil may also die. Tillage is most effective on annual weeds. It can result in the spread of perennial weeds that have rhizomes or bulbs, such as Johnson grass and wild onion. Common cultivation equipment includes discs, harrows, hand hoes, and sweeps or chisels on power tool bars and rotary tillers.

15-14. Hand hoeing is a form of mechanical weed control that is best used in small areas, such as gardens. (Courtesy, U.S. Department of Agriculture)

- Flame—Weeds can be controlled by burning and heat. Sterilizing soil and other medium used in greenhouses with heat destroys the embryo in the seed. Flame throwers are used to control some weeds. Carefully directed flames can be used in crops such as cotton to control small, tender weeds. Larger flames can be used on weeds in pastures, such as prickly pears—a kind of cactus.

CROPPING-PRACTICE METHODS

How a crop is grown helps manage weeds. The growth habits of crops are important in eliminating weeds. Several examples of how cropping practices manage weeds are included here.

- Mulching—**Mulching** is using a cover on the soil that smothers small weed plants. Shredded bark, grass clippings, straw, and sheets of paper or plastic are often used as mulch. Mulches maybe placed around shrubs or high-value fruits and vegetables. Some crops, such as strawberries and tomatoes, are planted through long strips of plastic. The plastic not only prevents weed growth, but also helps hold moisture in the soil.

15-15. Plastic mulch is used to control weeds that might compete with these tomatoes.

15-16. Flooding is used in a rice field to manage terrestrial weeds. (Courtesy, Spectra-Physics, Dayton, Ohio)

- Flooding—Flooding is primarily used in rice production to manage terrestrial weeds. The land where the rice is growing is covered with shallow water. Once the rice plants are up, the field is flooded for at least five weeks. Fields may be relatively weed free afterward except for any aquatic weeds that are present.

- Crop competition—Crops often grow fast if they have a good environment. Planting well-suited varieties in soil that provides the proper nutrients helps crop plants outgrow weed plants. Once crop plants get large enough, the ground is shaded and weeds are "choked-out." Placing fertilizer in a band near crop seed will help crop plants use the fertilizer before weeds develop.

- Crop rotation—Rotating crops helps keep weeds down. Some weeds are more prevalent in certain crops than in others. Crop rotation can be used to introduce a crop that is more successful in competing with the growth habits of certain weeds. Crop rotation can be used along with chemicals. Alternating crops that resist damage from chemicals permits the use of different chemicals on a field over a two-year period. This allows the opportunity to kill weeds that tend to be the same each year.

- Weeder animals—Animals that like to eat weeds, and not the crops in which the weeds grow, are used to manage weeds. This practice has limited use. One example is

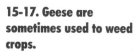

15-17. Geese are sometimes used to weed crops.

the use of geese—often known as weeder geese—in crops that geese do not like to eat with weeds the geese like to eat. The geese are to eat weeds and not crops!

CHEMICAL METHODS

Chemical weed management involves the use of herbicides. A herbicide is a chemical compound (pesticide) that kills plants. The chemical is often sprayed on the leaves, seeds, or stems of plants.

Herbicides are carefully developed to control specific weeds. They are manufactured to result in minimum damage to the environment. Of course, herbicides must be used properly and only as approved. Safety is an important concern in the use of any chemical material. (Chapter 12 has a section on safety in using chemicals.)

Herbicides are widely used in weed control. Many important facts must be considered. A good knowledge of the biology of the weed to be killed and the crop in which it is to be controlled is needed. Two important areas are which herbicides to use and when to use them. These are covered in the next sections of the chapter.

SELECTING THE HERBICIDE TO USE

Effective chemical weed management depends on selecting the best herbicide. Herbicides are classified according to the way they destroy weeds. Two areas are included: type of herbicide and selectivity.

TYPE OF HERBICIDE

Herbicides are classified by type based on how they kill plants. Three types of herbicides are used: contact, growth regulators, and soil sterilants.

Contact Herbicides

A *contact herbicide* is a weed chemical that kills plants by exposure. Only the parts of a plant in contact with the chemical are killed. The entire plant is killed only if all of it is exposed to the chemical. Root systems may not be exposed, but they may die if the top no longer provides food. For example, accidentally spilling gasoline or diesel fuel on a plant destroys the leaves, but not the roots.

15-18. Leaves on unwanted vegetation are showing symptoms of translocated herbicide action.

Contact herbicides work best on very young seedlings. They usually die shortly after exposure. Contact herbicides also work best on annuals. The roots of established perennials or biennials may not be killed. The leaves and stem may die, but new growth sprouts from the roots.

Growth Regulators

A *growth regulator herbicide* kills weeds by altering growth or metabolic processes. These herbicides are absorbed by the roots, stems, or leaves of a weed plant. Once inside, the chemical is moved through the xylem and phloem to all tissues of the plant. Response to the herbicide may require a week or more. Symptoms of action show as the plant wilts, twists, falls over, and dies. In the event that the treatment was inadequate to kill the weed, some signs may show, but the weed is not killed.

The movement of a herbicide inside a weed is known as *translocation*. Herbicides that only contact a few places on a plant are moved throughout the plant. Most growth regulators are translocated in plants.

Many herbicides used today are growth regulators designed to kill plants. They do so by disrupting normal functions of weed plants.

Soil Sterilants

A *soil sterilant* is a compound that prevents the growth of plants in the soil. They have limited use with row crops but are useful in greenhouses, along ditch banks, and with other special applications. Soil sterilants are not practical for use on a widespread basis. No plants

Table 15-1. Examples of Herbicides[1]

Common Name	Trade Name	Class[2]	Crop/Use
atrazine	Atrazine[3]	R	corn/broadleaf weeds and grasses
copper sulfate	Copper Sulfate	G	aquatic weeds and algae in ponds and tanks
glyphosate	Roundup[3]	G	cotton/broadleaf weeds and grasses rice/annual and perennial grasses and broadleaf weeds
methyl bromide[4]	Brom-O-Gas[3]	R	turf/dormant weed seeds and structures of perennial weeds, nematodes, and fungi
MSMA[5]	MSMA[3]	G	cotton/annual grasses and a few broadleaf weeds
paraquat	Gramoxone[3] Extra	R	soybeans/small annual weeds
picloram	Tordon[3] Grazon[3]	R	grain crops/annual weeds
trifluralin	Treflan[3]	G	cotton/winter annuals
2,4-D	2,4-D Amine 4	S	turf/broadleaf weeds including wild onion

[1] Only a few examples are included; no recommendations are implied; refer to labels for information on use; many restrictions may exist; changes occur frequently.

[2] Class refers to classification as follows: R—restricted use product by EPA; G—general use products not restricted by EPA; and S—state or regional restrictions.

[3] Trade names may vary depending on formulation and manufacturer.

[4] A fumigant that is no longer approved for widespread use.

[5] MSMA is monosodium monoarsenate.

can be growing in the area of application when they are used. Methyl bromide, as shown in Table 15-1, is an example that has been used but is no longer commonly available.

SELECTIVITY OF HERBICIDES

Some herbicides will kill any weeds; others will kill only a few selected weeds. A **selective herbicide** is a compound that kills only certain plant species and not others. This characteristic makes it possible to use herbicides on weeds in crops without injuring the crop. An example is using a herbicide that only kills broadleaf weeds in corn crops. The corn is not damaged, but the broadleaf weeds are killed. (Note the importance of plant structure. Corn is a monocot while the broadleaf weeds are dicots.)

A **non-selective herbicide** kills all plants regardless of species. These can be used only where all vegetation is to be destroyed. It should not be used in growing crops. Non-selective herbicides will destroy crops and weeds.

Improperly used, selective herbicides can damage crops or desired plants. Always read and follow instructions.

USING HERBICIDES

Since herbicides are compounds that have the power to kill, they must be carefully selected and properly used. The kind of weed, the crop it is in, and the nature of the herbicide are three important considerations. In getting the needed information, determine when and how applications are made.

15-19. A preplant application is being made to land that will be planted with corn. (Courtesy, Case Corporation)

15-20. A preemergence application is being made while planting. (Courtesy, National Cotton Council of America)

TIME OF APPLICATION

Time of application refers to when herbicides are applied relative to the stage of growth of the crop. Some herbicides can be used only at certain times. If used incorrectly, they may damage the crop or fail to control weeds effectively.

Herbicides may be applied at three times:

- Preplant applications—A **preplant application** is made before a crop is planted. The seedbed may be prepared and the herbicide applied. Planting is normally a few days after the preplant application. Sometimes, the preplant application may be made in the fall after the crop residue has been chopped and the land plowed.

- Preemergence applications—**Preemergence application** is the application of a herbicide before the crop comes up. It is usually made during the planting operation or after planting. Preemergence is also before any weeds are growing. Growers often like to use preemergence applications because the

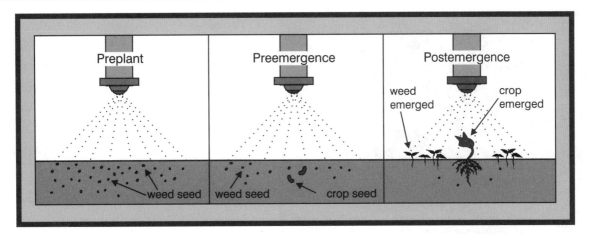

15-21. Comparison of preplant, preemergence, and postemergence applications.

equipment is attached to the planting equipment. Only one trip is needed over a field to both plant and apply herbicide. It also assures that the treatment is applied in a timely manner. With preplant treatments applied well ahead of time, some effect of the herbicide is lost. This does not happen with preemergence application. Preemergence is often used when planting corn, soybeans, cotton, and grain sorghum.

- Postemergence applications—A ***postemergence application*** is after the crops are growing and weeds are present. The best results are achieved when the weeds are small seedlings. Larger weeds are more difficult to manage and have likely cost some crop yield. In some crops, over-the-top methods of postemergence are used for tall weeds that have grown above the height of the crop. A wick-type applicator applies herbicide by touching the tops of the weeds. An example of over-the-top application is on tall grass in soybeans or cotton.

AREA OF APPLICATION

The area of application deals with the extent of coverage in a crop or turf. Applying herbicide to locations where weed control provides low returns is a waste of money. Herbicide should be applied where it reaches weeds in a cost-effective manner.

In general, four areas of application are used:

- Band—Band application is treating narrow strips with herbicide. The strips are centered on the seed drill, which is usually at the top of a row. A band is usually 10 to 12 inches wide. This is an efficient way to manage weeds by applying herbicides close to plants. Cultivation is used to destroy weeds in the middle between the rows. Cotton and corn are common crops where band application is used.

15-22. Broadcast application of herbicide. (Courtesy, John Deere)

- Broadcast—Broadcast application is treating the entire area of a field. No bands or strips are left untreated. Broadcast application is used with pastures, lawns, and grain crops that are not in rows. This type of application is used in places where cultivation cannot be used to destroy weeds.

- Directed—**Directed application** is treating only selected plants or target plants. Originally, this method was primarily used with trees. Small trees to be culled were injected with a herbicide just above the ground.

Some use of directed application has been made in row crops. Newer technologies with "smart" sprayers only apply herbicides where weeds are present. These sprayers use computers and photo detectors to determine the location of each weed. A tiny amount of herbicide is quickly sprayed on the weed. This approach saves the money it costs to spray areas in a field where there are no weeds.

- Spot—**Spot application** is treating only certain areas in a field or pasture. Locations where weeds are growing are targeted. This saves the cost of herbicide plus labor to apply it. It also protects the environment because no more herbicide is used than is necessary. Spot application is often used along fence rows, around barns, and in crops where weeds are in spots.

15-23. A "smart" sprayer applying herbicide in soybeans. (This sprayer uses a computer and photo detectors to detect the precise location of weed plants. A small amount of herbicide is applied directly to the weed. The silicon photo detectors know the difference between the crop and the weed.) (Courtesy, Agricultural Research Service, USDA)

APPLYING HERBICIDES

Knowing when and how to properly apply herbicides is important in efficient and effective weed management. Many factors are involved.

RECOMMENDATIONS

Only herbicides approved for a particular use should be applied. Further, the rate and timing of application should comply with the approved use.

Herbicide recommendations vary from one year to another. Most states annually release weed control guidelines. These are prepared by weed scientists from the state land-grant university, Department of Agriculture, or others, in conjunction with producers, herbicide manufacturers, and representatives of environmental protection agencies and groups. A copy of the recommendations is available from the land-grant university.

SOIL TYPE

The rate of herbicide to use with preplant and preemergence applications depends on the soil. Recommendations are based on soil type. Before buying and applying herbicide, determine the type of soil. Some herbicides are more effective in sandy soils than in clay and loam soils and vice versa. Soil type is not closely related to postemergence applications directed on weeds.

STAGE OF GROWTH

The size and age of both weed and crop plants must be considered. The best results will be achieved when the weeds are small—preferably in the seedling stage. Larger weeds are more difficult to control. Crop plants can be broken or otherwise damaged by ground application equipment if they are beyond 12 to 15 inches high. Over-the-top application requires that the weeds be taller than the crop plants.

FORM OF HERBICIDE

Herbicides are typically in a liquid form. Some are mixed with pellets of fertilizer or other materials for specific uses, such as in lawns. Most herbicides are bought in bottles or cans.

Table 15-2. Common Pesticide Formulations

Form	Abbreviated Symbol	Description
Dry Flowable	DF	finely ground, dry form mixed with liquid carrier and emulsifier
Dust	D	fine particles of active ingredient mixed with a dry carrier and applied in dry state
Emulsifiable Concentrate	EC or E	concentrate of one liquid suspended in another—usually oil in water
Granule	G	active ingredient attached to inert material, such as ground corncobs or clay
Slurry	SL	thin, watery mixture of finely ground dusts
Soluble Powder	SP	power that can be changed into a solution
Solution	S	liquid mixture of two or more substances
Wettable Powder	WP or W	any dust form that can be mixed with water—does not dissolve in water but is suspended in the solution

The amount bought should be no more than will be used. The material should be bought in the largest container practical to reduce problems with empty herbicide containers.

Liquid forms of herbicides may be in oil- or water-based forms. Knowing the form and how to prepare the herbicide for application is essential. In addition, most plant leaves have a waxy coating. The coating may result in spray bouncing off when applied.

Carriers

A **carrier** is the material in which the active ingredient of a herbicide is mixed. Most carriers are water, but some are oil. In effect, a carrier dilutes the herbicide and reduces toxicity. Carriers provide a form that allows maximum safety and effectiveness when a herbicide is used.

In mixing a herbicide in a carrier, the amount of active ingredient in the material determines how much to use. In general, smaller amounts of products with higher active ingredients need to be mixed with the carrier. Always read and follow directions.

Surfactant

A **surfactant** is a material added to herbicide mixes to assure that the applied material spreads over leaves and does not bounce off. Surfactants are like soap. They help solutions overcome differences between oil, water, and wax. Surface tension between the leaf and herbicide is reduced so the material spreads over a leaf's surface. Surfactants are sometimes known as wetting agents. In oil-based sprays, filming agents are used to help assure good coverage.

EQUIPMENT TO USE

The equipment used in applying a liquid herbicide must be "right for the job." Most liquid herbicides are applied with sprayers. Dust forms are applied with dusters, while granules may be applied with drills or fertilizer distributors. Select equipment based on the size of the area to be treated, time of treatment, size of the crop and weed plants, and environmental conditions.

Applicators

The most common liquid applicators are sprayers. Wick systems and other methods are sometimes used.

Sprayers vary in size, design, and method of application. Some are small hand sprayers that are useful in home lawns and other special applications. Other sprayers are large ground equipment or on airplanes for aerial application.

All sprayers have a tank for the herbicide carrier, a pump, and emitters (nozzles) connected by tubes to the tank. All parts must be in good operating condition. The nozzles must be designed and adjusted to meet the purpose of an application. Proper calibration is essential.

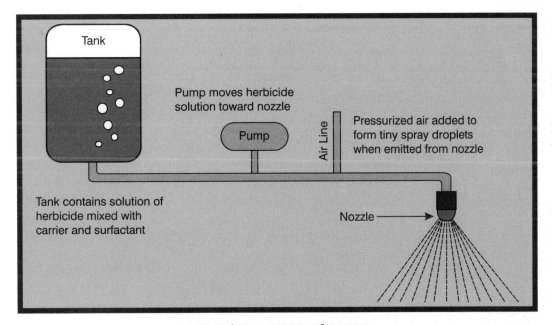

15-24. Major components of a sprayer.

Calibration is setting equipment to meter the exact amount of herbicide needed. Several factors are involved. The size of nozzle openings determines how much is sprayed. The pressure in the tank determines the rate and force of application. The speed at which the application equipment moves over the ground determines the amount applied on an area basis. Nozzle size, pressure, and speed must be factored together in making equipment adjustments.

Method of Application

Two general methods of herbicide application are used: ground and air.

Ground application is treating for weeds using equipment that travels through the field. It is used primarily with preplant and preemergence applications. It can also be used in some postemergence treatments. Weather conditions and the stage of growth of the crop are big factors in determining if ground application is best.

Aerial application is using airplanes or helicopters to apply herbicide. It is used when:

* crops are large and ground equipment would cause physical damage to the plants,

* the soil is too wet to support the weight of ground equipment, or

* speed of application is important (aerial application covers more area in less time than ground application).

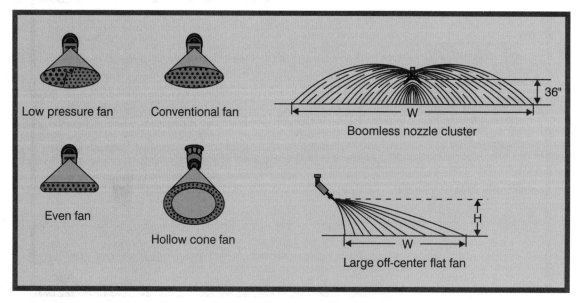

15-25. Common spray nozzle patterns.

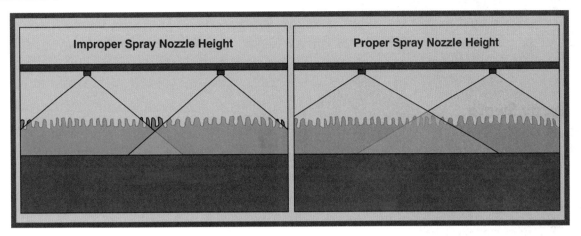

15-26. Spray nozzle height should give uniform coverage of all weeds.

Aerial application may pose additional hazards. Environmental safety is a major concern. Herbicides applied by air are more likely to drift to other crops or into streams and lakes. The pilot of the aerial aircraft should know the exact area to be treated and be informed of nearby areas to avoid. Spraying equipment should be turned off when flying over areas not being treated. Especially avoid spraying into water and onto animals.

The same equipment should not be used for applying insecticides and herbicides. Some residue may remain in the equipment. This might result in applying a weak herbicide when treating with an insecticide. Crop plants could be damaged in the process.

Nozzle Patterns and Settings

A *nozzle* is the device that dispenses spray. Proper adjustment is essential for accurate coverage. Nozzles wear with use and the openings become larger. Larger openings spray more herbicide and in different patterns. Use quality nozzles and replace them when they begin to wear. Equipment calibration changes with worn nozzles.

The design of the opening in a nozzle determines the pattern of the spray. Most nozzles provide hollow or fan type patterns. A nozzle should release equal coverage throughout the spray pattern. Worn nozzles fail to do so.

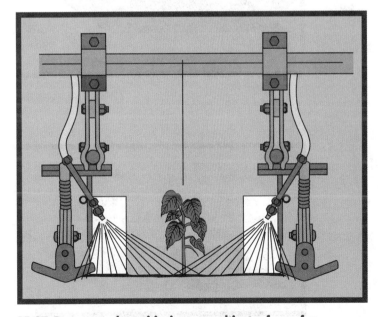

15-27. Proper nozzle positioning on a cultivator frame for postemergence in a row crop. (Flat fan nozzles direct the spray down and back to provide the desired overlap and band width. A shield keeps spray off the leaves of plants.)

Nozzles should be properly adjusted to give the desired coverage. The spray should be directed in the best pattern. With postemergence applications, nozzles must be adjusted to avoid spraying crop leaves but give maximum coverage of weeds.

Spray Swath

Spray swath is the width of area covered by a sprayer. The entire area should receive an equal and proper amount of herbicide. Skipping areas in an orchard or a field results in weeds not being killed. Operators of spray equipment need to carefully observe the travel pattern of the rig. Some overlap is needed, but no more should be used than is required for uniform coverage.

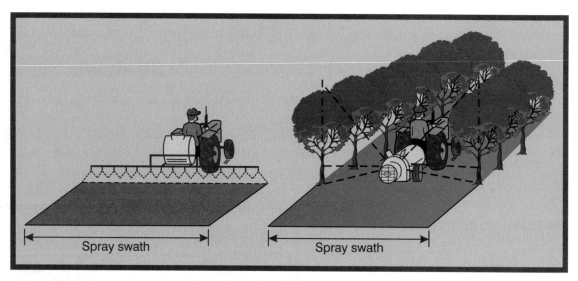

15-28. Two examples of spray swath.

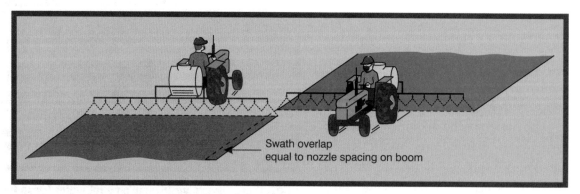

15-29. Swath overlap is needed to assure proper coverage of the edge of a swath.

Safety

Herbicides are poisons. Safe application and handling are essential. Always read the label. Applicator training may be required. Here are a few safety tips:

- All herbicides must be used only as approved.

- Do not treat food crops that are near harvest.

- Properly store herbicides.

- Keep herbicides out of the reach of children or other people who are not responsible.

- Dispose of empty containers properly.

- Flush skin and eyes with water if contact is made with herbicides.

- Know the emergency telephone number in case of an accident. Take a cell telephone in a protective container.

- Protect wildlife and domestic animals.

- Wear proper personal protection, including clothes, respirators, hats, gloves, and boots. (Refer to Chapter 12 for more details on personal protection.)

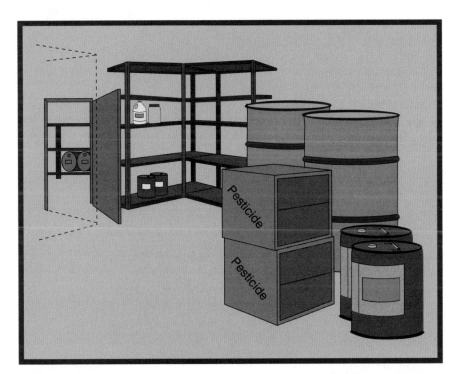

15-30. Proper storage of pesticides in original containers with labels attached is essential.

REVIEWING

MAIN IDEAS

A weed is a plant that is growing where it is not wanted. Noxious weeds are species that cause particular problems in crops. Weeds cause big losses to crop producers, including lowering yields and contaminating harvested crops.

Weeds can be classified in many ways, such as by life cycle and seed structures and growth. Weeds can also be classified by the crop in which they are a pest and methods of control.

Weeds are spread naturally and by human activity. Practices can be followed to minimize weed spread. Weed management is an important part of crop production. Common methods of weed management include mechanical, cropping-practice, and chemical approaches.

Selecting the herbicide to use is important. Select the right type of herbicide. Time the application properly. Use the most efficient area of coverage. Always follow manufacturers' recommendations. Use equipment that is properly calibrated in making applications.

QUESTIONS

Answer the following questions using complete sentences and correct spelling.

1. When is a plant a weed? Noxious weed?

2. Can a crop plant be a weed? Explain.

3. How do weeds cause losses?

4. What are the types of weeds based on life cycle? Give an example of each.

5. How are weeds dispersed?

6. What methods of weed control are commonly used?

7. What are the types of herbicides?

8. What is herbicide selectivity?

9. What are the three common times of application? Distinguish between them.

10. What is area of application? Name and explain four types.

11. What equipment is needed to apply herbicides?

12. What important safety practices should be followed in using herbicides?

EVALUATING

Match the terms with the definitions. Write the letter by the term in the blank that is provided.

a. weed   e. mulching   i. preemergence application
b. noxious weed   f. herbicide   j. nozzle
c. annual weed   g. translocation
d. dispersal   h. selective herbicide

1. _____ applying a herbicide before a crop grows

2. _____ device that dispenses spray

3. _____ a plant growing where it is not wanted

4. _____ spreading of weeds

5. _____ a particularly pesky weed

6. _____ controlling weeds with a ground cover that smothers weeds

7. _____ weed that completes its life cycle in one growing season

8. _____ a chemical compound that kills only certain plant species

9. _____ a chemical compound that kills plants

10. _____ movement of a herbicide inside the vascular system of a weed

EXPLORING

1. Identify a common weed in the lawns or crops in your community. Select a method of control. Use weed control recommendations to identify a herbicide. Determine the rate of active ingredient to use and the best method of application. Prepare a written report on your study. Give an oral report in class.

2. Visit a local garden center. Make a list of the different herbicides that are for sale. Include the kinds of weeds controlled and the rate and method of application. Summarize your findings in a written report.

3. Help calibrate a sprayer. Use information that was provided by the manufacturer or that is found in the weed control recommendations for your state.

CHAPTER 16

Plants and Their Environment

OBJECTIVES

This chapter covers important areas of plant environment, including how changes in the environment affect plants. It has the following objectives:

1 Explain the meaning of plant environment.

2 Describe the role of climate and weather.

3 Explain growing season and related concepts.

4 Describe the edaphic environment.

5 Explain the effect of pollution on plants.

TERMS

abiotic environment
acid rain
air quality
biotic environment
ecosystem
edaphic environment
forcing

frost
global warming
greenhouse effect
growing season
hardiness zones
permafrost
photoperiod

plant environment
pollution
soil contamination
soil degradation
weather

16-1. Interior plantscaping is used to promote plant health and assure an appealing hotel lobby.

PLANTS need good environments to grow. One role of a grower is to help provide a good environment. Growers do much better with plants if they know the needs of plants.

If you have an ornamental plant in your home, you are aware of and meet its environmental needs—light, water, nutrients, and others. Forgetting to water a plant for a few days can have unfortunate results—wilted and dead!

Crops vary in the environment they need. Some crop plants do not grow well when moved to places where they are not adapted. Knowing about the environmental needs of weed plants can help in controlling them. Many factors are a part of establishing a good environment.

PLANTS AND ENVIRONMENT

16-2. A horticulturist strives to provide a good environment for plants.

Plant environment is the above and below ground surroundings in which a plant grows. Desired plants and weeds grow in the plant environment. This environment is made of many different living and nonliving factors. These factors must provide nurturing conditions for plants.

PLANT NEEDS

For plants to thrive, distinct needs must be met. These are often provided by nature. Producers also have important roles in helping assure that the needs of plants are met.

For a plant to grow well, five environmental factors must be supplied:

* Light—Light provides energy for photosynthesis to take place in plants. Some plants require full sun, while others can exist in partial sun and limited light. Most crop plants, such as corn and wheat, require full sun. For crop plants, shade impairs productivity. Some ornamental plants grow satisfactorily in limited sunlight, including Saint Augustine grass and azaleas.

* Temperature—Most plant species are suited to a particular temperature range. Extremes beyond the range result in little growth and perhaps the death of the plant. Producers plant crops based on temperature requirements. Tomatoes are very sensitive to cold weather, but winter wheat can withstand considerable cold weather.

* Water—Plants require moisture (water) for growth. Without adequate moisture, plants wilt and

16-3. Colorful azaleas are well suited to their environment in the Biltmore Estates gardens in North Carolina.

are less productive. In some cases, plants will die of lack of moisture. The water needs of plants are met by natural precipitation as well as irrigation.

- Air—Plants use oxygen from the air in photosynthesis. Other factors in the air may promote or may damage plant growth. For example, pollution can reduce plant growth and, in extreme cases, kill plants.

- Nutrients—Plants use nutrients for growth. These are normally in the soil, or other medium, where roots are present. Nutrients are removed

16-4. Macadamia nut trees are provided a good environment in this grove.

with moisture by the roots and used by the plant in its life processes. Plants that do not have adequate nutrients will not grow well nor produce a crop. In extreme cases, the plants may die.

CAREER PROFILE

SOIL TECHNICIAN

Soil technicians prepare maps, take soil samples, and perform analyses in laboratories. The use of global positioning systems and computer equipment are common.

Soil technicians need to understand soils and crops. They need a good background in chemistry and the environment. Most positions require people with education beyond high school, though a few high school graduates with training in agriculture may be employed. Many soil technicians complete two-year associate degree programs in agronomy. Some have baccalaureate degrees in agronomy or related areas. On-the-job experience with crops and soils is important. (Courtesy, Joliet Junior College, Joliet, Illinois)

16-5. Trees in an Oregon old-growth forest live in an interesting ecosystem.

PLANT ECOSYSTEMS

An **ecosystem** consists of all the parts of the environment where a particular plant lives. Plants do not exist alone. Other living and nonliving factors are found in the environment. These form the biotic and abiotic environments of plants.

Biotic Plant Environment

The **biotic environment** of a plant consists of the living things that have an influence on its growth. Other plants, birds, insects, and microscopic organisms are in the biotic environment. The other plants include competing plants of the same species and weeds. Large animals can become environmental factors, with a good example being cattle in a pasture.

Some biotic factors damage plant growth, such as harmful insects. Other biotic factors support plant growth, such as microbial activity in the soil that helps provide nutrients. The producer must be able to help regulate biotic factors to support maximum plant growth. Cultural practices are used with crops to help provide a good biotic environment.

16-6. Fields of tomatoes in the foreground and cabbages in the background are surrounded by trees and other plants in the Blue Ridge Mountains to form a comprehensive biotic environment.

Abiotic Plant Environment

The **abiotic environment** includes all of the nonliving elements in a plant's environment. Nonliving elements consist of water, light, moisture, wind, temperature, soil, and others, including human-made changes. Plant growth requires an abiotic environment appropriate to its needs.

Producers can regulate some factors in the abiotic environment; other factors cannot be easily controlled, such as the weather. Of course, greenhouses and other plant growing structures can be used to help overcome abiotic environment adversities.

CLIMATE AND WEATHER

Climate and weather are important abiotic environmental factors. They set the crops that can be grown and the cultural practices needed to have good production.

CLIMATE

Climate is the average of the weather conditions that are found in a particular location. Some locations are warm and dry. Other locations are cool and damp. What happens on any single day with the weather has little to no impact on the overall climate.

Climate varies widely in the United States. Warm, tropical climates are found in Hawaii, south Florida, and the Southwestern states. Cold climates are found at higher elevations and areas of Alaska.

The climate is important in plant growth. Producers select crops and varieties of crops based on the climate situations where the crops will be grown. Some crops are not suited for locations where the climate does not provide for the needs of the plant. Crops that require warm weather will not grow in cold climates. An example is the orange. Cold weather kills orange trees.

16-7. The rock causes the tree to adapt to its abiotic environment.

16-8. The saguaro, a giant cactus, adapts to its environment in the Arizona desert. (This one is along the Apache Trail near Tortilla Flats, Arizona.)

16-9. Flowers on this dogwood have been damaged by unseasonably cold weather.

WEATHER

Weather is the current condition of the atmosphere. Major conditions include temperature, moisture, wind, and atmospheric pressure. The weather conditions on any single day can promote plant growth or damage plant growth.

Unseasonable weather can be a particular problem. Weather that is too cold or too wet can damage or destroy crops. Sudden cold snaps with unseasonable frost can kill tender crops.

Temperature

Temperature is a major factor in the abiotic environment of plants. Temperatures in North America range from tropical to arctic. Plants are adapted in nearly all of the temperature ranges. However, some plants will only grow in a very narrow temperature range.

Frost is frozen dew. Dew is water that has condensed on leaves, the roofs of buildings, and other surfaces. Tiny ice crystals are formed at about 32°F (0°C).

16-10. Prevailing winds on San Francisco Bay have resulted in this newly established tree losing all its limbs on one side.

16-11. Unseasonably low temperatures, including snow, have damaged this daffodil that was preparing to bloom.

16-13. Pineapples require a tropical climate.

16-12. Permafrost results in a very shallow soil in east central Alaska. Trees that would normally grow large trunks, many feet tall and 12 or more inches in diameter, are stunted and only a few feet tall and 3 to 4 inches in diameter. The shallow soil will not allow root growth.

Frost also refers to freezing and, in some cases, permanently frozen areas, known as permafrost. **Permafrost** occurs in the arctic and refers to permanently frozen subsoil. Few plants will grow in soil where permafrost is near the surface. Trees that typically grow to great heights are stunted and may only be a few feet (meters) tall.

Heat is also a factor of temperature. The weather can be too warm for crops. Hot dry spells are especially damaging. In hot weather, plants transpire. The water they lose keeps them cool. If the soil does not have adequate moisture for the plant to replace that lost by transpiration, the plant will suffer. Some crops also can withstand just so many days of heat above a certain degree level.

GROWING SEASON

Many crops require a specific number of days to reach maturity. The climate must provide an adequate growing season. If not, the crop is unsuited for growth in the area.

A **growing season** is the number of days from the average date of the last freeze in the spring to the average date of the first freeze in the fall. The growing season is also known as the frost-free period.

Crops grown must reach sufficient maturity during the growing season. In some cases, varieties have been developed that require fewer days to reach maturity. For example, some varieties of sweet corn require about 100 days to mature. Other varieties of corn take longer to mature.

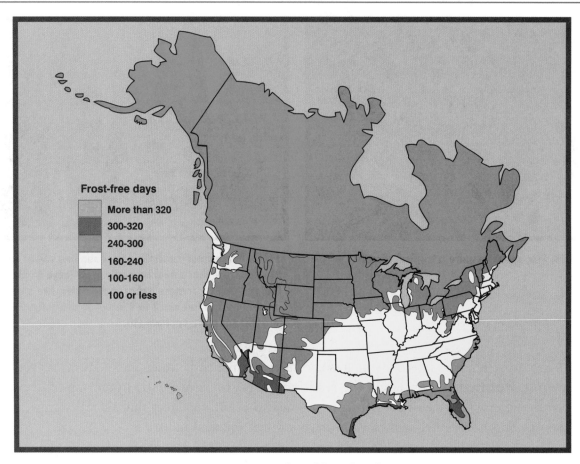

16-14. Average number of frost-free days.

Growing seasons are short in the northern parts of the United States. Some areas of Alaska, Canada, and elevations on the tallest mountains may have few or no frost-free days. Crops with short times to maturity are grown, such as radishes or cabbage.

PHOTOPERIOD

Photoperiod is the amount of light in a 24-hour day. The number of hours of light is greater in North America in the summer than in the winter. Some species of plants are sensitive to photoperiod; others are not.

A plant not affected by photoperiod is a day-neutral plant. Some plants can thrive with only a few hours of light each day. Other plants require more hours of light.

Photoperiod affects the stage of development of some plants. Flower formation may occur as photoperiods reach a specific length. By knowing how plants respond, growers can regulate environments. A good example is the production of chrysanthemums in greenhouses. Growers regulate the amount of light to force the chrysanthemums to bloom when flowers are desired.

Forcing is bringing a plant to a certain stage outside its normal season. This is usually accomplished in growing structures with artificial light. Greenhouses are the common structures for getting a plant to grow outside its natural cycle.

HARDINESS

Plants vary in response to temperature. Hardiness is the ability of a plant to withstand colder temperatures. Some plants, such as cotton and soybeans, are not hardy and are easily damaged by cold weather. Other plants are hardy and can

16-15. To meet market demands near the holiday season, these poinsettias have been forced.

withstand at least some cold weather. Wheat is an example of a plant with considerable hardiness. Cabbage, sweet peas, and mustard greens are examples of cool season crops that can tolerate some cooler temperatures. Plants that lack hardiness are known as tender. They are readily killed by frost.

Hardiness zones have been established. The zones, or divisions, are based on the average minimum temperatures that plants need. Eleven zones are used in North America. These are easily shown using a map, known as a plant hardiness zone map.

CONNECTION

CLIMATE CONTROL IN A VINEYARD

Small temperature changes in the air can destroy or protect plants. Viticulturists (grape growers) in Napa and Sonoma Valleys in California know that air layers drift in from the Pacific Ocean. A few degrees temperature difference in the layers helps protect grapes from frost.

Large smudge pot heaters can be used to warm the air layer next to the ground. The slight warming may keep the layer of cooler air from settling near the grapes and causing frost. This is not climate control, but it helps use information about the air layers to prevent frost damage.

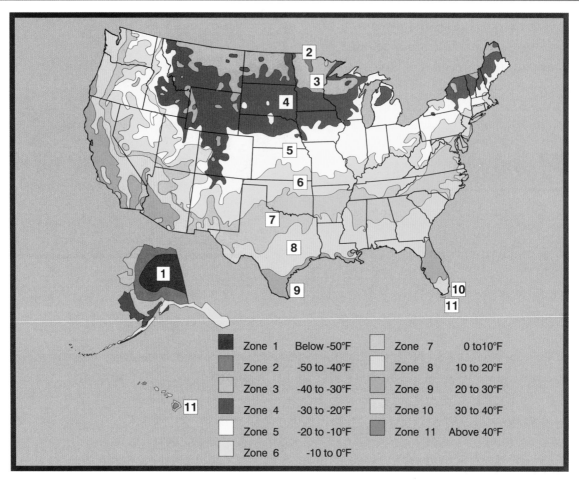

16-16. A plant hardiness zone map helps in selecting plants based on temperature tolerance. Developed by the U.S. Department of Agriculture, the map shows the average annual minimum temperatures of 11 zones.

HEAT TOLERANCE

Heat has a big impact on plant growth. The effects of heat damage may not be as evident as that of cold. Nevertheless, heat causes damage to plants and reduces yields and aesthetic appeal.

The American Horticultural Society (AHS) has coded some plants based on heat tolerance; many years will be required to code all plants. A map has been prepared that divides the United States into 12 zones. The zones are based on the number of days the temperature is above 86°F (30°C). Higher daytime temperatures lead to higher night temperatures, which are particularly damaging to plant productivity. For example, spruce and white pine trees do not grow well and may die in zones with high temperatures.

The web site for the AHS (www.ahs.org) has a wealth of information on the heat-zone map. It allows individuals to enter their zip codes for local information.

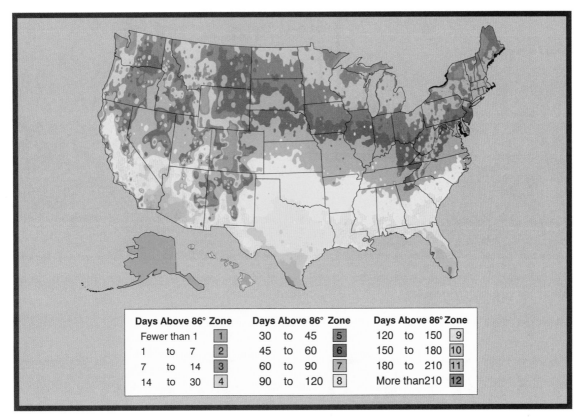

Days Above 86° Zone			Days Above 86° Zone			Days Above 86° Zone		
Fewer than 1		1	30	to	45	5	120	to	150	9
1	to	7	2	45	to	60	6	150	to	180	10
7	to	14	3	60	to	90	7	180	to	210	11
14	to	30	4	90	to	120	8	More than 210		12

16-17. Heat-zone map of the United States.

EDAPHIC ENVIRONMENT

In addition to an atmospheric environment, plants have an edaphic environment. The ***edaphic environment*** is the soil and the area where the roots are located. It deals with creating a good soil or media environment for plant roots and other structures that grow in the soil. Mulch and thatch on the soil are also part of the edaphic environment.

A good edaphic environment is essential for plant growth and productivity. Its importance is probably equal to the atmospheric environment. Without a good environment for the root system, a plant will not grow.

Plants that produce underground structures, such as potatoes and peanuts, must have an edaphic environment appropriate to the needs of the particular plant. These needs vary with different species.

Important edaphic environmental factors include soil moisture, temperature, aeration, pH, and salinity. These areas were covered in previous chapters on soils. Specific information for different crop areas is included in those chapters.

16-18. A soil thermometer is used to determine the temperature of the edaphic environment.

16-19. A good environment results in beautiful early spring blossoms on Spanish Bluebells.

PLANT ENVIRONMENT POLLUTION

The environment for plants can be damaged. The good atmospheric and edaphic environment that is naturally provided can be injured. Sometimes, the environment is damaged by the actions of people. At other times, natural factors cause damage to the quality of the environment for plant growth.

Plant environments may be polluted. ***Pollution*** is any substance or condition that damages the usefulness or productivity of the environment of plants. Some pollution may not do serious damage to plants. Other pollution may contaminate the environment, resulting in a condition that damages plant growth. In some cases, contamination can kill plants.

AIR QUALITY

Plants primarily use carbon dioxide and oxygen from the air. These must be present in appropriate amounts or the plant will not grow as it should. ***Air quality*** is the suitability of the air for use by living organisms. Air pollution occurs when substances get into the air and damage its quality.

Pollution can come from many sources. Gases released into the air by factories may damage plants. Automobile exhaust causes serious damage to the atmosphere. Burning crop residues and other activities that release gases or particulate into the air may cause damage to plants.

Three major concerns today are the destruction of the ozone layer, the increasing amount of carbon dioxide in the air, and acid rain. Each of these can influence plant growth.

Ozone

Ozone is a kind of oxygen. It is formed when three atoms of oxygen combine to form a molecule. Ozone is a concern because of its production and its destruction.

Ozone production is said to cause global warming. **Global warming** is the gradual warming of the earth's surface due to the greenhouse effect. Ozone production on the earth results from the use of certain products, including halon and fuel in engines. The ozone rises into the lower atmosphere where it combines with other gases creating a condition known as the **greenhouse effect**. This layer of ozone acts similar to a sheet of plastic or glass. It holds heat in place. Some scientists refer to this as global warming.

Ozone destruction is changing how the upper atmosphere protects the earth. In the upper atmosphere, a layer of ozone protects the surface of the earth from harmful radiation. As this layer gets thinner, plants will be exposed to increasing amounts of ultraviolet radiation. Crop yields and the ability to provide food could be threatened.

Carbon Dioxide

The amount of carbon dioxide (CO_2) in the air is gradually increasing. The effect this increase will have on plant life is unknown. The change is occurring so gradually that plants may be able to adapt. Scientists are studying the effects of increased carbon dioxide on plants. The air contains less than 1 percent carbon dioxide. It is used by plants in photosynthesis. Some early research has shown that the growth of plants is not damaged by slight increases of carbon dioxide. In fact, CO_2 is added in some greenhouses to speed plant growth.

16-20. The effect of global warming on pine tree growth is being studied with growth chambers that enclose only a part of a tree.

16-21. Scientific plant and atmospheric research (SPAR) units are used to study future environments. The carbon dioxide and temperature in this structure are carefully regulated in studying plant growth.

Acid Rain

Acid rain is any form of precipitation that contains acid. It results when acid-forming substances in the atmosphere get into precipitation. The exhaust from engines and the emissions from factories are major sources of nitrogen oxides, sulfur dioxide, and chloride in the air. These form acids when mixed with precipitation, such as nitric acid, hydrogen sulfide, and hydrochloric acid.

Scientists believe that many trees, crops, and other plants have been damaged by acid rain. Yellowing of leaves (known as chlorosis) may result from acid rain. Acid in rain leaches calcium, potassium, and magnesium from leaves. Acid rain will also lower the pH of soil.

Water quality

Water quality refers to the suitability of water for plants. Hazardous substances sometimes get into water. Water can dissolve hazardous substances and it readily transports solid materials and pathogens. Acids, heavy metals, and other materials may be in polluted wastewater. Some water problems are created by humans and some are natural.

Salt in water (salinity) is a major problem in areas where irrigation is practiced. Using irrigation water with salt builds up the salt content of the soil. Most plants do not grow well, if at all, in soil with high salinity. Irrigation water with high salt content can destroy the productivity of land.

16-22. Providing a good environment is the goal of plant producers.

Soil quality

Plants must have certain nutrients from the soil to grow and be productive. When the soil does not have these nutrients or when the soil has been damaged, plants are unable to grow. Sometimes, soil is polluted or otherwise degraded.

Soil degradation is any thing or process that lowers soil productivity. It results from contamination, erosion, and construction activities.

Soil contamination results when chemicals, oil, heavy metals, or other substances get into it. Some of these contaminants are from agricultural uses. Using pesticides improperly or applying excessive fertilizer can cause soil contamination. Dumping materials that damage soil often occurs. Hazardous materials are sometimes dumped on land. Contaminated water can degrade soil by overflowing onto the land or through irrigation.

Since most of the fertility of soil is in the top few inches, preventing erosion is very important. Good land management practices can go a long way in preventing erosion. (Previous chapters on soils contained information on erosion.)

Construction activities degrade soil in two ways: the land directly involved in the construction is changed and land adjacent to the construction site may be degraded. Most builders are careful to control how land is used and changed. Steps are taken to prevent degradation. Silt fences can be used to control soil particles washing into water runoff. Ground covers are also important in reducing soil loss and sediment in streams.

REVIEWING

MAIN IDEAS

Plants require a good environment to grow and be productive. Plants grow in an ecosystem. This ecosystem is made up of biotic and abiotic factors. The biotic factors are the interactions of the living organisms. The abiotic factors are the nonliving components.

Climate and weather are important to plant growth. Crops and varieties of crops are selected on the basis of climate. Species that are not adapted will not be productive. Within climates, the weather is the current condition of the atmosphere. Unseasonable weather can result in crop damage.

Understanding the climate and the characteristics of plants helps in making choices about what to plant. The growing season is the average number of days in a year that are frost-free. It runs from the average date of the last frost in the spring to the average date of the first frost in the fall. Photoperiod and plant hardiness are also important factors in gaining desired production.

The edaphic environment of a plant is the soil, including mulches on the soil. It is equally important to the atmospheric environment of plants.

Pollution threatens plants. Air, water, and soil pollution are major concerns. Depletion of the ozone layer, increasing levels of carbon dioxide, and acid rain are major air pollution concerns. Water and soil quality depend on preventing degradation of these natural resources.

QUESTIONS

Answer the following questions using complete sentences and correct spelling.

1. What is plant environment?

2. What five environmental factors must be met for a plant to grow?

3. Distinguish between the biotic and abiotic factors in a plant's ecosystem.

4. Why is the climate important in plant growth?

5. What weather factors affect plant growth?

6. What is a growing season? How is it distinguished from frost-free period?

7. What is hardiness?

8. What are the important factors in the edaphic environment?
9. What are three areas of concern in maintaining air quality for plant growth?
10. What is soil degradation?

EVALUATING

Match the term with the correct definition. Write the letter by the term in the blank that is provided.

a. ecosystem
b. biotic environment
c. abiotic environment
d. climate
e. weather
f. permafrost
g. photoperiod
h. frost
i. edaphic environment
j. acid rain

1. _____ any precipitation that contains acid
2. _____ frozen dew
3. _____ amount of light in a 24-hour day
4. _____ parts of the environment in which a plant is growing
5. _____ living things in an environment
6. _____ current condition of the atmosphere
7. _____ nonliving things in an environment
8. _____ average of weather conditions
9. _____ condition where subsoil or soil is permanently frozen
10. _____ environment provided to plants by the soil

EXPLORING

1. Prepare a description of the growing season in your area. Include recommended planting dates for selected crops. Information to help with this activity can be obtained from the local office of the Cooperative Extension Service. Give an oral report on your findings.

2. Investigate the biotic and abiotic environment of a tree or other plant near your home or school. Prepare a report that describes the major factors present. Assess how these factors influence the growth of the plant. (An alternative to using a plant that is growing outside is to use one in a greenhouse, or use an ornamental plant in a home or office.)

3. Observe a construction site in your community. Assess the impact of the construction on the soil. Determine the practices being followed to reduce soil degradation. Prepare a written report on your observations.

Part Five

Meeting Human Needs with Plants

HUMANS require many products from plants. A large number of plant species are used to get these products. These species require differing cultural conditions to be productive. Some crops have specific climate needs. Most all crops have specific cultural needs. Knowing these needs helps us to be better crop growers. It also helps us know which crops have potential where we live.

CHAPTER 17

Grain Crops

OBJECTIVES

This chapter provides basic information about the production of grain crops. It has the following objectives:

1 Name grain crops, grain uses, and leading grain production states.

2 Describe how to select grain crops and varieties.

3 Explain cultural requirements of major grain crops, including corn, wheat, rice, and grain sorghum.

4 Name and describe cultural practices with minor and emerging grain crops, including amaranth, barley, millet, oats, quinoa, rye, teff, triticale, and wild rice.

TERMS

acre-inch
dent corn
grain
grain length
grain marketing
groat

growing degree day (GDD)
kernel
kernel hardness
listing
lowland rice
plant population

shattering
spring wheat
upland rice
winter wheat

17-1. A teacher is reviewing cultural practices with a student in a corn test plot.

Grains are the most important food items worldwide. Most all of our meals have some kind of food from grain. This means that more land is given to grain crop production than any other crops.

What about the foods you eat? Over the last 24 hours, what foods have you eaten that contained grain—pizza, bread, etc.? Make a list of the grains including wheat, rice, corn, and others covered in this chapter. After you have made the list, suppose grain was not available. What would your food have been?

With some people, grain foods are the major foods in their diets. This means that good quality grain is needed. Poor grain means that the people will have poor quality food. Fortunately, quality grain crops can be produced.

17-2. Four grain species.

CROPS AND USE

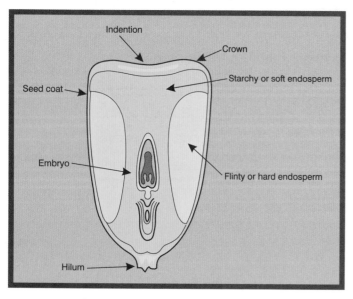

17-3. Major parts of a grain seed are shown with corn.

Grain has been used for food for thousands of years. Domestication of grain began about 9000 B.C. in the middle eastern areas of the earth. Corn was not one of these grains, however. Corn is thought to have originated in Mexico.

Grain is the seed of a cereal grain plant. Cereal grain plants are members of the *Gramineae* (grass) family. The grain kernels have a high starch content, which means that they are an excellent source of energy. A **kernel** is the part of an individual grain within the seed coat. Sometimes, the entire seed of grain is referred to as a kernel. The contents of the kernel provide the important nutrients for human food and live-stock feed.

GRAIN CROPS GROWN

The major grain crops in the United States are wheat, corn, and rice. Other important grain crops are oats, barley, rye, and grain sorghum. Grains of less national importance include amaranth, buckwheat, millet, quinoa, teff, triticale, and wild rice. Grain sorghum is primarily grown for feed, especially poultry and livestock. The other grains are used in food manufacturing as well as to feed animals.

Worldwide, the leading grain crops are wheat, corn, and rice, in that order. Corn is most important in the United States. Rice is most important in developing countries where it is a major human food.

Grain production is distributed throughout the United States. The largest concentration of production is in the mid-section of the United States. Kansas, North Dakota, Illinois, and Montana are major grain-producing states. Figure 17-4 shows the locations of the leading

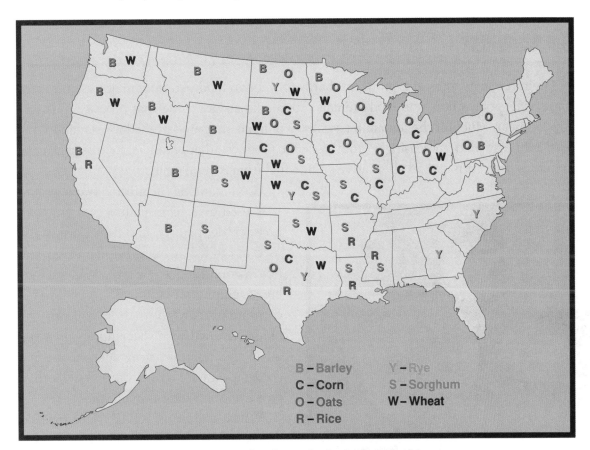

17-4. Major areas of grain production in the United States.

states in grain production. Table 17-1 shows the ranking of the top five states in production of the most important grain crops.

Table 17-1. Rank of Top Five States in Selected Grain Crop Production

| Grain Crop | Rank | | | | |
	One	Two	Three	Four	Five
wheat	Kansas	N. Dakota	Montana	Washington	Oklahoma
corn*	Iowa	Illinois	Nebraska	Minnesota	Indiana
oats**	Wisconsin	N. Dakota	Minnesota	Iowa	S. Dakota
barley**	N. Dakota	Montana	Idaho	Washington	Minnesota
rice	Arkansas	California	Louisiana	Texas	Mississippi
grain sorghum	Kansas	Texas	Nebraska	Missouri	Louisiana
rye	Texas	Georgia	N. Dakota	N. Carolina	Kansas

*harvested for grain and seed

**harvested for grain

Source: U.S. Department of Agriculture

GRAIN USE

Grain has many uses. It is grown primarily for human food. Another major use is as feed for service, companion, and food animals. Food animals are those that produce milk and eggs or meat. Much of the animal feed goes to animals that are fattened for slaughter and used as human food. Increasingly, grain is used to produce other products, such as packaging materials from corn and an alternative engine fuel from several grains.

Corn is among the most versatile grains. It is used as a high-quality feed grain for hogs, cattle, poultry, and fish. Corn is also used in producing many human food products. Ground and whole grain foods include bread, chips, and hominy. Special types of corn are also grown, such as popcorn and blue corn used with speciality foods. Corn oil is an important vegetable oil used in cooking and for other purposes. Corn is grown to some extent in all states.

Wheat is milled to make flour for baking, as well as used for animal feed. Rice is

17-5. Bread is a common product made from grain.

polished (husk removed) and used without grinding but is sometimes ground. Oats, barley, and rye are used in food production and animal feed. Grains of lesser importance include triticale, millet, and quinoa.

Some of the lesser grains are newer crops that may hold high future opportunities. One example is quinoa. Quinoa is a new grain in the United States, being introduced in the 1980s. It is grown at higher elevations in a few western states. Quinoa has long been a major food with the Inca Indians of South America.

17-6. Grain is milled into flour and other food products. Wheat is being checked after the first break in the milling stage. (Courtesy, Cargill, Minneapolis)

Table 17-2. Uses of Selected Grains

Crop	Examples of Use
corn	food—cooking oil, bread, breakfast cereal; feed—livestock, fish, poultry, and pets; others—fuel (gasohol), medicine, cosmetics
wheat	food—bread, pasta, breakfast cereal; feed—poultry and livestock; others—adhesives, alcohol, manufacturing
rice	food—boiled main or side dish, breakfast cereal; others—hull products, straw products
oats	food—breakfast cereal; feed—horses, cattle
barley	food—malting; feed—livestock
rye	food—bread; feed—livestock
grain sorghum	feed—livestock and poultry

SELECTING CROPS AND VARIETIES

All grain crops have common and scientific names. Most crops have numerous kinds and varieties. These are adapted to different climates and may meet different needs in the grain market. Some native kinds are maintained for research purposes. Information about the kinds and varieties adapted in a region is available from land-grant universities and local offices of the Cooperative Extension Service.

Selecting the grain crop to grow involves five important areas: climate, soil and water, market, technology, and personal skills and preferences.

Table 17-3. Names of Selected Grain Crops

Common Name	Scientific Name	Notes
amaranth	*Amaranthus cruentus*	a promising new crop
barley	*Hordeum vulgare*	vulgare subspecies
buckwheat	*Fagopyrum esculentum*	food and animal feed uses
canary grass	*Phalaris canariensis*	limited production for bird feed
corn (also known as maize)[1]	*Zea mays*	—[2]
grain sorghum	*Sorghum bicolor*	many races exist
millet	*Panicum miliaceum*	"Proso" type is only millet grown for grain in the United States (see Chapter 20, Forage and Turf Crops)
oats	*Avena sativa*	regional variations
quinoa	*Chenopodium quinoa*	pseudocereal
rice[1]	*Oryza sativa* *Oryza glaberrima*	68,000 cultivars 2,600 cultivars
rye	*Secale cereale*	grown for grain[3]
teff	*Eragrostis tef*	holds potential as grain
triticale	*Triticale hexaploide*	promising forage
wheat[1]	*Triticum aestivum* *Triticum turgidum*	aestivum subspecies (soft red winter) compactum subspecies (soft white winter)
wild rice	*Zizania palustris*	now becoming a domesticated crop

[1]Major grain crops.

[2]Corn is classified on the amount, quality, and arrangement of kernel endosperm into six types: dent (*indentata* subspecies), flint (*indurata* subspecies), floury (also known as Indian or blue corn) (*amylacea* subspecies), pop (*everta* subspecies), sweet (*saccharata* subspecies), and pod (*tunicata* subspecies). Pod corn has little value.

[3]Rye grown for cereal is distinguished from ryegrass. The two common species of ryegrass are *Lolium multiforum* (Italian ryegrass) and *Lolium perenne* (perennial ryegrass).

CLIMATE

The climate is the average weather condition over a long time. Grain crops vary in their climate requirement. With summer crops, the growing season must be long enough for the crop to mature and produce an abundance of grain. Planting must be after the danger of frost has passed. Maturity must be reached before frost in the fall. The weather conditions during the growing season must be what the crop needs. High temperature in the summer and adequate moisture levels are important with summer grain crops.

Select a variety to match the climate. Corn, for example, typically requires 110 to 125 days to grow to maturity. Some varieties require less than 100 days. The varieties with early maturity are best for climates with shorter growing seasons.

17-7. Irrigation is sometimes needed to promote good yields of wheat in the western United States.

Some varieties of wheat have been developed for summer production; others are planted in the fall and mature in the spring. Summer wheat crops are typically planted in the northern United States and southern Canada. Tender wheat plants would not likely survive the cold winter weather found there. Winter wheat can be planted in the fall and will survive winters that are not too harsh. Winter wheat is typically planted south of the northern tier of states in the United States.

Soil and Water

The available soil and water must be within the range of the requirements of the crop to be grown. Most grains require fertile soil, though requirements vary. Moisture requirements also vary widely with the crop. Rice and wheat have widely varying soil and moisture requirements.

Rice requires large amounts of water. Fields in which rice is planted are carefully planned for water management. The field is flooded a few inches (cm) for a large part of the growing season. The water for flooding is normally pumped from wells or natural sources, such as streams and lakes. Natural precipitation has some impact on the amount of water needed for rice. Rice cannot be successfully grown without soil with good water holding capacity and water for flooding. Low-lying areas in the South and in California are best for rice.

Wheat does not require large amounts of water as does rice. It can be grown in areas with only a few inches of rainfall a year. In very dry areas, some irrigation may be used. Wheat is often grown in the plains of the central part of the United States. It is adapted to a wide range of soil and water conditions.

17-8. Grain elevators are important in marketing grain crops.

MARKET

Grain crops are grown to be sold. Income is gained from the sale of the grain. Success depends on having a good market. A grain crop should never be produced unless a market is available.

Grain marketing includes all the processes that connect the grain producer with the consumer. It is much more than just the sale of grain. Most grain growers are involved with only a few steps in the process, such as growing a crop that consumers want and delivering it to the location where it can be sold, such as a local elevator.

In some cases, grain may be grown and fed to animals on the farm. In effect, the grain is marketed through animals. In that case, a good market must be available for the animals. Fewer grain growers are marketing grain by feeding it to animals. The grain is sold to others who are in the business of animal feeding.

Markets for grain should be convenient and provide returns to the grower. Hauling grain a long distance is expensive and reduces returns to the grower. It is best to plant grain crops for which a good market is available within a reasonable distance of the farm.

CAREER PROFILE

WHEAT PRODUCER

A wheat producer grows wheat. Sometimes known as a wheat farmer, a producer is responsible for all areas of wheat production. As an entrepreneur, the wheat producer must arrange financing, obtain land, select and plant seed, provide needed cultural practices, harvest, and market the grain. A wheat producer also hires and supervises employees and keeps careful financial records of the farming operation.

Most wheat farmers have considerable experience in wheat production. A high school education is needed, though many have associate and baccalaureate degrees. Regular participation in workshops and field days is required to stay up to date on the latest technology.

This shows a wheat producer in Washington examining wheat prior to harvesting. (Courtesy, U.S. Department of Agriculture)

Most grain is sold by cash marketing. The grower delivers the grain and is paid for it. In addition, marketing alternatives include futures markets and forward contracting. These often have advantages to grain producers who are interested in such approaches.

TECHNOLOGY

Grain technology is the use of science or the products of science in growing a grain crop. Equipment, fertilizer, improved seed, pest control, and other technology are used.

17-9. Equipment and other inputs are needed to produce a grain crop. Land is being harrowed in preparation for planting. (Courtesy, New Holland North America, Inc.)

Precision and variable rate technology are increasingly being used. Technology must be available in a local area for efficient grain production. If not, the potential grower runs the risk of trying to grow a crop without the benefits of technology.

Areas where grain is grown often have several providers of inputs needed to produce the crop. Equipment dealerships and repair services are available. Sources of fertilizer, seed, and chemicals are readily available. Elevators are prominent in the marketing process. Trucks and good roads are needed to haul the crop to the elevator. Crop consultants are often important.

SKILLS AND PREFERENCES

The personal skills and preferences of an individual should be considered. An individual with experience in growing a particular crop is more likely to be successful in producing that crop. Keeping up to date on new developments is a part of being successful.

Another consideration is the skill level of the available labor. Most growers hire tractor and equipment operators and other workers. The available labor supply needs to have individuals who have skill in the work.

Some individuals prefer one crop to another. As long as marketing and production factors are about equal, individuals should grow the crop they prefer. Overall, the ability to grow a grain crop that makes a profit is most important to the financial success of a farm business.

CORN PRODUCTION

Corn is much more important in the United States than elsewhere in the world. About half of the corn grown worldwide is in the United States. Worldwide, acreage planted to corn is about half that of wheat.

About 70 million acres are planted to corn in the United States each year. Nearly 85 percent of the planted acres are harvested for grain. The other 15 percent of the acreage is used for silage, popcorn, and other crops. (Additionally, 750,000 acres are planted to sweet corn each year.)

Corn yields per acre vary, with the average being about 123 bushels per acre in recent years. Silage yields usually average 12 or so tons per acre. With both grain and silage, yields depend on local weather conditions and cultural practices.

Most corn grains have about 65 percent carbohydrates, 10 percent protein, and 25 percent moisture, ash, and other materials. The high carbohydrate con-

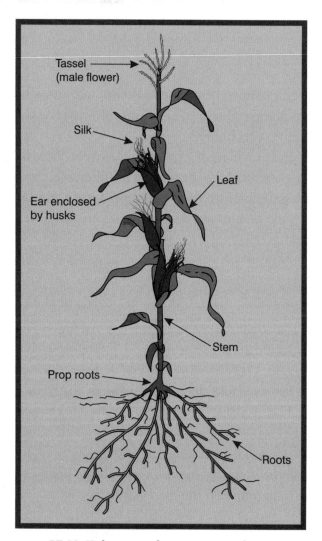

17-10. Major parts of a mature corn plant.

Tassel (male flower)

Silk

Ear enclosed by husks

Leaf

Stem

Prop roots

Roots

17-11. Prop roots help hold corn stalks erect.

tent is an excellent source of energy in an animal's diet. Some increase in protein would increase the use of corn as feed. Recent research has produced a line of corn with higher protein. Known as high lysine corn, the new lines are more nutritious, but they do not have the same high yield of grain.

TYPES GROWN

Seven types or subspecies of corn are found: dent, flint, floury, sweet, pop, decorative, and pod. Five have important markets, with pod corn being a curiosity with little market value. Each of the five types with economic value is briefly described here.

- Dent corn—**Dent corn** is the major type grown for grain. It is so widely planted that it is sometimes referred to as field corn. The grains are flat with an indention in the crown (end opposite the hilum). The outer covering is hard. The central core is soft. Dent corn is used for animal feed, human food products, and processing into oils, syrups, and other products. The grains may be yellow or white, depending on the variety. The best varieties produce two or three ears on each stalk. Genetically modified varieties are available to resist insect damage or provide for unique nutrient needs of some animal species.

17-12. Developing ears of corn are nearing maturity.

- Flint corn—Flint corn grains are more rounded than dent corn. Flint corn has a hard outer covering and less soft interior. It has a shorter time to maturity than dent corn, which makes it suited to growing in northern climates. Little acreage is planted to flint corn in the United States.

- Floury corn—Floury corn is soft and was the first corn grown in the United States. The grains may be white, blue, or shades of other colors. In recent years, blue corn has increased in popularity because of its use in chips and other specialty foods. Overall, acreage planted to floury corn is much less than dent corn.

- Sweet corn—Sweet corn is planted for human food. The ears are harvested while green when at the sweetest stage. The high sugar content and lower starch than the

17-13. Sweet corn is popular as roasting ears.

17-14. The wrinkled appearance of dried sweet corn kernels is obvious when compared to dent corn.

17-15. Attractive decorative corn shows a range of colors and genetic traits.

other types of corn makes sweet corn appealing. Sweet corn is widely used as roasting ears and for canning and freezing. Sweet corn should be cooked or processed as soon as possible after harvesting because of changes in the sugar and starch content. Sugar turns to starch rapidly after harvest, especially in a warm environment. Dried sweet corn grains have a wrinkled appearance. Other than for seed, sweet corn has little value in dried form. Yields are typically 80 to 100 cwt. of sweet corn per acre.

- Popcorn—Popcorn is a food that many people enjoy as a snack. The grains are smaller than other corn. Each stalk may grow several ears. Popcorn is harvested in the fall when mature and the leaves have dried. Moisture inside the grains turns to steam when heated causing the kernel to explode. Popcorn is grown commercially in a few states, with Illinois being a leader.

Cultural Practices

Cultural practices are the procedures used in producing a crop. They vary with the kind of crop and, with corn, from one type and variety to another. Cultural practices with corn include selecting the variety, planting, fertilization, pest control, and harvesting. In some cases, irrigation may be important. The primary focus in this section is on growing corn for grain.

Varieties

All varieties of corn sold for commercial grain and forage production in the United States are hybrids. Selecting the hybrid to plant is an important decision. Hundreds of hybrids are available. Many have been developed for use in local areas. They have been bred for local conditions. The land-grant university in your state or a local seed dealer can offer assistance with the variety to plant.

Corn varieties vary in height, number of ears per stalk, days to maturity, and other characteristics. Height ranges from about 2 feet to 20 feet (0.5 to 7.5 m), with 6 feet to 8 feet (1.8 to

17-16. Fast-growing corn plants indicate a variety that is obviously well suited to the land.

2.4 m) being most common. Number of ears per stalk ranges from one to three or more. In most cases, growers want each stalk to produce two or three well-developed ears. Number of days to maturity ranges from 50 to 330 days. Most hybrids mature in 110 to 120 days. The frost-free days should be at least 10 to 12 percent more than the days required for a variety to reach maturity. For example, if the growing season has 150 days, a corn hybrid with a maturity of 130 days or slightly less would be desirable. The variety planted should reach maturity at least 10 days before the first frost.

Planting

Planting should occur after the danger of frost. Planting dates range from February in south Texas to mid-May in the northern states. Temperature is an important consideration throughout the growing season. Timing should assure that the crop germinates and grows rapidly. The stages of growth of corn are related to growing degree days. In addition, seed germination is affected by the temperature of the soil. A general rule is to plant after the soil temperature is above 50°F (10°C) at a soil depth of 2 inches (5.08 cm).

Degree Days

A *growing degree day* (GDD) is a measure of the temperature requirements for best corn growth. It is the maximum temperature plus the minimum temperature in a day divided by 2 minus 50. The constant of 50 is subtracted because corn grows very little at 50°F (10°C) or below. Temperatures above 86°F (30°C) do not increase the rate of growth. Maximum temperatures over 86°F are counted as 86 in the formula. Here is an example: If the low temperature was 60°F and the high was 90°F, the GDD would be 60 + 86 = 146 ÷ 2 = 73 − 50 = 23 GDD.

Most hybrid corn has fairly specific GDD requirements. With a dent hybrid requiring a total of 2,450 GDD, 107 days of growing season would be needed.

Seedbed

Most corn is planted in a prepared seedbed. The seedbed is prepared with a chisel or moldboard plow and followed with a disk harrow or do-all. This procedure reduces clods and prepares a fine seedbed. Corn is usually planted in rows with a spacing of 20 to 40 inches (50.8 to 101.6 cm) apart. In general, yields per acre (hectare) increase with narrower rows. Many growers use 30-inch (76.2 cm) rows. The seed is typically planted 1 to 2 inches (2.54 to 5.1 cm) deep.

A planter is used to open the seed drill and properly place the seed. The planter should assure the desired plant population. *Plant population* is the number of plants growing in one acre of field. Seeds should be planted to assure a plant population of 24,000 to 32,000 plants per acre. Higher plant populations are sometimes used when extra attention is given to cultural practices, including fertilization and irrigation. Higher plant populations are also used with varieties that grow small stalks, such as sweet corn and popcorn. Lower plant populations may be used in dryland areas of the plains.

How far apart should seed be dropped (placed) to have the desired plant population? This is easy to calculate. Determine the row width and divide the width in feet

17-17. Initial seedbed preparation may be with a two-way moldboard plow. This implement makes it possible to throw all furrows the same direction. At the end of the field, the plow is raised and flipped over and the tractor is turned around to go back across the field.

17-18. A planter equipped with variable rate technology. (Courtesy, Case Corporation)

into 43,560 (square feet in an acre) to determine the total length of all rows in an acre. Divide the number of plants into the row length to determine the distance between seed. Here is an example: If the corn will be planted on 30-inch (2.5 feet) rows with a desired plant population of 32,000 plants, how far apart should the seed be dropped? 43,560 ÷ 2.5 = 17,424 ÷ 32,000 = 05.5 inches apart. (This example assumes 100 percent seed germination.)

No-Tillage Planting.

No-till planting involves different techniques and planting equipment from that of a prepared seedbed. Vegetation needs to be cut and controlled, often with chemicals. Planters

CONNECTION

QUALITY FOOD PRODUCTS FROM GRAIN

The linkage between the grain producer and consumers has many processes. Food quality can be no better than the quality of the grain that is harvested. Careful handling and processing are needed to assure quality food products.

Food products are tested to assure that they are wholesome. This shows a food scientist testing bread products for quality. Each batch of bread is carefully monitored to provide consistent and safe food products. (Courtesy, Cargill, Minneapolis)

17-19. Corn with a weed management problem.

17-20. A healthy tassel is a sign that pests have not destroyed the main bud of a corn plant.

need to be adjusted to place the seed at the proper depth. In most cases, seed planted no-till are placed slightly deeper but have less soil covering the seed.

Fertilization

Good corn yields require nutrients. Corn tends to need more fertilizer than most crops. Research has found that a yield of 150 bushels of corn per acre will require 170 pounds of N, 35 pounds of P_2O_5, and 175 pounds of K_2O. Soil with a pH of 5.0 to 8.0 is best.

Prior to seedbed preparation, a soil test should be made to determine the needed soil amendments. pH and nutrient needs can be assessed. These may be applied prior to planting, at the time of planting, and after the crop is growing.

Pest Control

Common corn pests include weeds, insects, nematodes, and some diseases. Recommendations on pest management should be obtained from the land-grant university in your state or a local agricultural consultant. IPM is increasingly used to manage pests.

Weeds are typically managed with cultivation and chemicals. Depth of cultivation is deeper when the plants are smaller and shallower as the plants grow. Cultivation should not be deep enough to damage the roots of growing corn plants. Only recommended herbicides should be used. These may be used preplant, preemergence, or as the corn is growing as postemergence.

Corn is particularly a target of bud worms and ear worms. These and other insect pests, nematodes, and diseases are managed with chemicals and by planting resistant varieties.

Harvesting

Harvesting corn for grain occurs at the end of the growing season when the kernel moisture content is 20 to 28 per-

17-21. Harvesting corn. (Courtesy, New Holland North America, Inc.)

cent. Higher moisture increases the need for artificial drying. As a general rule, corn with more than 15.5 percent moisture will need artificial drying if it is to be stored. In areas with high humidity, drying to 11 percent helps avoid aflatoxin buildup in stored grain.

Harvesting should be timed for maximum yield. Waiting too late results in brittle and lodged (fallen) stalks, resulting in grain loss. Corn is harvested with a picker or combine. Proper adjustment is essential to avoid grain loss.

Aflatoxins may be problems with improperly stored corn, other grains, and peanuts. An aflatoxin is a highly poisonous substance produced by fungi in grain. It is produced by *Aspergillus flavus* and a few related species of fungi in grain with a high moisture content. Grain that is below 12 percent moisture inhibits the growth of the fungus at any temperature. Using corn with aflatoxin in feed can result in the deaths of many animals. Grain is routinely inspected for the presence of aflatoxins. If excessive levels are found, it not sellable.

17-22. Examining corn for aflatoxin invasion during storage. (Courtesy, U.S. Department of Agriculture)

17-23. Quality ears of corn have been judged at a fair.

WHEAT PRODUCTION

17-24. Wheat ready for harvest.

Wheat is the most important cereal grain crop. It is grown almost as widely in the United States as corn. Wheat production is concentrated in the Plains states. The types grown and cultural requirements vary in different locations.

TYPES GROWN

The kind of wheat planted is classified based on time of planting, color, and kernel hardness. Various combinations of wheat characteristics may be selected and planted.

Season

Two main classes, based on time of planting, are spring and winter. **Spring wheat** is planted in the spring, grows during the summer, and matures in early fall. Spring wheat is planted in the northern states where the winters are too harsh for wheat to survive.

Winter wheat is planted in the fall, establishes itself over the winter, and grows rapidly in the spring. Winter wheat is typically grown in the central plains and southern states.

Color

Wheat is placed in two classes on the basis of color: red and white. The color of the seed coat determines its classification. Red-kernel wheat predominates in the United States. White-kernel wheat

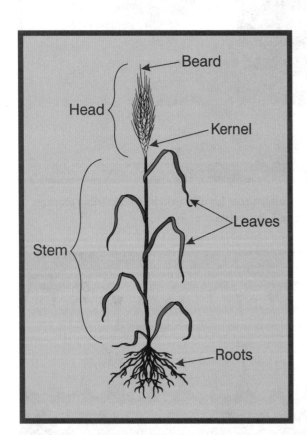

17-25. Major parts of a mature wheat plant.

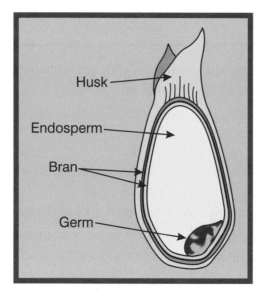

17-26. Parts of a wheat kernel.

17-27. Three kinds of wheat heads are found (from the left): common, which has a longer head and is bearded; club, which has a short head and no beard; and durum, which is bearded with hard kernels.

tends to produce slightly higher flour yields because less cleaning of red seed coat fragments is needed during milling.

Hardness

Kernel hardness is either soft or hard. Hardness is determined by the endosperm. The endosperm of hard wheat breaks along cell walls. Soft wheat will break across cells. Soft wheat yields slightly higher amounts of finely ground flour. Flour from hard wheat tends to be coarser. Bread is made with hard wheat flour; cakes and cookies are made with soft wheat flour.

Durum is a special kind of hard wheat. It is primarily used for making pasta. The flour from durum wheat is coarser than that from other hard wheat or soft wheat.

Kernel hardness may be used with color and season to have a wide range of types of wheat. Hard red winter wheat, hard red spring wheat, soft red winter wheat, and white winter wheat are examples.

17-28. Mature head of bearded white winter wheat, dry enough for harvesting. The moisture content in the kernel(s) is low. The wheat was grown in a dryland area of Arizona.

CULTURAL PRACTICES

Cultural practices tend to be similar with wheat but vary with local climates and other conditions. Information should be obtained locally on wheat production.

Varieties

Wheat varieties should be selected on the basis of climate adaptation and yield. Resistance to disease and pests is also a factor in selection.

Plant height may also be a consideration. Taller plants are more likely to lodge than shorter plants, particularly during a heavy thunderstorm if the heads are heavy with grain.

Winter hardiness is another consideration for growers of winter wheat in areas where there is considerable cold weather. Wheat tends to be more sensitive in the young growth stages shortly after planting. Even with the best winter wheat varieties, some damage results from cold weather. Both root and stem damage may occur. Some stems are damaged below 26°F (–3.3°C). Most wheat plants quickly overcome some cold weather damage when the weather warms in the spring.

Planting

Proper planting is a big part of successful wheat production. Three important factors to consider are planting date, seedbed preparation, and seeding rate and depth.

17-29. Wheat is often planted in a prepared seedbed. This shows a mulch-till chisel plow. (Courtesy, Case Corporation)

DATE. In general, planting dates are set by the season of the wheat, such as winter or spring.

Winter wheat is planted in the fall before winter begins. In the northern states, winter wheat is planted in September. In the southern states, winter wheat may be planted in October through early December. Planting should be early enough to allow young plants to become established before the onset of cold weather.

Spring wheat is planted in April and May in most locations. However, in southern areas, spring wheat may be planted in the winter months. Locations in the far north may plant in early June. Planting is somewhat related to attempts to control snow mold. Snow mold is a disease caused by a fungus that grows on leaves when they are pressed against the soil by the snow. Planting toward the end of the planting season results in smaller plants with leaves less likely to be mashed down by the snow.

SEEDBED. Most wheat is planted in prepared seedbeds. Preparation is similar to that for corn. Practices to conserve moisture are important in dryland areas. Disking land early, before planting, helps keep weed growth down.

SEEDING. Wheat seeds are planted using drills that provide a uniform plant population throughout a field. Each drill is about 6 inches (15 cm) apart. The number of wheat seeds in a pound varies from 12,000 to 20,000. The desired plant population varies with moisture and if the growing wheat will be grazed by livestock.

General recommendations are 20 to 100 pounds of seed per acre. The lower rates are used in dryland areas of the Plains states. The higher rates are used in the humid climates of the southeast. Wheat that will be grazed by livestock should be seeded at a heavier rate of 75 to 120 pounds per acre. A planter calibrated to drop 14 seeds per foot of drill based on 6-inch spacing will give about the right plant population.

Seeding rate influences yield. Not planting enough seed and not having a desired plant population results in wheat plants excessively tillering and in the growth of weeds. Having too many plants may result in lodging and decreased yield.

Wheat seed should be planted in moist soil, if possible. In areas with rain, planting depth is not as deep as in dry areas. A depth of 1 inch (2.5 cm) may be adequate in areas where rain is likely. A depth of 2 inches (5.0 cm) or more may be needed in dry areas where no rain may fall for some time after planting.

17-30. Wheat is usually seeded with a drill. This drill is specially designed for no-till wheat cropping. (Courtesy, AGCO Corporation, Duluth, Georgia)

17-31. A goal in wheat culture is to have a number of stems (tillers) on each plant. This compares a developing plant with good tillering (left) and one that has not tillered well. Proper fertilization promotes tillering.

Fertilization

As a kind of grass, wheat responds well to fertilizer. A wheat yield of 100 bushels per acre will require 120 pounds of N, 56 pounds of P_2O_5, and 31 pounds of K_2O. In addition, a 100-bushel-per-acre wheat yield requires 12 pounds of S, 13 pounds of Mg, and 5 pounds of Ca.

Soil tests are needed to determine the pH and nutrient condition of the soil. Wheat grows best in slightly acid soil. In addition, timing of nutrient applications is important, particularly in areas of high rainfall where leaching is more likely. Nitrogen should be incorporated into the seedbed prior to planting and during the mid-growing season prior to grain development. Applying fertilizer after a crop is growing is known as topdressing. It is usually not incorporated into the soil.

Pest Management

Wheat is subject to a number of pests. Insects and diseases are particular problems. Weeds can also be problems, but proper seeding helps manage weeds.

Diseases include snow mold, root rot, rust, powdery mildew, and scab. Snow molds are managed through cultural practices, such as delaying planting. Planting resistant vari-

17-32. No-till application of herbicide on an eastern Washington wheat field. (Courtesy, Agricultural Research Service, USDA)

eties, treating seed, rotating crops, and fungicides are among the approaches to disease management.

Many different insects attack wheat. These vary with the region and season of the year. The Hessian fly, fall armyworms, chinch bugs, false wireworms, grubs, and Russian wheat aphids are examples of wheat insect pests.

The Hessian fly is a major pest of wheat in some areas, particularly the South. The Hessian fly is managed by waiting to plant winter wheat until summer fly populations have died, often after the frost-free date in the southern states in the fall.

Cultural practices and insecticides are used to manage insects.

Harvesting

Harvesting should be after the heads of wheat are mature. The field will likely be a golden brown. Large combines cut the stalks with grain heads and separate the kernels from the stalks. Wheat should be harvested before lodging or shattering. **Shattering** is the point at which mature kernels fall from the wheat head. These kernels are lost grain. They fall to the ground and may benefit wildlife as feed.

The preferred moisture level for harvesting wheat is 12.5 percent. In some cases, harvested wheat may

17-33. Wheat is harvested with a combine. This shows harvested grain being moved from the combine to the wagon for hauling from the field. (Courtesy, AGCO Corporation, Duluth, Georgia)

have moisture levels as high as 20 percent. Artificial drying is needed above 12.5 percent. High-moisture wheat is penalized in selling, resulting in lower returns to the grower.

RICE PRODUCTION

Nearly half of the world's population depends on rice for food. It is especially important in Asia and Indonesia. The culture of rice differs from other cereal grain in that it is grown on land flooded with shallow water. The water provides moisture for the rice plant as well as helps control weeds. Over the last decade, acreage planted to rice has averaged nearly three million acres each year. The per acre yield has averaged 5,534 pounds.

17-34. Rice kernels before and after polishing. (Courtesy, U.S. Department of Agriculture)

Most rice is milled and polished into the white form common today. Milling involves removing trash and the outer hull. White rice may undergo additional polishing in rotating cylinders. Millers use great care to keep from breaking the rice kernels. People prefer whole kernels!

TYPES GROWN

Worldwide, nearly all rice grown today originated in Asia, though a small amount of is from Africa. In the United States, only Asian rice is produced. It is classified in two ways: grain length and cultural method.

- Grain length—**Grain length** is classifying rice by the length of the kernel. The following types are used: short-, medium-, and long-grain rice. Short-grain rice has kernels that are less that one-fifth inch (5 mm) long. Medium-grain rice is one-fifth to one-fourth inch (5 to 6 mm) long. Long-grain rice is one-fourth to five-sixteenth of an inch (6 to 8 mm) long. Long rice contains additional starch, which makes the rice fluffy and dry when cooked. Long-grain rices are grown in tropical climates. Short- and medium-grain rice varieties are grown in milder climates.

17-35. Lowland rice field.

- Cultural method—Cultural method refers to how the rice is grown. This is influenced by the terrain of the land on which it is grown. **Lowland rice** is grown in large, flat fields that are flooded by irrigation. The fields may be carefully formed using laser leveling systems. Small dikes are built around the field to direct the gradual flow of water. **Upland rice** is grown in small areas on hillsides. These locations usually depend on natural rainfall for water. These small fields are often referred to as rice paddies.

CULTURAL PRACTICES

In the United States, rice is grown on large farms that make considerable use of mechanization. This is far different from the small paddies found in developing countries.

Varieties

Rice varieties in the United States have been developed based on the length of the grain. The southern states grow long-grain varieties. California primarily grows medium- and short-grain varieties. Varieties are selected to assure production during the best part of the growing season.

17-36. A rice geneticist is checking a variety that releases a natural chemical that keeps weeds away. (Courtesy, U.S. Department of Agriculture)

Rice grain production formation is slow during very hot days. Planting so the grains are formed in late summer helps avoid this problem. Common varieties in the southern states include Newbonnet, Gulfmont, Rexmont, and Millie. Jasmine cultivars have been grown in Texas that compete with those produced in Thailand. The average days to maturity range from 117 to 130.

The California rice industry has developed a uniform naming system for rice based on grain length and days to maturity. Variety names are, for example, S-201 for an early-maturing short-grain rice to M-401 for a late-maturing, medium-grain rice. Days to maturity for the California rice range from 125 days for the early maturing to 165 for the late.

Growers should determine the specific varieties for their locations. The land-grant university or a local crop consultant can assist.

17-37. Jasmine 85 is a premium rice on the U.S. market. (Courtesy, U.S. Department of Agriculture)

17-38. The surfaces of rice fields are formed using a laser-guided system. (Courtesy, Spectra-Physics, Dayton, Ohio)

Planting

Rice is often planted in one of three ways: drilling into a dry seedbed, broadcasting on a dry seedbed, or broadcasting into standing water. When the seeds are placed on a dry seedbed, the field is flooded after planting.

Planting Date

The time of rice planting may depend more on air temperature than soil temperature. Water in shallow flooded areas will quickly warm when exposed to the air. In the southeast, an air temperature of at least 65°F (18.3°C) is recommended. In California, a slightly warmer air temperature is recommended—70°F (21.1°C). Dates of seeding are late March through late May in the southern states and mid-April to the first of June in California.

Seeding Rate

Seeding density should result in 15 to 20 plants per square foot. This is equal to 653,400 to 871,200 plants per acre. This plant population is obtained by water seeding 120 to 150 pounds per acre or drilling 90 to 100 pounds per acre. Overseeding is needed because as many as half of the seed do not produce plants that reach maturity. Birds, diseases, and other pests keep the rice plant population down.

Seedbed Preparation

How the seedbed is prepared depends on the method of seeding—dry seedbed versus water seeding. All seedbeds are formed to provide a near-level surface that will assure proper water dispersal throughout the field. Dry seedbeds should be tilled into a shallow, loose condition without large clods. Large rollers may be run over the land to firm the soil and decrease depressions. Seed should be planted about 1 inch (2.54 cm) deep.

With water seeding, the seedbed is prepared similarly to dry seeding, but some clods are left or grooves are made in the field. The grooves are made 2 inches (5.08 cm) deep and about 7 inches (17.78 cm) apart with specialized equipment. The clods or grooves are used to help assure an even dispersal of the seed over the field. Without the grooves, seed may drift to one side, resulting in areas without seed and other areas with too much seed. The rice seeds are often soaked in water before water seeding. Aerial equipment may be used to water seed rice.

Water Management

Unlike other grains, water management is necessary with rice. Soil characteristics influence the amount of water needed. Clay soils hold water, while sandy soils allow water to percolate and be lost to the rice crop. Water wells should be able to produce plenty of quality water.

Water in rice fields is measured in acre-inches. An **acre-inch** is an acre of land covered with water that is 1 inch deep. A field that has water 3

17-39. Water management in this rice uses contour levees to regulate water movement. (Courtesy, Spectra-Physics, Dayton, Ohio)

inches deep has 3 acre-inches per acre. Wells and pumps are used that provide the needed water. Pumps should be used that provide at least 1,000 gallons per minute. With a 40-acre field, 18 hours would be needed to cover the field with 1 acre-inch. The needs and pumping times of other field sizes can be easily determined. One acre-inch is 27,155 gallons of water. A 1,000 gpm pump requires 27 minutes to pump 1 acre-inch.

Fields of rice planted on dry seedbeds are flushed with water to moisten the soil and encourage seed germination. Flushing involves quickly covering the field with water and allowing it to drain off in about three days. When the plants are three to four weeks old, or before the plants are 6 inches (15.24 cm) tall, water is added. The field remains flooded for the growing season. Water depth should be 2 to 4 inches (5.08 to 10.16 cm) unless weeds develop. Additional water depth will help control most weeds. Also, water will need to be added to replace the amount lost to evaporation.

Less water is needed with continuous flood. Continuous flood is when the rice field is flooded throughout the growing season. Flushing requires more water and may result in more weeds.

Fertilization

Rice requires fertilizer to produce good yields. Soil testing should be done every three years to determine overall fertility levels. Nitrogen is the most important nutrient that is usually added. Rice requires 70 to 200 pounds of N per acre, depending on grain length. Long-grain rice requires a higher level of nitrogen. Phosphorus and potassium are added prior to planting on the basis of soil test analysis. Zinc is sometimes a limiting micronutrient. Fertilizer should be applied to dry fields prior to planting. Since rice involves an aquatic environment, nitrogen is often lost by bacterial processes in the soil.

17-40. Rice nearing maturity in a field that has been formed so the levees are straight. (Courtesy, James Lytle, Mississippi State University)

17-41. Rice is harvested with a combine that is equipped with oversize wheels or tracks to prevent bogging. (Courtesy, Case Corporation)

Pest Management

Rice has weed, insect, and disease pests. These tend to be different from other grain crops by virtue of the water environment. In some cases, small aquatic animals cause problems. Tadpole shrimp may cause problems with rice, especially in California. They chew off small rice plants and uproot plants by burrowing in the soil.

Common insect pests include rice weevils, armyworms, rice leaf miner, and grasshoppers. Diseases include a variety of molds, smuts, rots, and blights. Weed problems are mostly managed by flooding the fields. Using crop rotations, chemicals, and resistant varieties help in managing pests.

Harvesting

Water is removed from the field prior to harvesting. It should be off long enough for the soil to dry and support the weight of the combine. Some rice combines are equipped with oversize tires to prevent rutting. Combines fitted with tracks may also be used, especially in California.

The number of days water should be removed from a field prior to harvest depends on the soil composition.

In sandy soils, water can be drained off 10 days before harvest. In soils high in loam or clay, the water is drained off 15 days before harvest. This coincides with 20 to 25 days after heading by the rice.

Rice kernels should be harvested with 18 to 22 percent moisture. Harvesting with too much moisture results in kernels that are chalky rather than firm. Waiting until the moisture is below 18 percent may create fractures in the grain kernels in certain climate conditions. This results in lower-quality rice because the kernel will crack in milling.

GRAIN SORGHUM

Sorghum is the name of a group of grasses that originated in Africa and Asia. Of all grains, grain sorghum ranks fifth in worldwide grain production and follows corn and wheat in the United States. Over the last decade in the United States, an average of 13 million acres each year have been planted to grain sorghum (also known as milo). Some 90 percent was used for grain and the remainder was used for silage or other purposes. Grain sorghum is primarily used for animal feed in the United States. It is used as human food in Africa and Asia.

TYPES GROWN

Four general types of sorghum are grown in the United States: grain sorghum, sweet sorghum for forage, broom corn, and grass sorghums, such as Sudan grass and Johnson grass. Grain sorghum far outranks the other sorghums. Kansas, Texas, and Nebraska are the leading states in production. Little grain sorghum is grown north of Iowa.

17-42. Grain sorghum with well-developing grain heads.

Broom corn is used in manufacturing brooms. Little broom corn is produced in the United States today. Most is imported from Mexico and other countries. A severe shortage exists in the United States. Research is underway to develop varieties adapted to mechanical harvesting. The research is supported by the Association of Broom Manufacturers.

Grain sorghum is further classified on the basis of grain color. Four basic colors are found: yellow, white, brown, and mixed. Most sorghums grown in the United States are yellow hybrids, with some color variations including pink, red, and white-spotted. The feeding

17-43. Broom corn plants closely resemble grain sorghum. This broom corn is growing in North Carolina and obviously does not have a head of grain.

value of grain sorghum is somewhat less than corn, with the exception of white sorghum, which is equal to corn.

CULTURAL PRACTICES

Cultural practices with grain sorghum are somewhat different from the cereal grains. It is a warm-season crop. A wide range of tillage and other cultural practices are used with grain sorghum.

Varieties

Grain sorghum should be carefully selected to assure that the crop will meet market requirements. The variety chosen should mature in the growing season. To the extent possible, it should be resistant to the insects, diseases, and other pests in the local area. The variety should also resist lodging and shattering.

Purity of the produced grain is important with some markets. Any that is used for human consumption should be a white variety. In addition, the grain produced should have no more than 2 percent nonwhite grains.

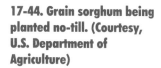

17-44. Grain sorghum being planted no-till. (Courtesy, U.S. Department of Agriculture)

Maturity requirements range from 60 to 150 days. The cultivars are often classified as early, medium, and late. The early varieties have fewer days to maturity, while the late varieties have the greater days to maturity. Most grain sorghum grown in the southern states requires 95 to 110 days to maturity. Varieties requiring 110 to 120 days tend to produce more grain.

Planting

As a summer annual, grain sorghum is planted after the frost season is over. The soil should have warmed to at least 55°F (12.8°C). Growers in northeast Texas plant when the early morning temperature is 55°F at 7:00 a.m. for three consecutive days. Many growers prefer to wait until the soil temperature is at least 65°F (13.5°C). Planting dates typically range from early April in the far south through May in the central plains.

Seedbed

Grain sorghum is typically planted in a prepared seedbed. However, it is increasingly being planted in no-till fields.

Traditional seedbed preparation begins with the harvest of the previous crop. Stalks, weeds, and other crop residue are disked under after harvest. Subsoiling is used if a hardpan exists. In the late winter or early spring, additional disking and harrowing are used to prepare the land.

Planting is often with a row planter, though some grain sorghum is planted with a drill. Row spacing is similar to that for corn—20 to 40 inches (50.8 to 101.6 cm). It may be planted on narrower rows, such as 18 inches (45.7 cm). Sorghum is often planted in shallow furrows to make better use of moisture in drier climates. This practice is known as *listing*.

Grain sorghum is often used in double-cropping rotations. Depending on the other crops grown in the area, it may be rotated with other small grains or with cotton. Because grain sorghum has a relatively short growing season, it fits well into a rotation system.

17-45. Preparing a seedbed for planting grain sorghum. (Courtesy, AGCO Corporation, Duluth, Georgia)

Plant Population

Plant populations tend to vary with available moisture. Populations may be as high as 100,000 or more plants per acre. Drier areas without irrigation may have 24,000 plants per acre. Depending on the cultivar, the number of seeds in a pound varies from 12,000 to 20,000. The seeding rate of 2 to 8 pounds per acre is used to get the desired plant population.

Some seeds do not germinate and seedlings may die before producing a crop. Most growers figure that only 65 to 75 percent of the seed will result in a mature plant. This means that some over-seeding is needed.

Proper planter calibration is important to achieve the desired plant population. The spacing of seed can be calculated based on the linear row length in an acre. An acre has 43,560 square feet. Total row length in one acre is found by dividing the width of a row in feet into 43,560. Here is an example for 24-inch (2 feet) rows: $43,560 \div 2 = 21,780$ linear row in an acre of field.

Distance between seed is calculated once the total seed requirement is known. A total seed requirement is found by adding 25 percent to the desired plant population to account for seeds that do not germinate or plants that fail to reach maturity. Once the total seed requirement is known, the number of seeds to drop in a foot of row is determined by dividing linear row length into the total number of seeds required. For an acre with a plant population of 50,000, an average of 2.87 seeds would be seeded in each foot of row. One seed would be dropped an average of 4.18 inches apart. (Here are the calculations: The number of additional seeds needed for plants failing to reach maturity is 12,500 ($50,000 \times .25$). The total seed requirement is 62,500 ($50,000 + 12,500$) per acre. The number of seeds to be planted per foot of row is 2.87 ($62,500 \div 21,780$).

17-46. Irrigation is being used with grain sorghum in Florida. (Courtesy, U.S. Department of Agriculture)

Moisture Management

The roots of grain sorghum make particularly good use of soil moisture. A crop will use up to 90 percent of the moisture for a depth of 35 inches (89 cm). This makes grain sorghum suited to areas with limited rainfall.

Grain sorghum requires considerable water to produce a crop. While blooming, grain sorghum may use 0.3 inch of water per day. Grain formation uses 0.2 inch per day. This need for water often results in good returns from irrigation in drier areas of the plains. Irrigation is not frequently used in the humid southern states.

Fertilization

The extensive root system of grain sorghum makes efficient use of existing nutrients in the soil. A general rule is that 2 pounds of N are needed for each 100 pounds of grain. Yields may reach 5,000 pounds per acre; therefore, 100 pounds of N should be applied per acre. Nitrogen is often applied in a split application, with about half at planting and the remainder 20 to 25 days after germination. Other nutrients should be used based on the results of soil testing. Both P and K will be needed.

A soil pH of 5.5 to 6.5 is best. Lime should be used to raise the pH if it is below 5.5.

Crops that follow grain sorghum in a rotation may need additional fertilizer. This is because grain sorghum tends to remove more nutrients from the soil than other crops. Some crops, such as soybeans, however, add nutrients which may add up to 60 pounds per acre. Using a grain sorghum and soybean rotation reduces the need for using commercial fertilizer high in nitrogen.

Pest Management

Grain sorghum is vulnerable to many pests. Insects, diseases, weeds, and birds and other animals cause losses. The insects in grain sorghum are similar to those that attack corn. Birds can cause large losses in mature grain.

Large flocks of birds often gather in fields. One bird will eat 3 ounces of grain sorghum a day. If 10,000 birds are in a flock, 185 pounds of grain can be lost each day. This number occurs over a period of several weeks. In two weeks, the flock of birds would eat well over a ton of grain. This robs the grower of high yields at harvest time! Gas-fired guns, balloons, and other kinds of "scare crows" are used to frighten birds away.

Insects attacking sorghum include the greenbug, sorghum webworm, and sorghum midge. Diseases include seed rot, damping-off, leaf blight, rust, and downy mildew. Insects and diseases are managed by planting disease resistant varieties and using pesticides. Crop rotation also lessens disease problems.

17-47. A developing sorghum head is being checked for signs of insect damage.

17-48. Pesticide is being applied to grain sorghum with a carefully calibrated sprayer.

Harvesting

Grain sorghum should be harvested as soon as possible after the heads have matured and the stalks begun to dry. Grain should have a moisture content of no more than 20 percent. Higher grades of grain sorghum will have lower moisture. For example, U.S. No. 1 grade has moisture content of 13 percent.

Harvesting should not be delayed. Grains are lost to shattering, lodging, birds, and the buildup of smut and weevils in humid climates. Harvested grain should also be free of small stones and weed seed.

OTHER GRAIN CROPS

The growth of other cereal grains is similar to wheat, except for sorghum. Grain sorghum is often treated separately from the cereal grains.

AMARANTH

Amaranth is a minor grain crop. It is sometimes considered a weed or herb. The species from which seed are collected are known as grain amaranth. Amaranth was the principle grain crop in South America hundreds of years ago. It is sometimes called the "mystical grain of the Aztec." Amaranth has higher protein than other grains (12 to 18 percent protein). It has a significantly higher lysine content. Some 60 species are in the *Amaranthus* genus.

Research on the potential and cultural needs of amaranth is underway in several states. Several states have issued producer guidelines, including Montana, Minnesota, Wisconsin, and Nebraska. Other states with some production underway include Tennessee, Arkansas, and Missouri.

The amaranth is a broadleaf plant that forms small seeds similar to grain sorghum. It is planted in a prepared seedbed similar to that for other grain crops. A soil pH of 6.0 or slightly above is preferred. Seed can be planted with standard grain drills or grain planters. Nutrient needs include N, P, and K. Harvesting is difficult because of the tiny size of the seed. Grain combines can be used to harvest amaranth. The seed will need to be dried to 12 percent or lower moisture for storage.

BARLEY

Barley is one of the oldest known grain crops. Little barley is used as food, though some is used in making malt and livestock feed. Barley is grown for grain and forage, primarily hay.

17-49. Developing amaranth head.

17-50. Head variations help distinguish between hay barley (left) and grain barley (right). Both have the same scientific name: *Hordeum vulgare.*

17-51. Young grain barley in a Minnesota field. (Courtesy, U.S. Department of Agriculture)

17-52. Barley nearing maturity in a California field.

17-53. Barley grain.

The growth and appearance of barley are similar to wheat. Most barley is grown in climates too cool for corn production. Barley can be easily substituted for corn in livestock rations.

Barley may be planted for winter or spring growth. Winter barley is planted in the fall and is hardier than oats. Winter barley is primarily grown in the southern states. Spring barley is planted as early as possible in the spring. Barley is seeded into prepared seedbeds. The rate of seeding is about 2 bushels (96 pounds) per acre. The rate may be reduced in drier climates to 12 to 24 pounds.

17-54. Barley stored outside at a feed mill for a feedlot where rainfall is rare.

Barley responds well to fertilizer, but care should be used not to apply too much N. Excessive N causes barley stems to be weak and easy to break over (lodge). Barley is harvested with a combine after the heads have dried.

MILLET

Millet is used for grain and summer forage. Several varieties of millet are grown. Proso millet is the only variety grown for grain in the United States. Four forage millets are commonly used: pearlmillet, brown top millet, foxtail millet and Japanese millet. Overall, millet production has declined in the United States. Millet continues as an important food crop in some African countries.

Proso millet has stout, erect stems that grow to about 4 feet (1 m) tall. Grain heads form similarly to grain sorghum. Ripened grains are ovate and rounded about 2 mm wide and 2.5 mm long. Most millet grain production in the United States is in the northern plains.

Pearlmillet may grow up to 15 feet (4.5 m) with a grain head similar to a cattail. Some species of foxtail millet are grown in the United States for birdseed.

Millet is planted after the danger of frost. It is more sensitive to frost than corn. Some producers recommend waiting three weeks after corn planting is initiated to plant millet. Millet is seeded at the rate of 25 to 30 pounds

17-55. Maturing pearlmillet head.

17-56. Using a drill to plant millet. (Courtesy, U.S. Department of Agriculture)

per acre in humid areas and 10 to 15 pounds per acre in drier areas. Most millet is planted in a prepared seed-bed similar to that for wheat. Grain drills or broadcast seeders are used. Additional N may be needed, as well as P and K. Soil testing and recommendations from local agricultural consultants are beneficial. Millet is harvested for hay while still in a growing stage. It is harvested for grain after the heads have matured. However, not all grains mature at the same time. To reduce shattering of grain, the crop is often mowed, raked in windrows, and combined.

17-57. Whole kernel (left) and rolled oats (right).

OATS

Overall production of oats has declined in recent years. Replacing draft animals with engine-powered tractors resulted in less demand for feed for horses and mules. The benefits of oats as human food have boosted the respect for the crop. Even with decline, the United States continues as the leading nation in oat production.

Oats are planted in the fall similarly to wheat. The seeds are drilled or broadcast on a prepared seedbed. Fertilizers similar to those for wheat are used. Soil analysis and following the

17-58. An oat field in Montana.

17-59. Planting oats in Minnesota. (Courtesy, U.S. Department of Agriculture)

recommendations of agronomists will result in increased yields. Topdressing with nitrogen in the early spring is particularly beneficial. Oats are harvested when dry with a combine.

Oats undergo processing somewhat different from other grains. The unground grain of oats is known as **groat**. The groats are roasted, separated from their hulls, and passed between large rollers. The rollers flatten the groats into rolled oats. With some additional processing, the rolled oats are used for breakfast cereal.

UINOA

Quinoa has long been grown in South America. It is viewed as a grain crop with some potential in the United States. Quinoa is sometimes called a pseudocereal because it is not in the grass family.

Quinoa is cold tolerant and grows in dry climates. The grain is similar to rice and has a high nutritional value. An oil similar to that from corn can be processed from quinoa seed. Some researchers have found the oil yield to be higher than corn.

Cultural conditions for quinoa in the United States have not been studied to a great extent. Best growth in South America occurs at altitudes of 8,000 or more feet. This means that the plant could be adapted to areas of the western United States.

RYE

Rye is a cereal grain crop similar to oats and barley. It is not as important as some of the other grains. About half of all rye is used for grain, with the remainder for pasture, hay, or as a cover crop. Rye grain has declined as preferences for white bread have increased. The yield is not as great as barley and wheat. The major use of the grain is for livestock feed though some is used to make flour for bread baking and in other foods in the United States. Some rye is plowed under as a green manure crop or killed with herbicide for planting no-till corn.

Rye is hardier than other small grains. It can be planted in the fall or spring, depending on the climate. Winter rye predominates. Growth is much like wheat. Planting is in prepared seedbeds with drills or broadcast. Fertilizer should include nitrogen topdressing in the early spring. Winter rye forms heads earlier than other grain crops in the spring. The grains shatter easily from the heads.

TEFF

Teff provides little grain in the United States, but it is an important food source in some nations. In Ethiopia, for example, teff provides two-thirds of the human nutrition. The grain is used for human food and as fodder for animal feed. The grain is white or reddish brown in color. It is smaller in size than most grains.

Seeds require a firm, moist seedbed. The seeds are planted less than a half inch (1.2 cm) deep. Planting often involves broadcasting the seed and running a packer over the seeded area. The seeds rapidly germinate, but initial growth is slow. Early growth is primarily to establish a good root system. Teff has few pest problems compared to other grain crops.

Some teff has been grown in South Dakota, Montana, and Oklahoma.

TRITICALE

Triticale is a cereal grain with potential for forage and grain for livestock. It was produced by crossbreeding rye and wheat. Triticale grain has a higher protein content than either rye or wheat. The cross resulted in a crop that has the high productivity of wheat and the climatic adaptations of rye. It is an important food crop in some countries not suited to wheat production. Triticale can grow in cold climates and is resistant to many diseases.

Triticale plants grow similar to wheat that are 18 to 41 inches (45 to 105 m) tall. It is currently being grown using similar cultural practices as with wheat and rye. Triticale can be planted in the fall or spring, depending on the climate. Experimental work has been carried out at a number of research stations in the United States. Alabama A&M University is a pioneering research station with triticale in the South.

WILD RICE

Wild rice is a crop that is quickly being domesticated. It is native to the Great Lakes region of the United States, where it grew wild in shallow lakes and rivers. Wild rice is now being cultured in Minnesota, Wisconsin, and California using diked, flooded fields. It matures in 120 days. Plants grow 24 to 28 inches (60 to 70 cm) tall and bear grain somewhat larger than traditional rice.

An early challenge that was met with wild rice was to develop varieties that were shatter resistant. Combines have been adapted to improve harvest efficiency. Most cultural practices are similar to other rice crops. Wild rice has several diseases and insect pests. Efforts are now underway to develop control measures. Crawfish, birds, raccoon, mink, skunk, deer, and other animals sometimes cause damage to the fields or dikes.

Because wild rice has been recognized as a gourmet food by consumers, it is in greater demand and sells for a higher price than most newly domesticated crops. The future for limited wild rice production looks good in the northern states.

REVIEWING

MAIN IDEAS

Grain is a major source of human food, animal feed, and raw material for manufacturing a wide range of products. The major grain crops grown in the United States are corn, wheat, and rice, with corn being the most important.

The grain crops to grow should be carefully selected. Areas to consider include climate, soil and water, market, technology, and skills and preferences. The geographical location should be compared with the requirements of the grain crops. A market should always be available. A good crop without a market has no value.

Several types of the major grains can be grown. These were developed to have varieties adapted to climates over widely differing geographical areas. Corn is grown primarily for grain though some is produced for animal forage and other uses. It is grown most widely in the United States.

Wheat is also grown in many locations. Some wheat is planted in the fall and is known as winter wheat. Other wheat is planted in the spring and harvested in the late summer. It is known as spring wheat.

Rice is the only major cereal grain that is produced in flooded fields. Rice is a major source of human nutrition in some countries. Techniques vary, depending on water management.

A number of minor grain crops are grown. In some locations, these crops are major agricultural enterprises. The most important minor crops are grain sorghum, barley, and rye. Promising new grain crops include triticale, quinoa, and wild rice.

QUESTIONS

Answer the following questions using complete sentences and correct spelling.

1. What are the leading grain crops worldwide? Compare this to the leading crops in the United States.

2. Name five important areas to consider in selecting the grain crop to grow. Briefly explain each.

3. Five economically important types of corn are found. Name and briefly describe each type.

4. What is a growing degree day? Calculate the GDD for a day with a low temperature of 62°F and a high of 88°F.

5. What is plant population? How is a crop planted to assure an adequate plant population?

6. What are the three ways wheat is classified? Briefly explain each.

7. What is shattering? Why is preventing shattering important?

8. What are the two major ways rice is classified? Distinguish between each type.

9. Why is water management important with rice? Briefly describe what water management involves.

10. Select one of the lesser grain crops and prepare a brief description of the crop. Is your selection an established or emerging grain?

EVALUATING

Match the term with the correct definition. Write the letter by the term in the blank that is provided.

a. acre-inch   e. plant population   i. growing degree day (GDD)
b. kernel   f. topdressing   j. shattering
c. grain marketing   g. lowland rice
d. cultural practices h. dent

1. _____ moving grain from the producer to the consumer
2. _____ most widely planted corn type
3. _____ procedures used in producing a crop
4. _____ measure of temperature requirements for best crop growth
5. _____ type grown in large, flat fields with irrigation
6. _____ one acre covered with 1 inch of water
7. _____ part of grain within the seed coat
8. _____ number of plants growing on an acre
9. _____ applying fertilizer after a crop is growing
10. _____ occurs when mature grains fall to the ground

EXPLORING

1. Prepare a report that describes the grain crops grown in the area near your home. The area can be the county or school district. Include the names of the crops, general cultural requirements, pest problems, and how they are marketed. Submit the written copy of your report to your teacher. Give an oral report in class.

2. Shadow a grain grower for a day or longer. Observe the preparation of land for planting and the equipment used to plant. Determine how the planter is calibrated to provide the desired plant population. Discuss other facets of production with the grower. Prepare a report on your findings.

3. Use the World Wide Web to explore grain production and use. A place to begin is the home page of DEKALB GENETICS, as follows:

http://www.dekalb.com/index.html

Print an example of your findings. Provide an oral report in class.

Sugar and Oil Crops

This chapter covers the basic production requirements of major sugar and oil crops. It has the following objectives:

1 Explain sugar and vegetable oil.

2 Identify major sources of sugar.

3 Describe the production of sugar crops.

4 Identify major sources of vegetable oil.

5 Describe the production of oil crops.

TERMS

beet pulp	cossettes	ratoon crop
bio-diesel	ethanol	sap
blackstrap molasses	granulated sugar	sucrose
brown sugar	harvest loss	sugar
confectioners' sugar	molasses	syrup
cooking oil	preharvest loss	vegetable oil

18-1. Many of the foods we enjoy involve oil and sugar. (Courtesy, U.S. Department of Agriculture)

SUGAR and vegetable oils make our foods enjoyable. Imagine what your day would be like without a sweet snack and potatoes or chicken cooked in vegetable oil! In addition, some of these products provide good health benefits.

It is amazing how our food products are manufactured. Who would think that we could get oil from a sunflower seed and sugar out of a beet? The modern agricultural industry works wonders! Sometimes, people take these for granted. They fail to remember the important cultural practices needed in the fertile fields of Earth. Technology has a big role. Without the use of modern science, we could not enjoy all the things that we do.

At your next meal, count all the foods with sugar and vegetable oil. Would you be willing to give them up? Most people want to have them.

SUGAR AND OIL: WHAT ARE THEY?

Sugars and vegetable oil are important food products. The crops grown for sugar and vegetable oil tend to be distinctly different, yet some are the same. Even some of the grain crops are used in making oils and sugars. Climate and cultural requirements also vary widely.

Sugars

Sugar is any food product used as sweetener. Several kinds of sugar are used. Common table sugar is known as **sucrose**. Other kinds of sugars include dextrose (which is corn sugar) and fructose (which is fruit sugar). **Brown sugar** is raw sugar with a small amount of molasses.

Syrup is a sweet liquid made from the watery juices of plants. Processes vary widely in how the syrup is produced. Sugar cane, maple trees, sweet sorghum, corn, and a few other plants are used to make syrup. Syrup made from sugar cane is known as **molasses**. Sweet sorghum (sorghum cane) is also used to make syrup (sometimes known as sorghum molasses). It is not as widely found as sugar cane molasses.

Both plants and animals are sources of sugar. Maple trees and corn are two plant sources. Honey is made by bees from the nectar of plant flowers. Milk from cattle or other mammals can be separated to produce lactose, or milk sugar.

The two major sources of sucrose vary widely in the part of the plant from which sugar is extracted and the locations where they are grown.

18-2. Sugar cane nearing maturity in Hawaii.

VEGETABLE OILS

Vegetable oil is a type of fat obtained from certain plants, usually the seeds of plants. The primary use of vegetable oil is in cooking foods. Many people prefer for their foods to be cooked in vegetable oil. They feel that it is better for their health than animal fat. The industrial uses of vegetable oil, such as making paint, are increasing in importance. Some vegetable oils are being used as fuel for engines.

The use of vegetable oil has increased considerably in recent years. This is because it is a substitute for cooking in animal fat—primarily lard. Health-conscious people want foods cooked in vegetable oil. It has little or no cholesterol, which is often considered damaging to human health.

Some vegetable oils are used as fuels for internal combustion engines. These oils can sometimes be substituted for petroleum oils in engines that use diesel fuel.

18-3. Both syrup and cooking oil are made from corn.

SUGAR SOURCES

Sugar is obtained from several sources, with plant sources being most important. The major sugar crops are sugar cane and sugar beets. Half the world's sugar is from sugar cane. Sugar beets account for about 40 percent of the sugar. The use of beet sugar has been increasing in recent years.

18-4. Granulated sugar.

Sucrose sugar is obtained from stored sugar in plants and is produced by the photosynthesis process. Good cultural conditions are needed to ensure that sugar is made and stored.

After harvest, sugar products are manufactured into solid and liquid forms. The solid form is granular, known as granulated sugar, and the liquid form is known as syrup.

Granulated sugar is made from crystals of raw sugar. As extracted from plants, raw sugar is brown and must be refined to produce the desired white sugar. Sugar is near 100 percent sucrose. Both sugar cane and sugar beets are used to make granulated sugar.

Confectioners' sugar is finely ground granulated (table) sugar. It is also known as powdered sugar. The manufacturing process may involve adding an anticaking ingredient. This makes it easier to use confectioners' sugar in baking and other food preparation activities.

CANE SUGAR MANUFACTURING

Sugar made from sugar cane is from the sap or juice in the stalk of the plant. The leaves and roots are not used. They are cut away from the stalk and left in the field. As a member of the grass family, the stalk is jointed and closely resembles a corn stalk. As a crop in tropical climates, sugar cane stalks may grow 7 to 15 feet (2 to 5 m) tall. In fields, the stalks often fall over and form thick mats of plants.

Sugar is manufactured at a refinery. The harvested stalks are crushed, squeezing out the sugary juice. This liquid is heated to drive the moisture out as vapor. Molasses is produced if the heating process is stopped before brown crystals begin to form or if the sweet liquid is removed during the process. The brown raw sugar undergoes additional processing. Some syrup remains after as much sugar as possible has been extracted. This syrup is known as **blackstrap molasses** and is used in cattle feed.

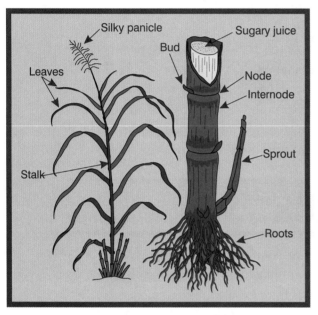

18-5. Structure of a sugar cane stalk.

18-6. Sugar refinery in Hawaii.

Beet Sugar Manufacturing

Sugar from sugar beets is made from the enlarged cone-shaped root of the plant. The tip of the root is a long tap root that may go 2 to 5 feet (0.6 to 1.5 m) into the ground. This long root gets moisture for the plant and makes the plant better suited to culture in drier climates than sugar cane.

18-7. Field of growing sugar beets nearing maturity.

CAREER PROFILE

SOYBEAN PRODUCER

A soybean producer is responsible for all activities in growing a profitable crop of soybeans. A producer must be able to make important decisions about varieties, land preparation, planting, cultural practices, and harvesting. These require a good knowledge of the ideal environment for soybeans.

The nature of the work varies with the season of the year. The work may be outside in fields operating equipment or inside keeping records and handling the payment of bills. Education and training needs vary. Some have a high school education; others have two-year and four-year college degrees. Experience working on a soybean farm is very important. Keeping up to date on current practices requires attending seminars and classes. Some use the World Wide Web to locate sources of information and talk about problems with other growers.

This shows a soybean producer adjusting a no-till planter. (Courtesy, U.S. Department of Agriculture)

18-8. A sugar beet infected with root-knot nematodes is being examined in a laboratory. (Courtesy, Agricultural Research Service, USDA)

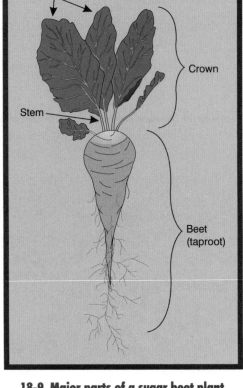

18-9. Major parts of a sugar beet plant.

Stems and leaves are not used. They are usually left in the field and plowed under to increase soil organic matter.

Whole beets are mechanically harvested and hauled to large beet collection areas. Since beets can be stored outside for several months without great loss in quality, processing continues nearly year round. The beets are cut into slices known as **cossettes** and soaked in water in large containers called diffusers. The water removes the sugar from the beet. The remaining solids are dried and used as **beet pulp**—a common feed ingredient. The sweet taste of beet pulp makes feed palatable to cattle. The remaining thin juice is processed in a manner similar to the processing of cane sugar.

WHERE GROWN

Sugar beets and sugar cane vary widely in climate requirements. The locations where they are grown reflect wide climatic differences. Sugar cane requires a near tropical climate. Long growing seasons of warm weather and plenty of moisture are needed. Sugar beets grow in cooler and drier climates than sugar cane. Sugar beets must often be irrigated to have adequate moisture.

Sugar beets are more widely grown than sugar cane in the United States. Minnesota is the leading state in acreage planted to sugar beets. It is followed by Idaho, North Dakota, and

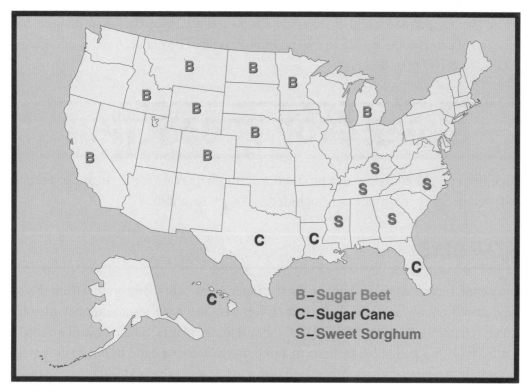

18-10. Map showing top states in sugar cane, sugar beet, and sweet sorghum production. Sweet sorghum is produced less intensively than sugar cane and sugar beets.

Michigan. These states are experiencing moderate increases each year. California has had some decline in sugar beet acreage. Yield per acre averages 20 tons of sugar beets. A ton of sugar beets yields about 290 pounds of refined sugar.

The leading states in acreage planted to sugar cane production are those with warm, tropical climates: Florida, Louisiana, and Hawaii. In Florida and Louisiana, the production is confined to the southernmost parts of the states. Texas also has some sugar cane produc-

Table 18-1. Leading Sugar Crop Production States

Rank	Sugar beets		Sugar Cane	
	State	Acreage	State	Acreage
1	Minnesota	434,000	Louisiana	490,000
2	North Dakota	231,000	Florida	454,000
3	Idaho	195,000	Texas	46,600
4	Michigan	175,000	Hawaii	35,400
5	California	96,000	—[1]	

[1]No other state has sufficient acreage to justify reporting.

Source: U.S. Department of Agriculture, 2000.

tion. The average per-acre yield of sugar cane is 33.2 tons. Since the sugar content of cane varies, a ton of cane produces 205 to 244 pounds of sugar. The per acre yields of sugar ranges from 6,800 to 8,100 pounds.

SUGAR CROP PRODUCTION

Since two crops dominate sugar production, this chapter includes sugar cane and sugar beets in detail. Some information is included on sweet sorghum.

SUGAR CANE

Sugar cane is considered one of the most efficient plants in using energy from the sun to produce stored energy as sugar. It thrives in hot climates with an abundance of sunlight. However, temperatures above 100°F (38°C) may damage plants. Sugar cane plants are damaged at 50°F (10°C) and killed by freezing temperatures. Sugar cane has been domesticated for thousands of years and was introduced into Louisiana in the mid-1700s.

Sugar cane is in the grass family. Its scientific name is *Saccharum officinarum*. Other species often used in plant breeding programs are *S. barberi* and *S. spontaneum*.

The large stalks will weigh 3 to 4 pounds without roots and leaves. Most of the weight is **sap**—a watery juice containing sugar. One average stalk will yield a little more than 4 ounces (0.15 kg) of refined sugar.

Sugar cane plants grow similar to corn. They have stems or stalks with blade-like leaves. A fibrous root system with shallow soil penetration may give way and the stalks lodge.

Sugar cane is propagated vegetatively. Nodes in the stalk

18-11. Section of a stalk of sugar cane showing nodes and internodes.

18-12. Sugar cane sprouting from a node of a partially covered seed stalk.

will sprout new growth. Mature stalks are carefully kept from the previous season to prevent killing the buds. The stalks are cut into sections that are 18 to 36 inches (46 to 91 cm) long. Sometimes, the entire stalk is used. These sections are laid end to end in a furrow and covered with soil 2 to 3 inches (5 to 7.5 cm) deep. The nodes grow tiny shoots that develop into mature stalks. Roots also grow from the nodes into the soil. Depending on the climate, 8 to 30 months are needed from planting until harvest.

At maturity, sugar cane produces seed in tropical climates. The seeds are formed in a silky panicle (similar to a corn tassel) at the top of the stalk. Most cane is harvested before this stage is reached. The few seeds produced are often infertile and will not germinate.

Sugar cane requires 45 to 51 inches (115 to 130 cm) of rainfall or irrigation during the growing season. A stalk will reach harvest size in one growing season or maturity in two years where the climate is warm year round. After harvest in tropical climates, the roots will sprout again and grow stalks. Up to three stubble crops are possible in about 8 years. The sugar cane produced by sprouting is known as a **ratoon crop**.

Sugar cane is harvested in the fall in the continental United States. Harvest extends nearly year round in Hawaii.

18-13. Sugar cane flower.

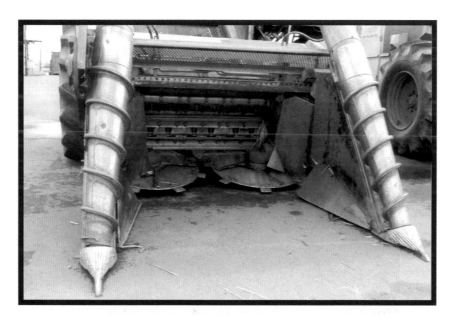

18-14. A mechanical sugar cane harvester has large rotary blades to cut and trim stalks.

18-15. Sugar cane is being harvested onto high dump trailers that will transfer the cane to trucks for hauling to a sugar mill. (Courtesy, U.S. Department of Agriculture)

The method of harvesting varies. Hand cutting and stripping of leaves have been replaced in most places with mechanical harvesters. Large harvesters may cut the stalks, remove the leaves, and load the stalks into trucks for hauling to a sugar refinery. In other cases, the field is burned to remove the leaves before cutting. The stalks are cut, raked into piles, and loaded on trucks.

SUGAR BEETS

The sugar beet (*Beta vulgaris*) is a biennial plant. Production for sugar involves cultural practices that treat it as an annual. Since sugar beets are planted from seed, producers of seed allow the plant to grow for two years so that seeds are formed.

Sugar beets grow in several regions in North America. Production ranges from the north central part of the United States to the West Coast. They grow best in average daily temperatures of 66 to 71°F (19 to 22°C). The preferred climate is one with warm days and cool nights.

The seeds are planted in the spring in a prepared seedbed using high-speed precision planting equipment. In one growing season, beets reach weights of 2.2 to 4.4 pounds (1 to 2

18-16. Sugar beets being planted. (Courtesy, Texas Department of Agriculture)

18-17. Harvesting begins by cutting off and shredding the tops of sugar beet plants. (The tops of beets can be seen sticking up in the foreground on this North Dakota farm.)

kg). Sugar formation increases rapidly in sugar beets in the late summer as the nights get cooler and nitrogen supplies are diminished in the soil. Vegetative growth slows and more sugar is produced. Harvesting is delayed as long as possible in the growing season to assure maximum sugar content.

Sugar beets are harvested using a machine that cuts off the tops, lifts the roots, and loads them onto a

18-18. Sugar beets being "lifted" from the soil and loaded into trucks.

18-19. After harvest, sugar beets are hauled to processing plants or piling stations.

truck. Harvested beets are hauled to processing plants or beet piling stations. Most sugar beets are grown under contract with a processor. The agreement specifies when they will be planted, the cultural practices used, and when to harvest.

18-20. Sweet sorghum plants growing in North Carolina.

SWEET SORGHUM

Sweet sorghum (*Sorghum bicolor*) is produced in several states on a relatively small scale. It has the same scientific name as grain sorghum. The sweet sorghum cultivar is distinguished by its "sweet juice" characteristic. Considerable local interest exists in the crop often known as sorghum cane. The sorghum is used for making syrup. The syrup is used as a sweetener. Most of the production is in the southeastern states. Kentucky and Tennessee have the largest acreages.

The goal is to get the greatest amount of the best possible syrup. As with other crops, attention to cultural practices is important.

Sweet sorghum is planted as seed (seeds are quite small) into fields or small plants are set. Soil temperature should be 65°F or higher for seeding. The rate of seeding is 3 to 4 seed per linear foot of row.

Beginning plants in a float-system bed (similar to that used with tobacco) results in getting a head start on the growing season, according to University of Kentucky

18-21. Sweet sorghum plants in a float-system bed in Mississippi nearing transplant size.

crops specialists. The young sorghum plants are transplanted to the field into a prepared seedbed with 36- to 40-inch rows when the soil temperature is at least 60°F. Maturity requires 90 to 120 days, depending on the variety grown. It is three weeks less for transplanted sorghum plants.

Stalks grow 6 to 12 feet tall and are 1 to 2 inches in diameter. Planting too thick results in small, spindly stalks.

Sweet sorghum prefers land low in organic matter and can be planted following corn, soybeans, and other crops. However, sweet sorghum should not be planted following a tobacco crop. Soils with a pH of 6.0 or slightly higher are best. A general fertilizer recommendation is 40 pounds per acre each of nitrogen, phosphate, and potash.

Weed control is often with cultivation, which is begun when the plants are 5 to 6 inches tall. No approved herbicides are available in some places.

Harvest involves stripping the leaves from mature stalks and clipping off the seed head. The stalk is cut close to the ground. The top is removed when it diminishes to one-half inch. The stalks are hauled to a roller mill for squeezing the juice. The juice is cooked in an evaporating pan. Impurities are removed and the syrup is canned or bottled.

18-22. Lower leaves have been stripped from sweet sorghum plants as part of harvesting.

VEGETABLE OIL SOURCES

The seeds of most plants contain at least a small quantity of oil. A jar of vegetable cooking oil represents remarkable production and processing activities. Most vegetable oils are liquids at room temperature.

Vegetable oil is used in making many other products. **Cooking oil** is fat in which foods can be fried or used in other ways in food preparation. The food you eat today will likely have several items with vegetable cooking oil. Mayonnaise, salad dressing, and shortening are three examples. Many meat and vegetable foods are very popular when fried in cooking oil.

18-23. A biofuel is used to power this vehicle.

Noncooking vegetable oil is used to make many products. Printing ink, soap, and leather tanning products are made from the oil. Some vegetable oil is used as fuel. The fuels are often blends of vegetable oil and petroleum. Gasoline blended with **ethanol** (a form of alcohol) is used on a limited basis. Vegetable oil may also be blended with diesel fuel for a fuel often referred to a **bio-diesel**.

CONNECTION

KNOWING THE LOCATION: GLOBAL POSITIONING

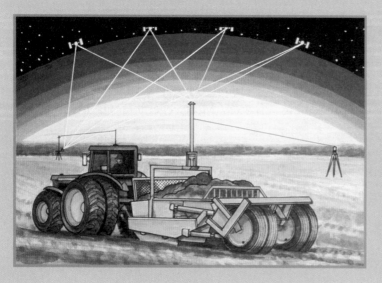

Sugar and oil crop production is increasingly carried out using global positioning systems (GPS). Mechanization and variable input technology require the ability to track all areas of the surface of a field. Mechanization requires carefully formed fields without low and high places. Uneven field surfaces result in large equipment not working the soil properly.

Global positioning involves using triangulation. The position of a location is determined using three satellites, with a fourth satellite used to correct for error. A station on the ground and on the operating machinery uses the information in the work. (Courtesy, Spectra-Physics, Inc., Dayton, Ohio)

Oil manufacturing results in several by-products. Starch by-products are used to make chewing gum, plywood, crayons, and degradable plastics. Meal by-products are used as animal feed, fertilizer, and for some industrial processes.

OIL CROPS

Vegetable oils may be obtained from seed or fruit. The most important source of vegetable oil in the United States is the soybean seed. Other important seed sources include canola, corn, cottonseed, and peanuts. Safflower seed, sunflower seed, sesame seed, flaxseed, tung seed, rapeseed, mustard seed and lesquerella seed are also used for oil. Olive and coconut oils are extracted from fruit.

Spearmint and peppermint are also grown for oil. The oil of these crops is not used for cooking but as flavoring in foods and snacks, such as chewing gum and jelly. Some spearmint oil is used in making medicine.

Some crops are grown almost exclusively for their oil value; others are grown primarily for products

18-24. Lesquerella growing in a pilot plot in Arizona. The oil is used for cosmetics manufacture and other purposes. (Courtesy, Agricultural Research Service, USDA)

18-25. Olive oil is one of the few vegetable oils extracted from fruit rather than seed.

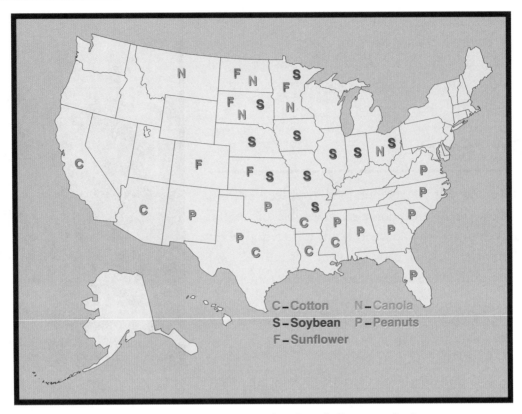

18-26. Locations of leading states in selected oil crop production.

besides oil. Corn and cotton, for example, are grown for grain and fiber, respectively. Soybeans are grown almost exclusively for oil, with meal and other by-products resulting.

Most vegetable oils are used as oils for preparing human foods. Tung and linseed oils are primarily used in manufacturing wood preservatives, especially paint.

Table 18-2. Leading States in Production of Major Crops Used for Vegetable Oil[1]

Rank[2]	Cottonseed	Soybeans	Sunflower Seeds	Canola	Peanuts
1	Texas	Iowa	N. Dakota	N. Dakota	Georgia
2	California	Illinois	S. Dakota	Minnesota	Texas
3	Georgia	Minnesota	Minnesota	S. Dakota	Alabama
4	Mississippi	Indiana	Kansas	Montana	N. Carolina
5	Arkansas	Ohio	Colorado	Ohio	Florida

[1] Corn not included; no information on peanuts used only for oil.

[2] Rank based on tons of cottonseed sold to mills, acreage planted to soybeans, and acreage planted to sunflowers used for oil.

Source: U.S. Department of Agriculture, 2000, North Dakota State University, and Ohio State University.

MANUFACTURING VEGETABLE OILS

Procedures used in manufacturing vegetable oils vary widely. Common methods include using solvents and presses. Solvents pull the oil out of the seed. The solvent itself evaporates and the solid materials are removed, leaving oil. Powerful presses may be used to squeeze the oil out of seed and fruit. Once extracted, the oils must undergo additional refining to prepare for use.

Vegetable oil production provides valuable by-products, such as soybean meal or cottonseed hulls. These products are used as animal feed. Some have a high protein content, such as cottonseed meal and soybean meal. These products are often mixed with grains in manufacturing feed.

OIL CROP PRODUCTION

Oil can be produced from a number of plant species. Several are widely used in producing vegetable oil. Most of the oil crops are used for other purposes. For example, corn is also used for grain and meal products. Cotton is also used for fiber. Peanuts are used for roasting and making peanut products.

Major crops used predominantly for products other than oil—corn and cotton—are covered in other chapters. Some crops used for oil are more widely grown outside the United States, such as sesame production in selected Asian countries. This chapter includes five crops with oil production being their major use: soybeans, canola, peanuts, sunflowers, and safflowers.

18-27. Sesame growing at a Florida research farm.

SOYBEANS

The soybean (*Glycine max*) is an annual legume. Plants grow 2 to 4 feet (61 to 122 cm) tall. Three or four beans are formed in pods during the growing season. Overall, soybeans are one of the most important crops in the United States.

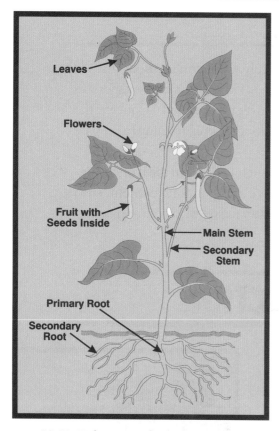

18-28. Major parts of a soybean plant.

18-29. Quality soybean seeds are needed.

Importance

More soybeans are grown in the United States than the rest of the world combined. One-third of the U.S. production is exported. Most exports are to Europe and Japan.

About 60 million acres have been planted to soybeans in the United States each year over the last decade. Annual production is 1.5 billion bushels (1 bushel weighs 60 pounds). This averages only a little over 25 bushels for each acre planted. Good growers typically have yields of 60 or more bushels per acre. In the last decade, some decrease in total acreage planted to soybeans has occurred.

Soybeans are good sources of protein and oil. A few soybeans are eaten as vegetable beans, but most are processed to yield oil and meal. Soybean meal has twice the protein of beef and fish. Soybean oil has many uses and contains no cholesterol. One bushel typically produces 11 pounds of oil, 43 pounds of meal, 4.2 pounds of hulls, and 1.8 pounds of other matter.

Cultural Requirements and Growth

The climate requirements for soybeans are similar to corn. Areas with high corn yields also have high soybean yields. Soybeans are planted in prepared seedbeds or in no-till fields when the soil temperature is at least 50°F (10°C). The ideal temperature for germination and growth is 86°F (30°C). Development occurs more slowly at higher or lower temperatures. In fact, temperatures below 75°F (24°C) delay flowering.

Many varieties of soybeans are planted. Varieties are typically adapted to fairly narrow climates and photoperiods. Growers should determine the varieties recommended in their areas. The variety selected should have resistance to the disease problems where it will be grown.

PLANTING. High-quality seeds should always be planted. These seeds are pure and have no soybean seed of another variety in them. The seeds should have good germination and vigor. They should be free of weed seeds and free of cracks and broken seeds.

Soybeans are planted in rows, drills, or broadcast. The general planting rate is 52 pounds per acre regardless of the method of planting. The seeds are planted 1 to 1.5 inches (2.5 to 3.8 cm) deep. Since moisture is required for germination and growth, seed may be planted slightly deeper in drier land.

Soybeans begin blooming six to eight weeks after planting. The first blooms produce mature beans in 12 to 15 weeks. Moisture is a major factor in blooming and bean development. A moisture deficiency at the time of blooming and bean development can greatly reduce yields.

PESTS. Soybeans are subject to damage from insects, diseases, and weeds. Insect pests include wireworms, white grubs, bean leaf beetles, grasshoppers, Mexican bean beetles, and fall armyworms. These are managed with a combination of biological, cultural, and chemical methods.

Most any weed present where soybeans are planted can be a pest. Weeds are managed by tilling before planting and when the plants are young. Herbicides are often applied as preemergence treatments. Weeds should be killed early and not allowed to compete with the crop. Weeds reduce crop yields. Roundup Ready® soybeans allow the use of a selected herbicide to control weeds that would kill soybeans that had not been genetically modified.

18-30. Soybeans are being planted in rows. (Courtesy, American Soybean Association, St. Louis, Missouri)

18-31. A maturing, genetically modified field of soybeans.

18-32. Holes in the leaves of these soybean plants are evidence of insect damage.

NUTRIENTS. Soybeans need essential nutrients to grow. As a legume, the plants form nitrogen nodules on their roots. Seeds must be inoculated at the time of planting to assure maximum nitrogen fixation. Inoculation is mixing nitrogen-fixing rhizobia bacteria with the seeds just before planting. These bacteria are carried with the seeds into the soil. Seeds planted on fields that have been previously planted to inoculated soybeans may not need to be inoculated each year.

Fertilizer requirements vary. Soil testing is needed to determine the amounts for individual fields. Local recommendations on fertilizer should be followed. Usually, soybeans do not consistently show increased yields when N is applied if the soil has adequate rhizobia bacteria. Soybeans require relatively large amounts of P and K. Other needed nutrients are often

18-33. Soybeans planted no-till. Dead weeds are evidence that a herbicide has reduced competition for the soybeans.

in the soil. pH should be 5.8 to 7.0. Raising the pH increases the availability of calcium and magnesium.

Harvesting

Soybeans are harvested with a combine. Harvesting begins when the beans are mature, the leaves have fallen off the plant, and the stalks and pods are golden brown.

Most beans are harvested at around 14 percent moisture content. Sometimes beans with higher levels of moisture are harvested to prevent preharvest losses. A **preharvest loss** is a loss that occurs before entering the field with a combine. This is also known as preharvest shatter. The bean pods pop open and release the beans which fall to the ground. Beans that have fallen to the ground cannot be harvested.

Harvest loss is caused by the harvesting process. Proper adjustment of a combine is essential to prevent losses. The combine must also be operated at an efficient speed to assure a maximum harvest yield. Beans with moisture levels over 14 percent must have some artificial drying. Moisture levels in harvested soybeans are related to relative humidity and storage conditions.

18-34. Mature soybeans ready for harvesting.

Peanuts

The peanut (*Arachis hypogaea*) is grown for oil, peanut butter, roasted peanuts, and other uses. It is the third most important oil crop worldwide though of less importance in the United States. Peanuts contain 40 to 50 percent oil. Once oil is pressed from the peanut, the remaining oil cake has 50 percent protein. This means that the cake is a very useful protein supplement in animal feeding.

The peanut originated in South America and is grown throughout warm climate regions worldwide. Little growth occurs below 56°F, with 86°F being optimal. The peanut plant is an annual legume and has nitrogen-fixing abilities. Small yellow flowers are self pollinated.

18-35. Peanut kernels are contained inside of pods. (Courtesy, U.S. Department of Agriculture)

18-36. Peanut flowers are bright yellow.

18-37. Following flowering the ovary elongates to form a peg that grows into the soil, where the pod is formed.

The fruit is a pod of one to five seeds (kernels) that forms at the end of a peg. A peg is an elongated part of the ovary from the flower that grows downward into the soil. Plants may have many pegs. Partially because of the growth of pegs, light-textured soils with low organic matter and good drainage are preferred.

Types Grown

Four types of peanuts are grown:

- Runner—The runner type accounts for 75 percent of the U.S. peanut production. The kernel is a good size and quite useful in manufacturing peanut butter. These peanuts are predominantly grown in Georgia, Alabama, Florida, Texas, and Oklahoma.

- Virginia—The Virginia type has the largest kernels and is used for roasting and in-shell sales. These peanuts are grown mostly in Virginia and North Carolina.

- Spanish—The Spanish peanut has smaller kernels that are covered with a reddish-brown skin. These peanuts are widely used in making candy. Some are sold as roasted peanuts and others are used in making peanut butter. Spanish peanuts are mostly grown in Texas and Oklahoma.

- Valencia—Valencia peanuts are known for having a larger number of kernels per shell. They are usually roasted or boiled in the shell. Valencias are mainly grown in New Mexico and account for about 1 percent of peanut production in the United States.

Cultural Practices

Seeds are planted in 18- to 30-inch rows in a prepared seedbed. Mechanical planters must carefully place the seed to avoid splitting them apart. Seed that split apart will not germinate. On 18-inch rows, the planting rate is 105,000 seeds per acre. Another rule-of-thumb is to plant four peanut seeds in hills 18 to 30 inches apart, with distance based on the spreading habits of the variety planted. Wider rows are seeded at a lower per-acre rate. The seeds are

18-38. Peanuts growing in Florida on 30-inch rows.

planted 1 to 2 inches deep after the danger of frost in May or June. Peanuts are typically fertilized similarly to soybeans. The optimal soil pH is 6.0 to 6.5.

Select a variety of the type to plant that is adapted to the local area. The length of the growing season is a factor. Spanish peanuts mature in 90 to 120 days. Valencia requires 90 to 110. The Runner and Virginia types require 130 to 150 days.

Weeds, insects, and other pests may need to be managed. Mechanical weed control is often used with young plants. Avoid cultivating near plants as they grow. Close cultivation can reduce flowering and the formation of pegs. Several herbicides may be approved for use in peanuts. Insects and diseases are also managed with approved pesticides, as needed. In some cases, soil-borne pests, such as nematodes, are problems. Keep deer and farm animals, such as goats, out of the fields to prevent damage to plants.

Harvesting

Peanuts are harvested to assure a maximum yield. Allowing peanuts to stay in the ground after maturity may result in sprouting if showers

18-39. A peanut plant that has been dug and shaken awaiting combining on an Alabama farm. (Courtesy, U.S. Department of Agriculture)

18-40. A peanut combine lifting windrows and removing the pods from the plants. (Courtesy, U.S. Department of Agriculture)

occur. In general, harvest when 75 percent of the pods show darkening on the inner surface of the hull. Harvesting after frost is okay, especially in northern climates.

Tops may be partially clipped to begin the harvesting process. Afterward, a digger-shaker-windrower is used to lift the plants with nuts attached. The digger should go deep enough to get the nuts and not break them off of the pegs. The windrows are harvested with combines while the peanuts pods are wet (green) or semidry. Waiting until the pods are dry may result in weather damage to peanuts. Leaves, stems, and other vegetative parts may be used as livestock feed or plowed back into the soil.

Once separated from the vines, the peanuts are cleaned and dried. A moisture content of 5 to 10 percent is safe. Artificial drying should be done immediately after harvest. This prevents mold and insect losses. Air drying is typically used. Heated dryers should not warm the peanuts above 95° F. Warm temperatures cause flavor and quality deterioration. Peanuts are stored in areas that protect them from insects, moisture, and contamination.

CANOLA

Canola has emerged as an important oil crop in recent years. The seeds are about 45 percent oil. The plant was developed from the rapeseed plant in 1974 by researchers in Canada. Rape is a flowering herb and a

18-41. Canola growing in North Dakota.

member of the mustard family. All varieties are members of one of two species: *Brassica napus* and *Brassica campestris*. Rape are annuals, planted in the spring and mature in late summer, or biennials, planted in the fall and mature in late spring.

The oil is considered to promote good health because it contains low levels of erucic acid, a substance that may cause heart disease. North Dakota and Minnesota lead the United States in canola production. Major areas of production are in western Canada.

Plants grow 2 to 6 feet tall. It has numerous branches with bluish-green leaves. Yellow flowers appear as the plants mature. The seeds are ripe when the plants turn the color of straw and the seeds are brown. Seeds are harvested with a combine. Proper timing and adjustment are essential to prevent loss of the small seed during harvest.

SUNFLOWERS

The sunflower (*Helianthus annuus*) is best known for having tall, showy yellow flowers. It is a fairly new seed crop in the United States. Some seeds are used as human snack food and birdfeed, but most go into vegetable oil processing. Sunflowers are also grown for ornamental uses. It is also a common and popular wildflower in the central United States.

The variety of sunflower planted depends on the use to be made of the seed. In general, two broad categories are used: oil and snack or birdfeed. The oil-type sunflower has smaller seed and was developed as a source of oil. The snack and birdfeed varieties are larger and were developed for appeal in human snack foods or in birdfeed.

Sunflowers grow from 3 to 10 feet (1 to 3 m) tall. They usually have one stem with a large single flower or several smaller flowers. Flowers may be 3 to 6 inches (8 to 15 cm) or more across, with some reaching 15 inches (38 cm). Seeds are formed in the middle of the flower,

18-42. Field of sunflowers.

18-43. Sunflower seed.

known as a sunflower head. One sunflower head may have as many as 1,000 seeds. People are always amazed at how sunflower heads turn during the day to face the sun.

Sunflowers rank third in importance in vegetable oil in the world. Seeds have 25 to 50 percent oil, which is similar to corn oil. They are also high in protein.

Because sunflowers are more tolerant of light frost in the seedling stage than corn, they can be planted in areas where corn may not be able to consistently produce high yields. Sunflowers mature in 90 to 120 days.

Sunflowers are also adapted to a wide range of soil types and require less moisture than other crops.

Sunflowers are planted in prepared seedbeds with planters similar to those used for corn. Rows are usually 24 to 30 inches (61 to 76 cm) apart. A plant population of 15,000 to 25,000 is wanted, with smaller populations used on drier land. Sunflowers require nutrients for growth but typically at a lower rate than other crops. Extensive root systems up to 7 feet (2.1 m) deep provide moisture and nutrients for the plants.

18-44. Mature, dry sunflower heads ready for harvesting on a North Dakota farm.

As with other crops, sunflowers are subject to damage by certain diseases and insects. Local recommendations are followed with pest management.

SAFFLOWER

The safflower (*Carthamus tinctorius*) has been cultured for many years in India and other Asian countries. It is now grown in Australia, Mexico, Spain, and the western United States. The seeds are similar to sunflower seed and provide important vegetable oil. Besides cooking oil and margarine, safflower oil is used in making paints and varnishes.

18-45. Field of safflower in full bloom on a California farm.

Safflower is planted as an annual crop in the spring. It is tolerant of cool weather, especially in the seedling stage, where it will survive temperatures as low as 20°F (–7°C). Safflower plants are less tolerant of cold weather as they mature. A growing season of 120 frost-free days is required for safflowers to grow.

Safflower plants resemble common thistles (they are in the thistle family) and may have small prickers that sting unprotected skin. The plants develop tap roots that reach into the soil for moisture and nutrients. Plant stems are 2 to 5 feet (60 to 150 cm) tall and form flowers and seed heads at the ends of the top branches. Flowers have a yellow-orange color. Oil content of the seed ranges from 32 to 40 percent. Protein content is 11 to 17 percent.

Safflowers will grow in areas with lower moisture levels than some other crops. They require about 18 inches (46 cm) of moisture but will grow better with 25 inches (63 cm). Above 25 inches (63 cm) of moisture, they are more subject to disease.

Safflower is planted with a grain drill or a row planter in a seedbed prepared similar to that for wheat. The seedbed should be firm at the time of planting. Drills are 6 to 15 inches (15 to 35 cm) apart. Seeding rates range from 15 to 40 pounds per acre. Seeds should be planted 1 to 2 inches (2.5 to 5 cm) deep. Safflowers are not affected by many insect pests. Blights, rusts, and head and root rot are sometimes problems.

Harvesting can begin when the plants are dry. Most safflower is harvested with a combine. Fields with green weeds are more difficult to harvest. Weedy fields are sometimes cut and windrowed for drying. A pick-up attachment is used on the combine to harvest from the windrows. Shattering does not cause major losses with safflowers.

REVIEWING

MAIN IDEAS

Sugar and oil crops are important sources of food. Sugar is any food product used as a sweetener. Sugar beets and sugar cane are the two most important sugar crops. Corn syrup, maple syrup, and a few other sources are used.

Oil crops are used to produce vegetable oil that is primarily used in cooking oil, margarine, and similar foods. Some oils are used for other purposes, such as paint and fuel. The leading oil crops are soybeans, peanuts, cotton seed, corn, canola, sunflowers, and safflowers.

Cultural requirements of sugar crops vary widely. Sugar cane requires a tropical climate with high moisture. Sugar beets grow in cooler climates with lower moisture levels.

The soybean is the most important oil crop. It will grow where corn grows but has some adaptations that make it preferable to corn as a crop for some producers. As a legume, fertilization is greatly different from corn and the other oil seed crops. Peanuts and canola are also important sources of vegetable oil.

Sunflowers and safflowers are also used for oil. They are more tolerant of drier climates and cooler weather than other oil crops.

QUESTIONS

Answer the following questions using complete sentences and correct spelling.

1. What is sugar? What kind is obtained from sugar cane and sugar beets?

2. What is vegetable oil? Why has the use of vegetable oil increased in recent years?

3. How is sugar made from sugar cane? Sugar beets?

4. What are the major crops grown for producing vegetable oil? Which is most important?

5. How is sugar cane propagated? Sugar beets?

6. Compare and contrast the climate requirements for sugar cane and sugar beets.

7. What are the important production considerations with soybeans?

8. Briefly describe the production of sunflowers and safflowers.

EVALUATING

Match the term with the correct definition. Write the letter by the term in the blank that is provided.

a. sucrose
b. preharvest loss
c. vegetable oil

d. syrup
e. blackstrap molasses
f. ratoon crop

g. harvest loss
h. granulated sugar

1. _____ sweet liquid made from plant juices

2. _____ fat obtained from plants

3. _____ table sugar

4. _____ loss of crop yield before harvest is begun

5. _____ product of raw sugar crystals

6. _____ syrup remaining after sugar has been processed

7. _____ sugar cane produced by sprouts from roots of harvested crop

8. _____ loss of crop yield as a part of harvesting

EXPLORING

1. Make a field trip to a farm that produces oil or sugar crops. Determine the kind of crop, variety planted, cultural practices followed, and how the crop is harvested. Prepare a report on your observations.

2. Obtain a sugar cane stalk. Cut it into 18- to 24-inch sections and plant it in the school greenhouse, garden, or field. Keep a log of events in the emergence of a sprout and growth of the plant.

3. Make an inventory of the products made from vegetable oils in the local supermarket. List the kind of product and the crop from which the vegetable oil was made.

Fiber Crops

TERMS

animal fiber	defoliant	module
boll	desiccant	plant fiber
bur	fiber	seed cotton
classing	ginning	square
cotton bale	lint	staple

19-1. Checking lost cotton in a field following harvest.

FLAX, cotton, and kenaf are examples of fiber crops. They provide us with some of the natural fibers that we use. What if we had no materials made from natural fibers? We would not have most of the clothes that we wear. Neither would we have towels, bed sheets, curtains, and many other things that we take for granted.

Fibers have been used in many ways for a long time. Cotton has been the leader among the fibers. Some 5,000 years ago, cotton was grown in Pakistan. Before that, it was harvested wild in East Africa. In addition to clothing, cotton was used in many other ways. One early use was to make harnesses for elephants—cotton is strong!

Fiber production—especially cotton—uses high technology. People have to know what they are doing to produce a good crop. Among crops, cotton requires more cultural attention than other common crops.

FIBER SOURCES

A ***fiber*** is a long threadlike strand of a substance that has many useful qualities. Most fibers are considerably longer than wide. Proportionate to size, many fibers are quite strong. Fibers are also flexible and can be easily bent and formed into products.

Fibers are of two major kinds: natural and manufactured.

19-2. A field of growing cotton. (Courtesy, Terra)

19-3. Young kenaf plants.

NATURAL FIBERS

Natural fibers are from two major sources: plants and animals. Both sources are widely used to meet the needs of humans.

Plant Fibers

A ***plant fiber*** is from plant origin. Common plant fibers include cotton, flax, hemp, jute, sisal, and kenaf. Cotton is by far the most widely used fiber. Flax has some special uses, particularly in making linen products. Hemp, jute, and sisal are course fibers used in making rope, burlap bags, and other products where softness is not required. Other plant fibers include kenaf and ramie.

Some flax is grown in North America for fiber. Linen, a quality cloth material, is made from flax fiber. Several Canadian provinces as well as North and South Dakota in the United States grow flax. Worldwide, most flax is grown in Europe and Asia. As a cool season crop, flax thrives at temperatures of 70 to 80°F

(21 to 27°C). In addition to linen, flax seed is used to make bread (reportedly quite nutritious and health-promoting) and various oil products. The oil from flax seed is known as linseed oil. Linseed oil is used in paint and similar products, such as an adhesive for particle board materials.

Hemp, jute, and sisal produce large, coarse fibers. Jute is second to cotton in worldwide production. The fibers are used in many ways where fine quality is not required. Next to the human skin, fibers from hemp, jute, and sisal feel scratchy and uncomfortable. India, China, and other Asian countries are the leading producers of these plant fibers.

Kenaf is a relatively new fiber crop being experimentally grown in some areas. It has been used to make fabric for clothing, but its greatest potential may be in paper manufacture. Kenaf fibers tend to be coarse and less flexible but stronger than other natural fibers.

Table 19-1. Names of Selected Plants Used for Fiber

Common Name	Scientific Name
cotton (upland)	Gossypium hirsutum
cotton (long or pima)	Gossypium barbadense
flax	Linum usitatissimum
hemp	Cannabis sativa
jute	Corchorus spp.
kenaf	Hibiscus cannabinus
ramie	Boehmeria nivea
sisal	Agave sisalana

Ramie, also known as china grass, produces very long fibers—nearly a foot long! The fibers are also strong. A major drawback is the difficulty of separating the fibers from surrounding plant tissues. Ramie production in the United States has not been successful. China, Japan, and Indonesia are major ramie producers.

Animal Fibers

An **animal fiber** is from animal origin. These fibers grow on, or are produced by, animals. Wool is the most widely used animal fiber. It is produced by some sheep and goats. Camel hair, mohair (from the angora goat), alpaca, and other animal hairs are sometimes used as fiber. Silk is the strongest of all natural fibers. Silkworms make cocoons, which are unwound to obtain individual fibers for cloth manufacture.

MANUFACTURED FIBERS

Manufactured fibers are also known as synthetic or human-made fibers. They are not natural fibers. Manufactured fibers include nylon, polyester, acrylic, and other kinds of fiber. More than 65 percent of the fiber used in textile mills in the United States is manufactured. In some cases, they are mixed with natural fibers for products such as elastic fibers. Many of the manufactured fibers are used in making carpet, upholstery, and similar goods.

COTTON PRODUCTION

Cotton is a warm season crop. Treated as an annual, the plants will live as perennials if protected from cold weather. Cotton is typically planted in the spring and harvested in the fall. It requires a growing season of at least 180 frost-free days. Some newer varieties require fewer than 180 days.

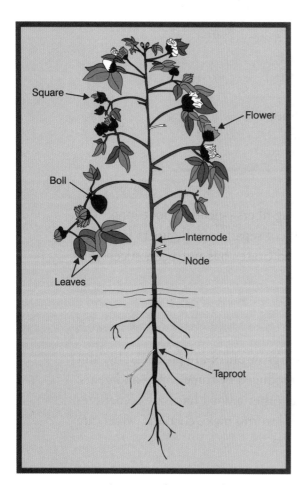

19-4. Major parts of a cotton plant.

Cotton, as picked from the field, is known as **seed cotton**. Seed cotton is **lint** (the fiber) and seed. The seed cotton is ginned to separate the fiber and the seed. The lint goes to textile mills for manufacture into cloth materials. The seed goes to mills where the oil is extracted. The outer part of the seed becomes seed hulls, which are used as cattle feed. A high-quality protein cotton seed meal remains after the oil has been extracted. Cotton seed meal is used as a source of protein in animal feeds.

THE PLANT

A cotton plant typically grows 3 to 5 feet (1 to 1.5 m) tall. Height depends on the species as well as the available moisture. The cotton grown in the plains of Texas may be only 18 to 24 inches (45 to 60 cm) tall. Cotton grown in the Delta area of the South grows much taller sometimes reaching 6 feet (2 m). In warm climates, cotton plants may live as perennials and grow taller.

The plant grows upright with branches that spread in all directions. The broad leaves may have three or five lobes. The tap root system may extend to a depth that is twice the height of the plant. About six weeks after emergence from the soil, flower buds, known as squares, begin to form. A **square** opens into a flower that normally lasts two days. Color changes with the age of the flower. Flowers are white early on the first day and turn pink to red after pollination on the second day. Each plant may have several flowers a day. The first flowers are on the lower limbs while the plant continues to grow vertically. As the season progresses, flowers will open on the upper limbs. A mature plant will have 18 to 24 stem nodes where branches develop.

19-5. Cotton flowers. (Color is a function of age. White flowers are in the first day of bloom. Red or pink are in the second day of bloom.)

After pollination, the flowers dry up and a **boll** is formed. A boll is the fruit of the cotton plant that contains both seeds and fiber. Several weeks are required for a boll to mature and open. Once open, the boll is dry and known as a **bur**. Bolls are divided into several sections containing locks of cotton. The sections of the bur have sharp points that can stick the human skin. Experienced hand cotton pickers know how to work fast and avoid being stuck by the bur. Most cotton is picked by machine.

Cotton is subject to damage by insects at any stage of development. Squares can be punctured and bolls can be attacked. Some pests, especially the boll weevil, lay eggs inside the squares. Boll worms may eat holes into young bolls and destroy them before any fibers have formed.

19-6. Developing cotton bolls (left) and open boll (right).

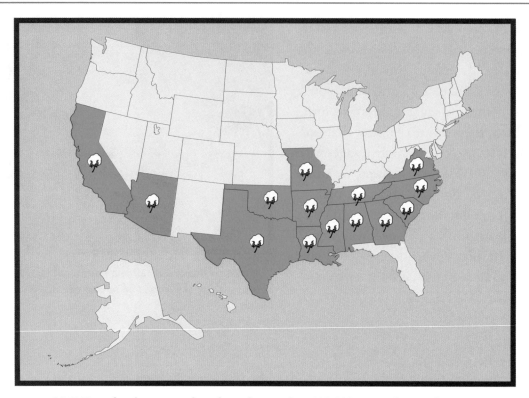

19-7. Map showing states that planted more than 100,000 acres of cotton in 2000.

Table 19-2. Upland Cotton Production by State

State[1]	Rank	Number of Acres Planted	Number of Bales Produced[2]
Texas	1	4,800,000	4,300,000
Georgia	2	1,300,000	1,610,000
Mississippi	3	1,280,000	1,750,000
Arkansas	4	950,000	1,460,000
North Carolina	5	930,000	1,450,000
California	6	765,000	2,050,000[2]
Louisiana	7	700,000	930,000
Tennessee	8	570,000	720,000
Alabama	9	540,000	550,000

Note: Data are for the Year 2000. Source: U.S. Department of Agriculture.

[1]Other states with reported cotton acreage: Kansas, Missouri, Florida, New Mexico, North Carolina, South Carolina, Oklahoma, and Virginia.

[2]Net weight of a bale is 480 pounds (500 pounds with bagging—wrapping around the bale to hold it together); data for 2000 are estimated.

[3]California has the highest per-acre yield at 2.68 bales.

KINDS OF COTTON

Four kinds of cotton are grown worldwide. The distinction is primarily based on the length of the fiber, known as **staple**. Kinds of cotton:

- Upland cotton—This kind of cotton predominates in the United States and is grown in many other countries. It is capable of producing a good yield under a range of climate conditions. The fibers can be used for a wide range of products including fine quality clothing as well as heavy canvas materials. Plants range from 1 to 7 feet (30 cm to 2.1 m) tall, with height largely depending on water supply and the variety planted. Staple length is typically $7/8$ to $1^{1}/_{4}$ inches (22 to 32 mm). Texas is the leading state in upland cotton production, though the per acre yields are lower than California and other states.

Table 19-3. American-Pima Cotton Production by State

State	Rank	Number of Acres Planted	Number of Bales Produced[1]
California	1	144,000	370,000
Texas	2	16,000	27,000
Arizona	3	6,000	10,300
New Mexico	4	6,000	8,500

[1]New weight of bale is 480 pounds.
Source: U.S. Department of Agriculture, 2000.

- Pima (Egyptian) cotton—Known for exceptionally long and strong fibers, Pima cotton was developed from Egyptian cotton to better suit growing conditions in some areas. Some Pima cotton is grown in the United States but not nearly to the extent as upland cotton. The staple of Egyptian cotton may reach 1.5 inches (38 mm). Pima cotton tends to have slightly shorter staple. Pima and Egyptian cotton are used for fine quality clothing and other products where staple is important. Egyptian cotton fibers have a light tan color.

- Sea-island cotton—Sea-island cotton is grown in the West Indies though it originated in the islands off the coast of South Carolina, Georgia,

19-8. Label in a shirt made of sea-island cotton.

and northern Florida. It produces very high-quality fibers but has low yields and small bolls. The silky fibers reach $1^3/_4$ inches (44 mm) in length. Sea-island cotton is imported to the United States. Most clothing made from this cotton is specially labeled and demands a premium price.

- Asiatic cotton—Asiatic cotton is declining in production and importance. It is primarily grown in China, India, and Pakistan. The staple is short and the fibers are coarse. The cotton is often used as padding or for blankets. Upland cotton is replacing Asiatic cotton. No Asiatic cotton is grown in the United States, though U.S. cotton competes with it on the world market.

VARIETIES, LAND PREPARATION, AND PLANTING

Many varieties of cotton are available. A grower should select a variety that has been proven in the local area. Many are using genetically modified varieties that resist pests, such as Bt cotton. Reputable seed companies and agronomists with land-grant universities can assist in variety selection. In selecting a variety, consider the market that is available for the cotton. In some cases, all growers in an area grow nearly identical varieties to assure uniform fiber quality in ginning.

Cotton is planted in a carefully prepared seedbed. The land is prepared by chopping stalks from the previous crop, plowing, disking, and harrowing. Growing no-till cotton has met with mixed results. Yields have been low and the fiber produced is of reduced quality.

CAREER PROFILE

COTTON GIN OPERATOR

A cotton gin operator oversees the operation of a cotton gin. Duties include hiring and supervising employees, as well as dealing with growers when seed cotton is delivered. The work involves mechanical skills, record keeping, and human relations skills.

Cotton gin operators need practical experience in the operation of gins. They begin as workers in the gin and advance as they learn the operation of the equipment. Some operators have associate degrees in gin management. Others have high school diplomas or baccalaureate degrees.

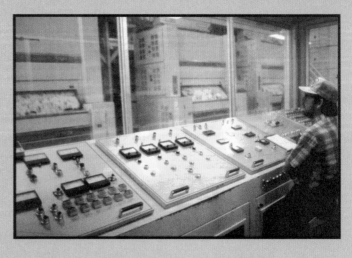

This photograph shows a gin operator monitoring the ginning process from a control panel. (Courtesy, National Cotton Council)

19-9. Primary tillage initiates land preparation for planting cotton. (Courtesy, Case Corporation)

Planting is with a high-speed, multiple-row planter. Seeds are delinted and treated to resist disease. Seeds from cotton normally have a few fibers attached. These fibers can clog up equipment and cause planting problems. A carefully managed chemical process is used to remove the fibers by seed processors. Cotton is planted in raised rows (beds) spaced 30 to 36 inches (76 to 91 cm)

19-10. Comparison of cotton seed, with the normal ginned seed on the right and chemically delinted seed on the left.

19-11. Cotton being planted. (Courtesy, The National Cotton Council of America)

19-12. Growers plant quality seed. (Courtesy, Terra)

apart. In some cases, a skip-row planting pattern is used. This involves planting two to four rows and skipping two to four rows.

Seeding rates are relatively consistent throughout the entire Cotton Belt in the United States. Plant population levels are 50,000 to 60,000 per acre, though some producers want only 40,000. Most growers want to have three or four plants per linear foot of row. To achieve this number of plants, the planter is calibrated to drop one additional seed per foot over the desired plant population. Planting more seeds compensates for the seed that do not germinate. Cotton seed germination is often 80 percent. Cotton seeds are planted about 2 inches (5.1 cm) deep.

Cotton is planted in the spring after the soil temperature has reached at least 65°F (13°C) at planting depth for three consecutive days. Germination is optimum if the temperature is 85°F (35°C). Weather forecasts should include warm, sunny days following planting to get the best germination. The seeds sprout and emerge from the soil in five to seven days.

Once up, the plant may remain in the cotyledon stage of development for seven to nine days. This stage of the plant is particularly vulnerable to insect damage, especially by thrips if the weather is not warm. Cotton planting begins in south Texas in late February and concludes in the Texas Plains in early June. The vast majority of the cotton in the Delta area is planted in May. Arizona and California cotton may be planted in April.

19-13. Young cotton plants in the cotyledon stage. Note that the seeds have been planted in hills, with four to five plants per hill.

19-14. Examples of skip-row cotton are shown here. On the left, two rows have been planted and one skipped in the Mississippi Delta. On the right, four rows have been planted and four skipped to start an almond orchard in California.

PEST MANAGEMENT

Cotton is subject to a wide range of insect, disease, and weed pests.

Insect management begins following harvest of the previous crop. Stalks are chopped to destroy hiding places for insects to overwinter. Cold weather will kill some pests, especially if they do not have a protective place to hide. Many growers use IPM programs in insect control. Repeated spraying with insecticides has often been used to manage cotton pests. Scouting is an important management tool for cotton producers. The transgenic insect-resistant varieties are reducing the need to apply insecticides. The major insect pests of cotton are listed in Table 19-4.

Most producers use a preemergence herbicide to keep down weeds from the beginning. Postemergence herbicides are used on weeds that emerge after the cotton is growing. Over-the-top applications may be made on larger cotton to kill tall weeds, such as Johnson grass and cockleburs. Keeping weeds out of cotton is especially important in producing a trash-free harvested fiber product.

Cultivation is sometimes used to keep down weeds and provide a soil mulch. Cotton may be cultivated sev-

19-15. Cotton being damaged by a worm pest. (Courtesy, James Lytle, Mississippi State University)

Table 19-4. Major Insect Pests of Cotton

Pest	Damage	Management[1]
boll weevil	feeds on and deposits eggs in squares and bolls	chemical; cultural; plant resistant varieties
bollworm	bores into squares and bolls	chemical; cultural; Bt varieties
tobacco budworm	bores into squares and bolls	chemical; Bt varieties
pink bollworm	bores into bolls and eats developing seed	chemical; quarantine Bt varieties
plant bug	feeds on and destroys squares	chemical
fleahopper	feeds on squares	chemical
thrips	sucks sap from young plants	chemical
spider mites	destroys leaves by creating massive webs on underside of leaves	chemical; biological
aphids	feeds on young leaves and squares	beneficial insects; chemical
whitefly	feeds on underside of leaves, reducing plant growth efficiency	chemical

[1]Use only recommended and approved methods.

19-16. Considerable herbicide has been used on this land to prepare it for no-till cotton.

eral times during a growing season. The plows should be adjusted for maximum weed destruction with minimum damage to the cotton plants. Plows that cut too deeply destroy cotton roots and reduce yields.

Growers should get the recommendations for insect, disease, and weed management from the land-grant university in their state. These recommendations have been carefully prepared to meet environmental guidelines and provide effective pest control.

FERTILIZATION AND IRRIGATION

Cotton requires fertile soil with adequate moisture. As a crop, only the seed cotton is removed from the field. Stalks, leaves, and other plant parts are plowed back into the land.

Each bale of cotton removes 40 pounds of nitrogen (N), 15 pounds of potassium (K), 20 pounds of phosphorus (P), 1 pound of calcium (Ca), 5 pounds of magnesium (Mg), and 2 pounds of sulfur (S) from the soil.

Cotton also uses minor amounts of the following trace elements: boron (B), manganese (Mn), zinc (Zn), chlorine (Cl), copper (Cu), iron (Fe), and molybdenum (Mo). These nutrients must be replaced in the soil or the level of production will decline. Soil testing is often used to determine the amounts of nutrients needed. Growers also estimate nutrient needs on the basis of previous experience and desired future production.

19-17. Comparison of a normal cotton leaf (center) with those showing potassium deficiency. (Courtesy, Potash and Phosphate Institute)

Over-fertilization can create problems with cotton. Too much nitrogen results in excessive stem and leaf growth. The stalks will grow taller and produce fewer squares and bolls.

Cotton prefers a soil pH of 5.7 to 7.0. It will grow satisfactorily in a pH range of 4.5 to 8.5. Some trace elements, such as boron and iron, are less available in soils below 6.0 pH.

Fertilizer is applied before planting, as well as after the crop is growing. The preplant fertilization program is normally based on complete nutrient needs of cotton, including nitro-

19-18. Furrow flood irrigation system on a California cotton farm.

gen, phosphorus, potassium, and the secondary and micronutrients. Side dressing is used to apply additional nitrogen after the crop is underway.

Irrigation is used as a primary source of water or to supplement rainfall during brief drought intervals. In the dry plains and southwestern United States, irrigation may be the major source of water. In the Delta states, irrigation is used only as needed to supplement rainfall.

HARVESTING

Three methods of harvesting are used–hand, picker, and stripper. Very little cotton is hand picked, though the quality of hand-picked lint is cleaner than that of machine-picked lint. Pickers are used with much of the cotton in the United States. Specially made spindles remove the cotton from the burs. The equipment must be carefully adjusted and properly operated to harvest a high proportion of the cotton. Seed cotton left on stalks in the field is lost income!

Strippers pull all limbs, bolls, burs, and lint from the cotton plant. Strippers are used in the plains areas, such as West Texas, where cotton stalks are small. The ginning process must clean the stems and other plant debris from the lint.

Prior to harvesting, a defoliant or desiccant is applied to the cotton plants. A **defoliant** is a chemical applied to a plant that causes the leaves to fall off. The bolls remain on the plant. This results in cleaner seed cotton because trash from the

19-19. A close-up of spindles on a cotton picker. (Courtesy, Marco Nicovich, Mississippi State University)

19-20. Picker harvesting cotton. (Notice that there are no leaves on the cotton plants—the result of defoliation.) (Courtesy, Case Corporation)

leaves does not stick in the lint. Nearly all cotton harvested with a picker is defoliated. In some cases, cotton will regrow some of its leaves if the defoliant is applied too early or if the frost is later than usual in the fall. A *desiccant* is a chemical that causes the entire plant to dry up. Desiccants are most often used prior to harvesting with a stripper.

Harvesting begins when the bolls are open. Some growers wait until all bolls are open and make one trip through the field with a picker. Other growers harvest two times. The first harvest removes the open bolls and leaves the green bolls time to fully mature. If a stripper is used, all cotton should be open because the stalk is destroyed.

19-21. Picker dumping into a wagon. (Courtesy, Case Corporation)

Harvested cotton may be hauled to a gin or stored in the field in modules. A *module* is a specially compacted unit of seed cotton that can be easily transported. In effect, seed cotton can be harvested and stored in the field until ginning.

CONNECTION

BOLL WEEVIL ERADICATION

The boll weevil has been a major cotton pest for many years. It punctures squares and bolls and lays eggs inside. The feeding larvae destroy the square and boll. The square and boll fall off the plant and do not make any cotton.

A major initiative has led to eradication of the boll weevil in most cotton-producing locations. One procedure has involved raising millions of male boll weevils in carefully controlled laboratories. The adult males are sterilized and released into cotton areas to mate with the females. The mating does not result in fertile eggs. Since a female mates only once, she will lay infertile eggs throughout her lifetime. This results in no offspring from her.

Another step has been to outlaw ornamental cotton. Homeowners are not allowed to have cotton plants in their yards. These become places where boll weevils can live and reproduce. (Courtesy, Marco Nicovich, Mississippi State University)

19-22. Covered module with a module builder.

GINNING AND CLASSING

Ginning is the process of separating the lint and seed in harvested seed cotton. It also includes drying the cotton to reduce moisture content and cleaning the lint to remove trash. The ginned cotton is pressed into a *cotton bale*. One bale of lint cotton requires 1,100 to 1,200 pounds of seed cotton. The weight of seed cotton is more than one-half seed along with some moisture and trash that will be removed.

19-23. Cotton moving through a gin stand as part of cleaning and separating lint and seed. (Courtesy, Marco Nicovich, Mississippi State University)

9-24. A ginned, compressed bale of cotton. (Courtesy, Marco Nicovich, Mississippi State University)

19-25. Fibers can be carefully pulled between the fingers to assess the staple.

The standard weight of a bale is 480 pounds net lint cotton. The weight may vary from less than 480 to more than 600 pounds, though efforts are being made to standardize all bales. A 480-pound bale will be wrapped with burlap or synthetic bagging and bound with six steel ties. The weight of the lint, bagging, and ties will be 500 pounds.

Classing is the process of grading cotton. Every bale is classed. A small sample is typically removed from the bale. This sample is graded for length of staple, color, and the presence of trash. The diameter of fibers may also be measured with a micronaire—a kind of caliper for measuring very small distances. Growers are paid by the pound on the basis of the grade of the cotton. Grades allow all individuals involved in cotton trading to understand the characteristics of a given bale of cotton.

19-26. Sophisticated technology is used in classing cotton. (Courtesy, National Cotton Council)

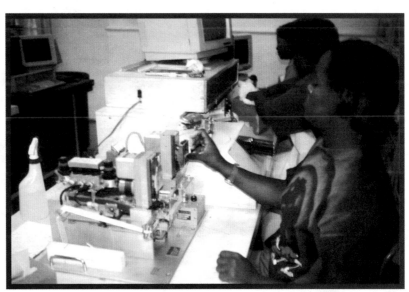

REVIEWING

MAIN IDEAS

Natural fibers are derived from plant and animal sources. The major plant source is cotton. Some experimental work is underway on kenaf. Other plant fibers include hemp, jute, sisal, and ramie.

Cotton requires a long, warm growing season. Upland cotton is the predominant kind planted, though some Pima cotton is grown in the United States. Sea-island and Asiatic kinds of cotton are also grown.

Most cotton is planted in a prepared seedbed. Chemically delinted seed of locally recommended varieties should be planted. Seeding rates vary, but desired plant populations are usually 50,000 to 60,000 cotton plants per acre.

Cotton can be damaged by insect, disease, and weed pests. Over the years, considerable use of chemicals has kept cotton production levels high. In recent years, more attention has been given to IPM.

Harvesting is usually with mechanical pickers though strippers are used in some places, especially where the stalks do not grow very tall. Seed cotton is ginned and classed in preparation for manufacture at textile mills.

QUESTIONS

Answer the following questions using complete sentences and correct spelling.

1. What are the sources of plant fibers?

2. Distinguish between seed cotton and lint.

3. What is a square? Boll?

4. What kinds of cotton are planted? Which is the leading kind in the United States?

5. What soil conditions are needed to plant cotton?

6. What are the major insect pests of cotton?

7. What are the fertilizer needs of cotton?

8. How is cotton harvested?

9. What is ginning? Classing?

10. What is the weight of a bale of lint cotton? How much seed cotton is needed to produce a bale?

EVALUATING

Match the term with the correct definition. Write the letter by the term in the blank that is provided.

a. classing
b. defoliant
c. desiccant
d. seed cotton

e. lint
f. square
g. boll
h. staple

i. module
j. ginning

1. _____ a compacted unit of seed cotton

2. _____ process of separating seed and lint

3. _____ chemical applied to cotton to cause the leaves to fall off

4. _____ picked cotton containing fiber (lint) and seed

5. _____ chemical applied to cotton that causes the entire plant to dry up

6. _____ cotton fibers from which the seeds have been removed

7. _____ a cotton bud

8. _____ a cotton fruit

9. _____ length of cotton fibers

10. _____ grading cotton on the basis of color, fiber length, and presence of trash

EXPLORING

1. Visit a department store that sells a wide range of clothing. Review the labels on the clothing to determine the different plant and animal fibers, as well as manufactured fibers, used in manufacturing the clothing. Prepare a report on your observations.

2. Tour a cotton plantation and observe the cultural practices that are used. Interview the manager about cotton production. Determine the varieties planted, when planted, the equipment used, pests and control, and harvesting. Take a camera and make photographs of your observations. Prepare a poster that depicts the steps in cotton production.

3. Accompany a cotton scout while scouting cotton. Ask the scout to explain what to look for in scouting. Practice locating insects, evidence of insect damage, weeds, and other pests. Review pest control recommendation guidelines for your area in cotton production.

Forage and Turf Crops

20-1. Turf is important in having attractive landscaped areas.

FORAGE and turf crops are important to our quality of life. Some provide food; others provide a pleasing environment. Much of the land area of North America is used for forage and turf. Some of the plants are native plants that grow without human intervention. Others require careful management by humans to get them to live and grow.

Ranges, pastures, and lawn areas are covered with plants. Most are grasses and others are legume plants. Some are used to support the livestock industry. Others are used to prevent erosion and conserve the natural resources. Fine-textured plants may be established for their beauty and personal appeal, such as on golf courses. Some will grow in dryland areas, while others require considerable moisture.

USES OF FORAGE AND TURF CROPS

20-2. Cattle, as ruminants, convert grasses and legumes into animal protein.

20-3. Hay is often in large round bales. (Courtesy, Case Corporation)

Forage and turf are much alike, but there are differences. How the plants are used helps to distinguish between forage and turf.

FORAGE

Forage is vegetation fed to livestock. It may be in fresh, dried, or ensiled forms. Fresh forms include pasture grasses and other plants that the animals graze. Dried forms include hay and plant materials remaining on the land after the growing season. Ensiled plant materials refer to silage.

Pasture is improved or unimproved plant materials on land areas where animals graze. Pastures are mostly of grass plants, though others, including legumes, can be used for pasture. Most pastures are fenced and maintained to improve the quality of the forage. Large open areas are referred to as **range**. Most range areas have native plants and may be relatively unimproved over the natural conditions.

Pastures may be permanent or temporary. Permanent pastures are planted to annual grasses and legumes. The plants live from one year to the next. Common permanent pasture plants include fescue grass, Bermudagrass, and white clover. Both pastures and ranges usually have perennial plants. Temporary pastures are established to meet summer or winter grazing needs. Wheat, oats, and other winter small grains may be grazed during the winter and early spring. Temporary summer pastures use annuals, such as millet or sorghum.

Hay is green plant material that has been cut and dried for use as livestock feed. Hay may be made from any of several grasses or other plants, such as alfalfa. Careful attention to cutting and drying are needed to maintain nutrients in the hay. Allowing hay to get rained on lowers its quality. Excessive moisture can result in the hay rotting. Rotten hay is not suitable for animal feed. Hay is often baled for easy handling. It may be stored in the field or a barn. All hay in areas with rainfall should be protected.

Silage is chopped plant material that has fermented. Silage may contain leaves, stems, grain heads, and other aboveground plant parts. Most silage is made from green, growing crops, such as corn, grass, or sorghum. It usually contains 60 to 70 percent moisture and undergoes a fermentation process. The fermentation produces an acid that prevents spoilage of the silage. Silage is stored in a silo. A *silo* is an upright or horizontal facility that maintains the quality of the silage. Some silos are trenches constructed of concrete, while others are large cylinders.

Silage that contains less than 50 percent moisture is *haylage*. Haylage is produced similarly to silage, except that it is drier. It may be harvested at a slightly later stage of plant development and less water is added during the ensiling process.

Both silage and haylage retain more nutrients than hay. This is because the plant materials have not been dried. Fresh green plants contain more nutrients than dried plant materials.

Forages are important because of their efficient use as feed by ruminants. Land that would not produce

20-4. Forage harvester chopping corn silage. (Courtesy, AGCO Corporation, Duluth, Georgia)

20-5. Bunker silos used for storing silage. The insert shows properly cured corn silage.

20-6. Maintenance is important in having good turf. This shows water aeration being used on a putting green.

crops for any other purpose can be used for grazing, such as the vast range areas of the western United States.

TURF

Turf is short-growing, matted grasses or other plants maintained for aesthetic, recreational, or functional purposes. Lawns and areas around homes and businesses and in parks are made attractive with quality turf. Playing fields are more enjoyable when the turf is attractive.

The functional purpose of turf is to protect the soil and prevent erosion. Turf includes the aboveground and below ground area where the stems and roots are growing.

Most turf undergoes careful establishment and considerable mowing, fertilization, watering, and other practices to keep it in good condition. Turf is sometimes considered to be a part of ornamental horticulture. This is because turf is often established and maintained as a part of an overall landscape plan for a home or business.

CAREER PROFILE

TURFGRASS SPECIALIST

Turfgrass specialists study problems and help people establish and improve turfgrass areas. They often collect soil samples, recommend kinds of turf, and consult on turf establishment and maintenance.

Turfgrass specialists need practical experience in working with turf. They typically have college degrees in agronomy or a related area. Advancement is often based on having masters or doctors degrees in a closely related area.

This photograph shows a turf specialist examining sod. (Courtesy, Agricultural Research Service, USDA)

Part of the value of turf is based on its personal appeal to people. Everyone likes to see well-kept lawns. Turf increases the value of property and encourages people to go to a particular location. Who would want to play golf on a course without good turf?

FORAGE CROPS

Many different plants are used as forage. These include grasses as well as legumes.

GRASSES

A **grass** is a plant that typically has leaves with parallel veins and stems that are hollow or solid. Grasses rarely have woody stems, except for the hollow bamboo. Height and other growth characteristics vary widely. Some prefer cool seasons, while others are warm-season species. Grasses also vary as to life cycles.

Grasses are members of the Gramineae family. Many of the members of this family are used as cereal grain crops, such as wheat and barley. Sugar cane and related plants are also members of the grass family. Some 1,400 different species of grasses are found in the United States.

Parts of a Grass Plant

Grasses have roots, stems, and leaves. In addition, they have flowers and form fruit or seed. The root systems are fibrous and do not grow deeply into the soil. The leaf of a grass is called a **blade** and the stem is known as a **culm**.

Some grasses have **rhizomes**, which are ways the plant has of reproducing by sending shoots out below the ground. Other plants have stolons. A **stolon** is an aboveground creeping stem also used to reproduce the plant.

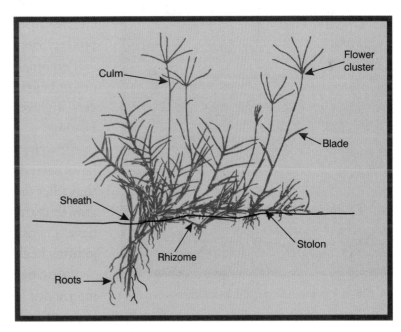

20-7. Major parts of a Bermudagrass plant.

20-8. Week-old fescue grass with dew on the tips of the blades.

20-9. Pasture with established Bahia grass.

How Grasses Grow

Grasses grow in a variety of ways. Growth varies both horizontally and vertically.

Some grasses bunch, while others form sod. Bunch grasses tend to grow in circular patterns. Sod grasses are more aggressive in sending stolons and rhizomes. Sod grasses form a thick mat of stems, leaves, and roots.

Grasses vary in vertical growth between being short and tall. The shorter grasses are more tolerant of close mowing and grazing.

Most grasses have shallow, fibrous root systems. They do not have long taproots as do the legumes and other plants. This means that grasses do not store as much food nor reach as deeply into moist areas of dry soil.

Grasses are either annuals or perennials. Annual grasses are planted each year. Few annual grasses are used, except those as temporary pastures, such as ryegrass. Most grasses are established for long-term productivity.

Perennial grasses regrow each year from buds at, or just below, the surface of the soil. The growth process involves three phases: tillering, jointing, and heading.

Tillering is the growth of shoots from buds that have been dormant. Leaves grow rapidly. The buds that tiller are sometimes known as tiller buds. This growth is the first growth of the year. These shoots have a consistent distance between the nodes until jointing begins.

Jointing is the growth phase in which the internodes begin to elongate. The plant grows rapidly vertically during jointing. The stage after jointing at the end of stem elongation is the **boot stage**.

No heads have appeared to produce seed. No new basal buds are evident at the ground level. Heading follows jointing.

The phase in which seed heads form on the plant is called *heading*. Shoots begin to grow from the base of the plant to replace the mature shoots with heads. The heading stage is a good time to cut forage grasses because they have basal buds ready to repeat the growth process. The jointing stage is not a good time to cut forages because this is a time of rapid growth and no new ground-level buds have formed.

Establishing Forages

Selecting the forage to establish depends on the use to be made of it and how well forages are adapted to the climate. Local recommendations from the land-grant university in your state or a reputable seed dealer will be helpful.

Forages may be established by seeding, sprigging, or in other specialized ways. Seeding may be in a prepared seedbed or by planting into existing vegetation. In some cases, existing grasses and legumes may be used. Additional forage crops are planted into the existing turf with minor disturbance of the soil. This helps protect land susceptible to erosion. Thorough pasture renovation often involves using a prepared seedbed. Plowing and disking are used to kill existing vegetation. In other cases, herbicides may be used to kill vegetation.

CONNECTION

LYSIMETER RESEARCH TO IMPROVE TURF SOIL

What makes the best soil for turf? Agronomists are trying to find an answer through research. The focus here is on the media used in putting greens.

Lysimeters are instruments for measuring the amount of water-soluble substances in the soil. These rectangular box lysimeters are 20 inches wide, 78 inches long, and 16 inches deep. All water and anything dissolved in the water that passes through the lysimeter is collected and analyzed.

The findings of this research could result in changing how putting greens are made. Substituting vitrified clay for sand shows improved soil moisture retention and reduces the leaching of nitrogen. More clay and less sand may be used if the research findings continue to show good results. (Courtesy, Jeff Krans, Mike Goatley, and Jac Varco, Mississippi State University)

Table 20-1. Examples of Forage Grasses

Common Name	Scientific Name	Description/Notation
Bahia grass	*Paspalum notatum*	perennial, warm-season sod grass; deep root system; used for pasture, hay, and turfgrass; adapted to sandy soil; grows 1 to 3 feet tall; grows best in warm climates
Bermudagrass (Common)	*Cynodon dactylon*	perennial, warm-season sod grass; good grazing grass that withstands some drought; grows 4 inches to 1 foot tall; pasture and sometimes hay
Brome grass (several species)	*Bromus* spp.	perennial, warm-season sod adapted to warm or cool, dry climates; popular in northern plains; sod grass; grows 3 to 4 feet tall; pasture
Dallis grass	*Paspalum dilatatum*	perennial, warm-season bunch grass; good grazing grass on bottomland; grows 1 to 4 feet tall; used for pasture
Fescue, Red (several species)	*Festuca rubra*	perennial, cool-season sod forming grass; use for pasture and turf; grows 1 to 2 feet tall
Hairy Grama	*Bouteloua hirsuta*	perennial bunch grass; grows on rocky hills and plains; 0.5 to 1.5 feet tall; used for pasture
Kentucky Blue Grass	*Poa pratensis*	perennial sod grass; slow to establish from seed; shallow root system; widely grown in Midwest and northeastern states; grows 1 to 2.5 feet tall
Ryegrass* (Perennial)	*Lolium perenne*	annual or short-lived perennial; cool-season grass that may be planted in the fall for winter grazing in climates that do not have severe cold; grows 2 to 3 feet tall; not a permanent forage; used for pasture or hay
Timothy	*Phleum pratense*	short-lived perennial; erect growing bunch grass; prefers clay or loam soil; planted in spring for hay harvest in late summer; grows 24 to 40 inches tall in upper Midwest and northeastern United States
Weeping Love Grass	*Eragrostis curvula*	perennial, warm-season dense bunch grass; best in sandy soil; grows 2 to 4 feet tall

*Ryegrass is to be distinguished from rye cultivated for grain (see Chapter 11). Also, two kinds of ryegrass are grown in the United States: perennial and Italian. Italian ryegrass (*Lolium multiflorum*) is an annual grown for hay, pasture, and silage west of the Cascade Mountains and in the southern states.

Temporary forage crops may be planted into permanent pasture. For example, winter grazing crops can be seeded into summer permanent pasture.

Crops grown intensively for hay or silage are often planted on a prepared seedbed. Corn to be used for silage, for example, is planted much as it would be planted for grain.

Soil testing is used to determine the nutrients needed and the pH of the soil. The nutrient requirements of the proposed crop should be considered. Adjustments of pH are best made when a forage is being established in a prepared seedbed. Fertilizer can be applied in a prepared seedbed or as a top dressing on existing forages.

20-10. Ryegrass used as winter grazing that has gone to seed.

Pasture Maintenance

Permanent pasture needs regular maintenance each year. In some cases, fertilizer will be needed to encourage growth, especially with hay crops where several cuttings will be made. Nitrogen is particularly needed for hay grasses.

20-11. Using a flail cutter in pasture maintenance. (Courtesy, Bush Hog Division of Allied Products Corporation, Selma, Alabama)

Weeds and other pests should be controlled. Herbicides may be used to kill weeds. Mowers may be used to cut weeds and, in general, give the pasture a neat appearance. Insects and diseases may be problems with some grasses. Fungus diseases can destroy areas within a pasture.

Irrigation may be needed during short drought periods. It is best to avoid irrigating permanent pasture by selecting drought-tolerant varieties.

20-12. Alfalfa is a high-quality legume forage.

LEGUMES

Legume forage plants fix nitrogen from the air in the soil. They are often planted in mixed grass forages or as pure stands of legumes.

The legumes used as forage plants are known as forbs. A **forb** is a flowering, broad-leaved plant that has a soft stem. Forbs do not have woody stems. Forbs make good plants for pasture, hay, and silage.

Since forage legumes are similar to soybeans and other peas, their growth is much like these crops. (Refer to Chapter 18 for information on the growth of soybeans for more details.)

20-13. White clover (left) and crimson clover (right) may be seeded with grasses in pastures.

Table 20-2. Examples of Forage Legumes

Common Name	Scientific Name	Description/Notations
Alfalfa	*Medicago sativa*	often considered the best forage crop for hay or pasture; grown in Midwest, northeast, and western United States; prefers neutral to alkaline soil; responds to irrigation in dry climates; responds well to fertilizer, especially nitrogen; several cuttings of hay can be made each season with good alfalfa; grows 2 to 3 feet tall; high protein content makes excellent hay
Birdsfoot Trefoil	*Lotus corniculatus*	perennial used for pasture and hay; grown in the upper Midwest and northeastern U.S.; similar to alfalfa in culture, except will tolerate a slightly lower pH; grows 24 to 36 inches tall

Clovers
  Alsike
  Crimson
  Red
  White
  Sweet
	
	Trifolium hybridum
	T. incarnatum
	T. pratense
	T. repens
	Melatotus officianalis
		considerable variation exists among clover species: perennial, cool-season forage; clover prefers fertile soil; can be seeded with grass or in monoculture stands; grows 1 foot tall grown in upper Midwest; grows 2 or more feet tall; colorful yellow flowers; forage uses

Lespedeza (common)	*Kummerowia striata*	annual, warm-season legume for hay and pasture; mostly in the southeastern U.S., though some is grown in Illinois and Iowa; may be grown in rotations with other crops; grows 16 to 20 inches tall; lespedeza will grow reasonably well on eroded, acidic soils; with proper management, lespedeza should reseed itself (Note: Sericea lespedeza is a perennial legume lacking some of the qualities of common lespedeza.)
Vetch, Hairy	*Vicia villosa*	winter cold-hardy annual grown for green manure and forage; tolerant of poorly drained soil; grows 24 to 70 inches tall; flowers and forms obvious seed pods in the spring

Establishing Legume Forages

A well-suited legume forage should be selected. It should be adapted to the climate and to the land conditions. The forage should also serve the intended purpose well, such as grazing or hay. Several excellent forage legumes are available.

Some legumes are planted in prepared seedbeds. Others are seeded into existing grass forages. Alfalfa, in particular, is often grown in pure stands, especially in irrigated areas.

Maximum benefits require that the soil contain adequate rhizobia bacteria. This assures that nitrogen is fixed and that nodules are formed on the roots. Existing soil may have ade-

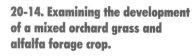

20-14. Examining the development of a mixed orchard grass and alfalfa forage crop.

20-15. Hairy vetch, a winter annual, may be used in temporary pasture and as a cover crop.

quate rhizobia. If no legumes have been on land for a few years, the seed can be inoculated with bacteria prior to planting.

Maintaining Legume Forages

Maintaining legumes requires routine attention to the growing conditions. Weeds and other pests need to be managed. Irrigation may be needed. Harvesting and grazing should be regulated to assure that the annual crops reseed themselves. Mowing before the seeds have matured may prevent reseeding.

20-16. Cages for leaf cutter bees may be used in alfalfa and sweet clover fields to promote pollination. This cage is in Montana. Pollination by bees is essential with some crops.

Some legumes are more tolerant of other plants. Weed control is important with those that are not, such as alfalfa. Regular fertilization is needed to assure good growth. Some require additional nitrogen, though nitrogen is fixed by the legumes.

HAY MAKING

Hay provides an economical source of nutrients for livestock. Hay making is cutting, curing, and storing hay crops for high-quality, nutritious feed. Good-quality hay is green, fine stemmed, free of weeds, and cut before full maturity. Hay should be cured to preserve as much green color as possible in the leaves and stems of the hay plants. Good-quality hay requires good-quality forage—hay can be no better than the crop that is harvested.

Most any of the legume and grass forages can be used for hay. In addition, soybeans and cereal grains are also used for hay. Mixtures of grasses and legumes are commonly used for hay. A few species are kept as pure as possible, such as alfalfa.

Hay should be cut at the growth stage of the plant to produce the largest amount of the

20-17. Quality hay with a good color and tender stems and leaves.

20-18. A combination mower and conditioner leaves cut hay in windrows to facilitate baling. (Courtesy, New Holland North America, Inc.)

20-19. Hay may be formed into large square bales. (Courtesy, Case Corporation)

highest-quality hay. Very young, tender plants may be the most nutritious hay, but the amount produced is low. In general, hay is cut in the bloom stage, or shortly thereafter, before the plant has begun to dry. With some crops, such as Bermudagrass and alfalfa, several cuttings may be made a year.

Curing is the process of drying hay to a moisture level that makes it safe for storage. Hay that is stored with too much moisture can mildew or create a fire by spontaneous combustion. Hay should not be cut soon after a rain. The hay should be baled or stored after curing while it has a sweet aroma. Hay with mildew has a sour smell. The curing and baling process should keep as many leaves as possible. Leaves are high in nutrients.

Cut hay is conditioned to speed curing. This involves crushing stems to allow moisture to dry out more rapidly. About 30 hours of sunshine are needed to cure hay. In some cases, artificial drying is used and this shortens the amount of time in the field. Many growers now use machines called haybines® that cut, condition, and windrow with one pass over the field.

Cured hay may be packed into small or large bales. These bales should be stored so they do not get rained on. In the dry, southwestern area of the United States, hay can be stored without a cover. Elsewhere, protecting hay from rain greatly increases its nutritional quality.

TURF

Turf includes some of the same species of grasses as the forage crops. Special turf varieties are often used. Several of the important turfgrasses are not used as forage crops. A good visual characteristic is essential with turf.

ESTABLISHING TURF

Good turf requires careful attention in establishment. Turf may be seeded or vegetatively propagated. Some species and varieties do not seed well. Sodding or sprigging is needed to get the turf established.

20-20. Sod being installed on the grounds of the Washington Monument with the U.S. Capitol Building in the background.

Sod is the layer of the turf including the stems, leaves, and roots of the plants. It is often used in propagating new turf areas. Sod is dug in rolls or squares. In some cases, it may be produced in small plugs that are set out and allowed to grow to cover the lawn area.

Soil Preparation

Preparing soil to plant turf includes testing the soil for nutrients and pH, adding amendments, and tilling the soil for planting.

Soil samples should be taken and analyzed by a reputable laboratory. Fertilizer and lime may be required. These should be applied as part of tilling the soil.

20-21. Fescue may be seeded in the fall or spring in some parts of the United States.

Soil amendments include materials worked into the soil to improve its tilth and make it more productive. In many cases, materials high in organic matter are needed. Composted cow manure, leaf mold, and saw dust are examples of products used. Sand may be added to soil high in clay.

Plows, rotary tillers, and harrows can be used for tilling. The existing grass cover will need to be crushed and broken apart. In some cases, the roots will need to be allowed to dry and die before continuing with planting. Also a part of tilling is smoothing the surface and establishing contours needed to give aesthetic appeal and manage water.

20-22. A rotary tiller may be used to prepare the seedbed for turf. (Courtesy, Bush Hog Division of Allied Products Corporation, Selma, Alabama)

Seeds are normally broadcast over the area. In most cases, watering is used to cover the seed. Raking or harrowing may result in the seed being covered too deeply. Seedbed preparation is needed with sodding and sprigging to assure a smooth surface and that the grass has a good environment for root growth.

Laser leveling may be used to properly shape athletic fields. Good drainage is a part of preparation. This includes using tile drainage systems and soil amendments, such as sand, to promote excess water removal.

20-23. A laser-guided leveling system is being used to prepare the surface of a soccer field.

20-24. A layer of coarse river sand has been placed on top of the base soil. The sand promotes internal and surface drainage for the athletic field turf that is being established.

20-25. A layer of burlap over soil is being fastened with a large staple. Sod is installed on top of the burlap.

20-26. A large roll of sod is being installed on top of burlap using low-compaction crawler equipment developed by Woerner Turf farms in Alabama.

Variety Selection

Variety selection is choosing the grass that will be planted. The grass chosen should be adapted to the general climate in the area and the specific site where it will be grown. For example, if the lawn area has trees and is shaded, a variety tolerant of shade will be needed.

20-27. A pallet of cut fescue sod is ready for installation.

20-28. Cut sod is being placed on prepared soil.

Table 20-3. Examples of Common Turfgrasses

Common Name	Scientific Name	Description/Notations
Grasses for warmer climates:		

Bermudagrasses
  Common
  Bradley
  Megennis
  Others
	
	Cynodon dactylon
	C. bradleyi
	C. magennisii
		specialized varieties and hybrids make it popular; fine-textured; established by seed or sod, depending on the kind; often used on athletic fields and home lawns; prefers full sun; mow at a height of 0.5 to 1.5 inches, except for varieties on golf greens that can be mowed shorter

Centipedegrass	*Eremochloa ophiuroides*	medium-textured grass; adapted to a wide range of soil conditions; minimal maintenance needed; mow 1 to 2 inches high
Saint Augustine	*Stenotaphrum secundatum*	coarse-textured and aggressive growing; prefers well-drained sandy soil; not cold tolerant; will tolerate shade; subject to fungus diseases; easy to propagate with sprigs or runners; mow at a height of 2.5 to 3.5 inches
Zoysia	*Zoysia japonica*	medium-textured grass; grows slowly; will tolerate some cold; mow with a reel mower at heights of 0.5 to 1 inch
Grasses for cooler climates:		
Bluegrass Kentucky	*Poa pratensis*	medium-textured grass; often established on lawns with sod; prefers moist, temperate climates; mow at a height of 2.5 to 3.5 inches
Bent grass	*Agrostis palustris*	common on closely cut areas, such as putting greens; grows by creeping along the ground; not a good home lawn grass

Fescue
  Fine
  Tall
	
	Festuca rubra
	F. arundinacea
		popular grasses; fine textured; tolerant of some shade; mowing height depends on variety, with fine fescue mowed 1.5 to 2.5 inches high, and tall fescue never mowed below 3 inches high

Perennial Ryegrass	*Lolium perenne*	widely used in home lawns; difficult to have good appearance after mowing; mow at a height of 1.5 to 2.5 inches

Information about turf varieties is available from the land-grant university in your state or a local nursery.

MAINTAINING TURF

Common maintenance activities include mowing, watering, fertilizing, and managing pests. Dethatching, aerating, and rolling are needed on some lawns.

Mowing

Mowing controls the height of the grass. Always mow at the correct height for the particular species. Mowing too short can injure or kill the grass. Most lawns perform better if mowed at a height of 2 to 3 inches. Rotary or reel mowers may be used. The blade should always be sharp. The mower should be safely operated.

In some cases, mowing involves collecting and removing clippings. **Clippings** are the pieces of grass blade that are cut off. Clippings left on a lawn may encourage disease and detract from the quality of the recently cut lawn. Clippings that remain on the lawn eventually filter to the soil level, decay, and return fertility to the soil. Clippings can be caught in a large bag and composted for future use.

20-29. Mowing maintains the desired height of turf. A vacuum attachment collects clippings.

20-30. An improperly mowed Bermudagrass lawn shows damage to the turf and an unattractive appearance.

Irrigation

Lawn grasses require much water to have an attractive appearance. In dry locations, irrigation is essential. It is a good idea to select a variety of grass that is compatible with the natural precipitation in the area if a person does not plan to irrigate.

Water should be used only when needed. Most lawns are irrigated with sprinkler systems. The soil should be soaked to a depth of 6 inches every few days. Watering lightly each day probably does more harm than good.

20-31. Irrigation is important in maintaining turf in dry climates.

20-32. Bermudagrass damaged by cold weather.

Fertilizing

Once a lawn is established, periodic applications of fertilizer are needed. This will keep the grass healthy and growing. Nitrogen is often needed to produce growth and a nice green color.

Recommendations on lawn fertilization should be followed. In many cases, a light application of a complete fertilizer is used in the fall on perennial grasses. Heavier applications and increased nitrogen are used in the spring. With winter grasses, fall fertilization and late winter fertilization are used.

All fertilizer should be evenly broadcast over the entire area of the lawn.

Managing Pests

Weeds, insects, diseases, and other pests may need to be managed. Lawns should be regularly checked for any problems. Weeds should be controlled before they become established in a lawn. A selective herbicide can be used in many grasses on broadleaf weeds. Insects and diseases should be properly managed using approved pesticides.

Small animals are sometimes pests in lawns. Moles, squirrels, crawfish, and other animals may burrow or dig into the soil. They may leave small holes or mounds that interfere with mowing and appearance. These can be managed by trapping or using eradication chemicals. In some cases, people prefer to see the squirrels and do not feel that the damage is a problem.

Other Practices

Dethatching, aerating, and rolling are needed only occasionally in some lawns.

Thatch is the accumulated grass stems, leaves, and roots in the turf. Dethatching is used to remove these materials. Large rakes or dethatching machines are used to remove thatch.

20-33. Aerating a lawn improves soil oxygen and moisture percolation.

Aeration is making holes or openings in the soil to allow air to enter. It is sometimes used to help control thatch. The holes are about 0.5 inch in diameter and 2 to 10 inches deep. Long tines push into the soil and remove cores from the holes. The cores are left on the surface and broken up to return to the soil.

Rolling is used in the spring on lawns in areas where the soil often freezes and thaws during the winter. Rolling smooths the surface. It should be used wisely to prevent unwanted soil compaction.

20-34. Small cores are cut and lifted from turf during aeration.

REVIEWING

MAIN IDEAS

Forage and turf are used in important ways. Forage is vegetation used as feed for animals. Pasture, hay, and silage are forages. Turf is short-growing sod used for functional and aesthetic purposes.

Forages may be grasses or legumes. Understanding how grasses grow is important in management. Cutting or performing other activities at the wrong stage of growth can result in serious damage to the forage crop. Many pastures are mixes of grasses and legumes, such as Bermudagrass and white clover. Legumes have the advantage of producing feedstuffs with higher protein content.

Forages should be carefully made into hay. Proper curing and storage are needed to retain important nutrients. Hay that gets rained on rapidly deteriorates in quality.

Turfgrasses are carefully maintained for appearance and function. Athletic fields, lawns, and similar locations are made useful and attractive with good turf. Sod is the layer of turf including the roots of plants. Many improvements have been made in turf species and varieties. These should be carefully selected and established to meet use and climate needs. Once in place, maintenance is needed to assure good growth and appearance.

QUESTIONS

Answer the following questions using complete sentences and correct spelling.

1. What is forage? What forms are used?

2. What are the major parts of a grass plant? (Use a drawing to label the parts.)

3. Why are tillering, jointing, and heading important in forage and turf maintenance?

4. Name and briefly describe three grass and three legume forages.

5. How is hay made?

6. What are the major steps in establishing turf?

7. What are the important activities in maintaining turf?

8. How should grass clippings be handled?

EVALUATING

Match the term with the correct definition. Write the letter by the term in the blank that is provided.

a. forage   e. grass   i. sod
b. turf   f. blade   j. thatch
c. hay   g. heading
d. pasture   h. curing

1. _____ plant material that has been cut and dried for feed

2. _____ accumulated grass stems, leaves, and roots in turf

3. _____ formation of seed heads on a grass plant

4. _____ layer of turf including roots

5. _____ plant with parallel veins and hollow or solid stems

6. _____ drying hay to a moisture level safe for storage

7. _____ vegetation fed to livestock

8. _____ plant materials growing on land where cattle graze

9. _____ grass leaf

10. _____ short-growing matted grasses maintained for aesthetic, recreational, or functional purposes

EXPLORING

1. Take a field trip to a sod farm. Determine the kind of sod produced and the cultural practices that are followed. Observe the digging of sod and how it is transported. Make photographs or digital images of all phases of the sod farm operation. Prepare a bulletin board or computer-based presentation on your observations.

2. Select a lawn or playing field and analyze the procedures needed to improve the turf. Determine if major renovation is needed. Recommend practices to improve the turf. Prepare a written report on your findings.

3. Obtain brochures or other materials from the land-grant university in your state on recommendations for forage and turf varieties, fertilization, pest control, and maintenance. Summarize your findings into a written report.

CHAPTER 21

Vegetable, Fruit, and Nut Crops

OBJECTIVES

This chapter covers the fundamentals of vegetable, fruit, and nut crop production. It has the following objectives:

1 Name important vegetable, fruit, and nut crops and classify by use and climate requirements.

2 Describe how to select vegetable, fruit, and nut crops.

3 Explain important cultural requirements for vegetable, fruit, and nut crops.

4 Describe the role of technology in vegetable, fruit, and nut production.

5 Explain harvesting and postharvest management of vegetable, fruit, and nut crops.

TERMS

cane fruit
cool-season vegetable
dwarf
nut
orchard

pomologist
pruning
semi-dwarf
standard
tree fruit

vegetable
vine fruit
warm-season vegetable

21-1. Many kinds of vegetables, fruits, and nuts are consumed in North America. (Courtesy, Agricultural Research Service, USDA)

"**E**AT your vegetables!" "Why don't you have a piece of fruit for dessert?" This advice seems to come naturally for parents and it is supported by the United States Department of Agriculture's Food Guide Pyramid. Vegetables, fruits, and nuts are good for you. In fact, the pyramid recommends that everyone eat three to five servings of vegetables and two to four servings of fruit every day.

When you consider that the population of the United States is more than 284 million people, that is a tremendous demand for vegetables and fruits. The value of vegetables grown each year in the United States alone totals more than $10 billion.

VEGETABLES, FRUITS, AND NUTS DEFINED

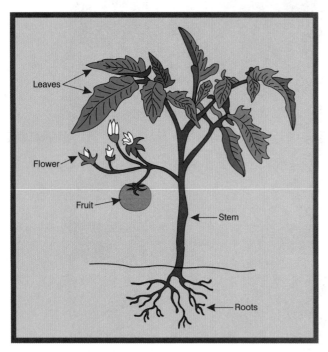

21-2. Major parts of a tomato plant.

People often have difficulty when trying to define the differences of vegetables, fruits, and nuts. Do you get confused on this issue? Is a tomato a fruit? Is a strawberry a vegetable? Is a walnut a fruit? What about green beans, kiwi, pears, muskmelons, peanuts, or sweet corn? Technically, these are all fruits. You may recall from Chapter 3 that a fruit is a ripened ovary. The ovary also contains the plant's seeds. However, in industry, vegetables, fruits, and nuts are defined differently. Some accepted definitions follow:

- ***Vegetable***—an herbaceous plant or plant part that is eaten raw or cooked during the principal part of the meal rather than as a dessert (USDA).

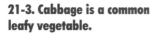
21-3. Cabbage is a common leafy vegetable.

21-5. The walnut is a popular nut crop.

21-4. The apple is an important fruit crop.

- Fruit—a fleshy, ripened ovary of a tree, shrub, or woody vine eaten raw or cooked.

- **Nut**—a hard, bony, one-seeded fruit of a woody plant.

VEGETABLES

Production of vegetables has sustained populations in North America for thousands of years. Native Americans grew squash, beans, and corn to supplement their diets. Colonists maintained vegetable gardens as a source of fresh food. Home vegetable gardening remains one of the most popular leisure activities in the United States. Early commercial vegetable farms grew many types of vegetables. As the United States developed, individual producers became more specialized and grew fewer types of vegetable crops. Today, vegetable production is a multibillion dollar industry.

The types of vegetables number in the hundreds. Each type of vegetable plant has evolved certain characteristics and has become adapted to grow in specific environmental conditions. Because of the diversity of vegetables, there is a need to classify vegetables. They are often classified by botanical family, edible parts, hardiness, or life cycle. People who classify vegetables and study vegetable production work in a field called olericulture.

Today, specific kinds of vegetables are grown commercially where the climate, water requirements, and soil are ideal for their growth. Areas of California have ideal conditions

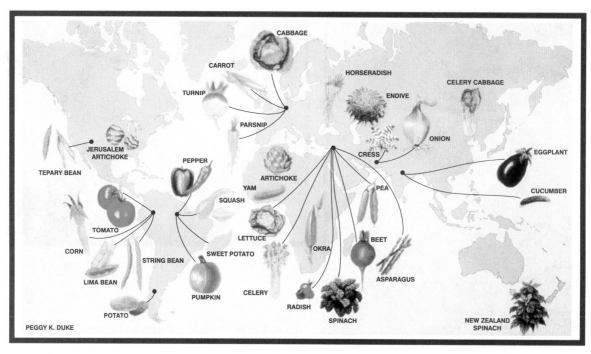

21-6. Common vegetables originated in many parts of the world. (Courtesy, Agricultural Research Service)

for many crops. As a result, California produces roughly half of all the vegetables grown in the United States. Other states that produce large quantities of vegetables are Florida, Arizona, Texas, and Oregon.

21-7. Supermarkets display numerous fresh vegetables.

VEGETABLE CROPS

Vegetables are grown for fresh market use or for processing. Supermarkets display an abundant variety of fresh vegetables. One can also find processed vegetables in cans and jars or in the frozen food section.

Eleven vegetables are grown mainly for the fresh market. These include Brussels sprouts, celery, eggplants, escarole, garlic, honeydews, lettuce, muskmelons, onions, green peppers, and watermelons.

In commercial production, twelve crops are commonly used for both the

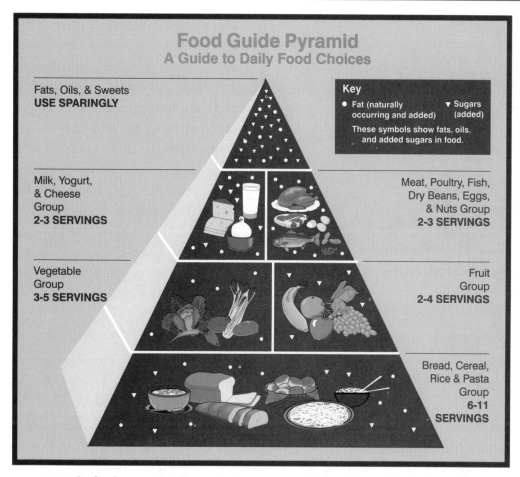

21-8. The food pyramid indicates the importance of vegetables and fruits in our diet.

fresh market and processing. They are asparagus, broccoli, cabbage, carrots, cauliflower, cucumbers, potatoes, snap beans, spinach, sweet corn, sweet potatoes, and tomatoes.

Three other vegetables are almost always grown with processing (canning or freezing) in mind. These are lima beans, beets, and green beans.

Vegetables are an essential part of a person's balanced diet. They provide us with vitamins, minerals, and roughage that keep us healthy. It is interesting to consider the different parts of a plant

21-9. A field of potatoes, cool-season vegetables, is being harvested in Colorado. (Courtesy, U.S. Department of Agriculture)

21-10. The watermelon is a popular vegetable fruit.

that we eat. A vegetable can be found for every organ of a plant. Carrots, radishes, beets, sweet potatoes, and horseradish are roots. Asparagus, potatoes (tubers), and yams (tubers) are stems. Lettuce, celery (petiole), spinach, chives, Brussels sprouts (buds), and onions (bulbs) are technically leaves. Broccoli, cauliflower, and artichokes are immature flowers. Cucumbers, eggplants, peas (seeds), snap beans, and sweet corn (seeds) are immature fruits. Mature fruits we eat include muskmelons, peppers, tomatoes, and watermelons.

VEGETABLE PRODUCTION

Vegetable production is greatly influenced by climate. Temperature is probably the most important climatic factor for vegetables. In fact, seasonal temperatures really determine the location in which most vegetables are grown. Broccoli, lettuce, onions, potatoes, asparagus, and spinach perform best in cool growing conditions. Vegetable plants that can withstand a frost are often called *cool-season vegetables*. On the other hand, tomatoes, peppers, melons, and squash like warm temperatures and would be killed by a frost. These are sometimes referred to as *warm-season vegetables*.

21-11. Pumpkins are warm-season vegetables. The vines die with the first heavy frost.

Rainfall is extremely important for vegetable production. It is particularly important in areas where irrigation is not commonly used, such as the Northeast. In the arid southwestern states, irrigation is required. Without irrigation the plants would not grow. The cost of irrigation in the Southwest is offset by high yields.

A third factor that influences which vegetables can be grown in a specific area is soil. Loose, well-drained, loam soils with 3 to 5 percent organic matter content are the best for most vegetables. High fertility is also desirable since vegetable crops are fast-growing plants with limited root systems. Some vegetables have specific soil requirements. Soils with a very high percentage of organic matter, or those with a high peat content, are required for onions, celery, and lettuce. Loose soils are essential for vegetable crops produced below the soil level, such as potatoes, carrots, and sweet potatoes. Heavy clay soils that compact easily and restrict air exchange limit vegetable growth.

21-12. Green beans are produced commercially and in home gardens. (Courtesy, U.S. Department of Agriculture)

Propagation

Most vegetable plants are propagated by seed. The seeds of some plants are directly sown in the field. Other vegetable plant seeds are started in a cold frame or greenhouse. Plants started in a cold frame or greenhouse are transplanted to the field when weather conditions are right and the plants have shown sufficient growth. Some vegetable plants, such as asparagus, are propagated asexually through means such as division.

CAREER PROFILE

VEGETABLE PRODUCER

Vegetable producers are involved in growing vegetable crops. They may be specialists in one type of plant, or they may grow many different types of vegetables in one season. Vegetable producers must be

knowledgeable in areas including business management, computer science, plant science, soil science, and integrated pest management.

A vegetable producer usually has formal education beyond the high school level. A college degree in agronomy, horticulture, or a related field is useful. Job opportunities as a vegetable producer are somewhat limited. Family-run businesses provide employment. Other opportunities are found on vegetable operations owned and operated by large corporations.

This photograph shows a tomato producer checking his plants. (Courtesy, U.S. Department of Agriculture)

Vegetable Production Has Changed

The number and size of vegetable farms have changed over the past 40 years. There has been a drop in the number of farms producing vegetables. At the same time, the size of vegetable farms has increased. The result is that there are fewer farms with about the same number of acres under production and yields have increased.

Advancements in the industry have lead to an increase in production. A number of factors account for the increase:

1. Plant breeders and genetic engineers have introduced new varieties of plants that perform better than older varieties.

2. The expanded use of fertilizers and improved methods of application have been beneficial.

3. Integrated pest management (IPM) has reduced losses caused by insect, weed, and disease pests.

4. Machines used for planting, maintaining, and harvesting vegetable crops have been improved to permit better timing of tasks and less damage to the crops.

5. Irrigation practices have improved the health and yields of vegetable plants.

6. An improved transportation system has allowed vegetables to be grown in parts of the country where the climate and soil are best suited for production.

FRUITS

We are all familiar with fruit crops. Apples, grapes, peaches, strawberries, and bananas are part of our everyday lives. Some people have made it their career to grow fruits, classify fruits, or conduct research on fruits. These people are **pomologists**. The field in which they work is called pomology.

Fruits are grown on different types of woody plants. The type of plant on which fruit is produced is used to classify fruits. Three main categories are tree fruits, small bush or cane fruits, and vine fruits. Numerous kinds of fruit can be found in each category.

*T*REE FRUITS

Some of the most common **tree fruits** are apples, oranges, lemons, grapefruit, cherries, pears, apricots, peaches, and plums. They are eaten fresh or processed for use as juices, jams,

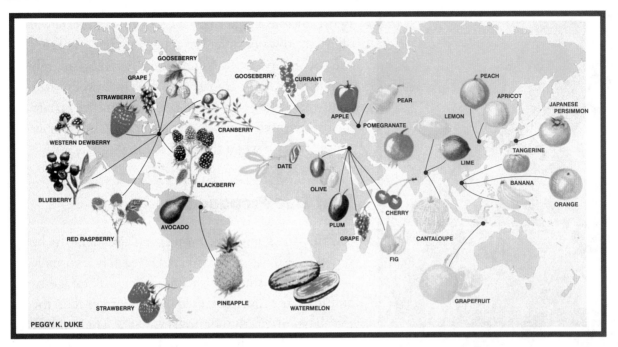

21-13. The origin of fruits can be traced to places around the world. (Courtesy Agricultural Research Service, USDA)

jellies, dried fruit, or canned fruit. The producers of tree fruits grow many trees in an area called an **orchard**.

Where Fruits Are Grown

Where tree fruits are grown depends largely on the climate. Fruit tree flowers are easily damaged by late frosts in the spring. Damaged flowers result in a poor yield of fruit. Peaches, cherries, and apricots are particularly sensitive to late frosts that can damage the flower buds or flowers. Some fruit trees, such as the citrus crops, including oranges, lemons, and grapefruits, can only be grown in the warm southern states.

Pollination of Flowers

Some trees must be cross-pollinated with pollen from another tree to produce abundant fruit. For example, the Red Delicious apple and the Jonathon apple are poor self-pollinators. Golden Delicious apple trees, on the other hand, are excellent pollinators for Jonathon and

21-14. Apples and pears populate these orchards in Yakima, Washington. (Courtesy, Agricultural Research Service, USDA)

21-15. Cherries are tree fruits.

Red Delicious apple trees. Therefore, producers will alternate rows of different apple varieties to ensure the necessary cross-pollination. Pomologists rely heavily on bees to transfer the pollen between their fruit trees. They often keep bee hives in the orchard for this purpose or pay to have someone bring hives to the orchard when the trees are in flower.

Fruit Tree Propagation

Fruit trees must be propagated asexually to be sure the desired varieties are obtained. This is usually accomplished through grafting. One advantage to grafting is the rootstock used can be hardier than the root system of the desired variety. The hardier rootstock allows the growth of the tree in colder climates with less chance of damage.

The rootstocks can also be selected to dwarf the growth of the tree. A normal size tree is called a *standard*. By grafting the variety on certain rootstocks, the growth of the tree is slowed. Grafting on selected rootstocks can produce *semi-dwarf* trees, about one-half the size of a standard. Other rootstocks produce *dwarf* trees less than one-third

21-16. Figs grow best in climates with hot dry summers and moist cool winters.

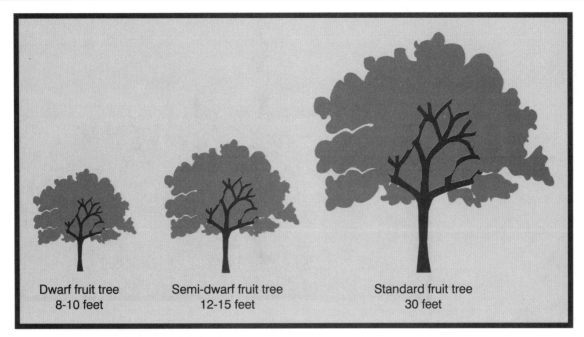

Dwarf fruit tree
8-10 feet

Semi-dwarf fruit tree
12-15 feet

Standard fruit tree
30 feet

21-17. The size of the fruit tree can be limited by grafting onto dwarf rootstocks.

the size of a standard. Of course, the advantage of the dwarfed trees is the fruit is easier to harvest. Dwarf fruit trees are very popular with homeowners with small yards.

Care and Maintenance of Trees

Each type of fruit tree has its own requirements for care and mainte-nance. In general, fruit trees perform well when planted in a good, well-drained soil. At planting, the trees are evenly spaced. Even spac-ing reduces competition among trees for sunlight, moisture, and minerals. It also makes it easier for the pro-ducer to care for the trees and to harvest fruit.

As the young trees mature, they are pruned to develop a strong framework of branches. **Pruning** is a selective removal of branches. The purpose is to develop a framework of

21-18. Fruit trees are evenly spaced to improve yields, care for the trees, and harvest the fruit.

21-19. These trees have been severely pruned to promote fruiting.

21-20. The branches of these young apple trees are being trained to improve the branching structure.

strong branches that will support the heavy load of ripening fruit. Well-spaced branches also allow easier harvest of fruit and reduce some disease problems. Each type of fruit tree is pruned in a different manner.

Young fruits are often thinned or pruned before they develop. This practice allows the remaining fruit to benefit from the sugars produced by the plant. The result is larger individual fruits. Depending on the type of fruit tree, thinning is accomplished by spraying plant hormones, or an insecticide called carbaryl, on the tree or by physically removing young fruit.

Pest Management

People have come to expect high-quality, unblemished fruit in grocery stores. The expected quality is only possible if pests are limited. Any of a number of pests can seriously damage individual fruit or entire crops. Insect pests attack and damage fruit, eat tree leaves, suck sap from leaves, spread diseases, and burrow into the wood. Bacterial, fungal, and viral diseases cause plant stress and disfigure fruit. Rabbits, deer, and mice can also inflict damage to the trees by chewing bark and eating buds.

Producers use an integrated pest management (IPM) plan to control pests. Managing pests is both time consuming and costly to the producer. While implementing IPM, care is taken to properly use pesticides to eliminate dangers to the consumer. Precautions are also taken when using insecticides to protect the bee populations. Another recommended practice is to keep the area around the tree trunk free from grass and weeds to reduce the damage from mice.

Harvest and Storage of Fruit

When fruit reaches maturity on the tree, it is time to harvest. Harvests are done by hand or by machines, depending on the type of fruit and the producer. During the process, care is taken not to bruise or damage the fruit. Harvested fruit is stored in coolers that slow the life processes in the cells of the fruit. In effect, the slowing of the processes slows the ripening and ultimate death of the fruit. Any damaged or overripe fruit is removed from the

21-21. Breeding programs can improve disease resistance, such as this fireblight resistant pear. (Courtesy, Agricultural Research Service, USDA)

CONNECTION

NEW EQUIPMENT IMPROVES HARVESTS

Harvesting abundant fruit, vegetables, or nuts can be very labor intensive. New equipment has been designed by agricultural engineers to mechanically harvest crops. With the equipment, crops are harvested more quickly, and less labor is required. Shown in the photograph are two people examining citrus fruit harvested by a canopy shaker. The machine shakes citrus from the tree branches.

Continued improvements in technology will result in improved harvest systems. This helps consumers have better quality products at reasonable costs. (Courtesy, Agricultural Research Service, USDA)

21-22. Strawberries (left), blueberries (lower right), and blackberries (top) fall in the category of cane fruits or small fruits. (Courtesy, Agricultural Research Service, USDA)

healthy fruit to limit the level of ethylene gas that speeds ripening.

BUSH OR CANE FRUITS

Bush or **cane fruits** are often referred to as small fruit crops because the fruit is produced on small perennial plants, never trees. The plants grow either low to the ground or 4 to 6 feet in height. They require less care than tree fruits, and they start producing fruit in a much shorter period of time. Some of the most common bush fruits are blueberries, blackberries, currants, gooseberries, raspberries, cranberries, and strawberries.

Growing Conditions

As with other crops, climate and soil conditions dictate which small fruits can be grown successfully in a certain area. A loose, well-drained, organic soil is best for these fruits. All of the small fruits grow well in a soil with a pH of 5.5 to 7.5, with the exception of blueberries. Blueberries require a pH of 4.2 to 5.2 for best

21-23. These raspberries are on a U-pick operation.

21-24. Strawberries produce runners that are easily propagated.

21-25. Because they are easily damaged, blackberries are harvested by hand. (Courtesy, Agricultural Research Service, USDA)

growth. States along the Pacific Coast and the East Coast, and some northern states, are home to the largest commercial producers of small fruits. However, the size of the plants and the relative ease of growing has made small fruits popular with home-owners.

Propagation

Small fruits are propagated asexually by using vegetative portions of a plant. Asexual propagation ensures that the exact same variety or cultivar will be produced. Blackberries and gooseberries are propagated by layering. Blueberries and currants are generally propagated by hardwood cuttings. Raspberries are usually propagated by division. Strawberries produce runners with a small plant that is removed and planted.

21-26. Machines are used in the harvest of these New Jersey cranberries. (Courtesy, Agricultural Research Service, USDA)

Vine Fruits

Vine fruits grow on a woody plant that requires support. The most important vine fruit is the grape. They have been in cultivation for thousands of years. Grapes are grown throughout the United States. Three main classes of grapes are grown. Each is adapted to growing conditions in different parts of the country. From them, we get juices and wines, jellies and jams, and raisins. Grapes are grown by large commercial growers and homeowners.

Grapes require room to grow and they require special care. In general, grapes are adaptable to many types of soil. Because they are a vine crop, they must be provided some type of support. They need to be heavily pruned every year for the best production. They also need to be fertilized and pests need to be controlled.

21-27. Grapes are the most common vine fruit. (Courtesy, Agricultural Research Service, USDA)

NUTS

Nut-bearing trees and shrubs are grown across the United States. Some of the species from which nuts are harvested are walnut, pecan, chestnut, almond, filbert, hazelnut, macadamia, and coconut. Most of these are commercially produced in certain parts of the country in large orchards. Some nut-bearing plants are used in the home landscape, but protecting the ripening nuts from wildlife is difficult. Peanuts are actually like a pea and grow on an herbaceous plant.

Climate

As with vegetable and fruit crops, the climate and soil type determine where nut culture occurs. Winter cold, spring frosts, early fall freezes, droughts, wet humid weather, and the length of a frost-free growing season affect nut production.

PROPAGATION

Propagation of nut crops is usually done by seed or by grafting. In orchards, the plants are evenly spaced in rows. The spacing is often quite far apart as trees, such as the walnut, hickory, and pecan, can grow very large. Different tree varieties are interplanted to ensure good cross-pollination. As the plants grow, they are pruned to develop a desirable shape and branching structure.

PESTS

The nut crops need to be protected from a multitude of pests. Insect and disease pests can be very damaging and methods of control must be applied. Birds and squirrels also reduce nut harvests.

HARVEST AND STORAGE

Nuts are harvested by hand from the plant, picked up from the ground, or harvested with the aid of machinery. Machines are used with some nut trees to shake the nuts loose from the limbs. Once fallen, the nuts are collected. They are then allowed to air dry before being placed in storage. Storage involves refrigeration.

21-28. Pecan flowers are elongated structures known as catkins.

21-29. Coconuts grow in tropical climates. (Courtesy, Agricultural Research Service, USDA)

REVIEWING

MAIN IDEAS

Vegetables, fruits, and nuts are important parts of a good diet. These foods are produced across the United States as part of a huge multibillion dollar industry. Where a crop is grown is based mostly on the climate of a region. Soil type is also an important consideration.

The difference between vegetables, fruits, and nuts is sometimes confusing. However, a simple definition can be given to each. Vegetables are herbaceous plants or plant parts eaten raw or cooked. Fruits are ripened ovaries of woody plants eaten raw or cooked. Nuts are hard, bony, one-seeded fruits of woody plants.

The production or study of vegetables is called olericulture. Producers grow vegetables for sale as a fresh product or processed food. Advancements in the industry have greatly increased the production of vegetables. There are better varieties, improved fertilizer practices, better pest control, more efficient machinery, elaborate irrigation systems, and superior roads. Propagation is usually by seed.

Tree fruits are grown in orchards. Flowers must be pollinated to produce fruit, and bees do much of the pollinating. Young fruit trees are pruned to develop a strong framework of branches that will hold fruit. Bush or cane fruits grow low to the ground. Fruit-bearing plants are generally propagated by grafting, cuttings, or division.

Nuts are produced on woody plants. Most nuts produced for commercial purposes are harvested in orchards. Cross-pollination is necessary, and pruning advisable to develop a strong branching structure.

QUESTIONS

Answer the following questions using complete sentences and correct spelling.

1. What are the differences in vegetables, fruits, and nuts?
2. What determines where crops are grown?
3. What plant organs are eaten as vegetables? Give examples for each.
4. What advancements have caused an increase in vegetable production?
5. What are the major tree fruits?
6. Why is pollination of fruit trees important?
7. How are fruit trees propagated?
8. How do tree fruits differ from bush or cane fruits and vine fruits?
9. What are the major nut crops grown in the United States?
10. Why are fruit and nut trees pruned?

EVALUATING

Match the term with the correct definition. Place the letter by the term in the blank that is provided.

a. nut
b. orchard
c. pruning
d. semi-dwarf

e. standard
f. olericulture
g. vegetable
h. pomology

i. fruit
j. cool-season vegetable

1. _____ a ripened ovary

2. _____ selective removal of branches

3. _____ an acreage of fruit or nut trees

4. _____ the study of vegetables or the production of vegetables

5. _____ an herbaceous plant or plant part eaten raw or cooked

6. _____ vegetables that can withstand a frost

7. _____ the study or production of fruit

8. _____ a normal-size tree

9. _____ a hard, bony, one-seeded fruit of a woody plant

10. _____ trees about one-half the normal size

EXPLORING

1. Look at the fruits and vegetables in your home. Identify the different plant structures that we eat. Go to a grocery store where there is a wide selection of vegetables and do the same.

2. Plan and grow a vegetable garden at home or on the school grounds as a class project. In the planning stage, poll people as to what kinds of vegetables they like. If it is to be a school garden, poll all the students. Determine if your climate and soil will support the growth of all the vegetables mentioned in the survey.

CHAPTER 22

Ornamental Crops

OBJECTIVES

This chapter covers the fundamentals of vegetable, fruit, and nut crop production. It has the following objectives:

1. Explain ornamental horticulture.
2. Describe floriculture and how it is carried out.
3. Explain landscape horticulture and identify major areas in landscaping.

TERMS

balled and burlapped (B&B)
bare root (BR)
bedding plant
containerized
cut flower
foliage plant
greenhouse
ground cover
hardscapes
nursery
potted flowering plant
shrub
transplanting
tree
vine

22-1. Throughout history, people have enjoyed sitting in pleasant landscapes.

PLANTS may be grown simply for the pleasure they give. People enjoy a beautiful flower, a stately 200-year-old oak, and soft green grass beneath their feet. This area of agriculture that involves the growing of plants for their appearance is called ornamental horticulture.

Plants are part of our everyday lives. We tend to surround our homes with plants in the form of landscaping. We grow house plants indoors. We give flowers for special occasions. Enough people like plants that gardening is considered the number one hobby in America.

Ornamental horticulture is also one of the fastest growing fields in the agriculture industry. Several factors account for this. An increasing number of people have a desire to "get in touch" with nature. Our society has given people a greater amount of leisure time. A third factor is the additional money people have and are willing to spend on plants. The result is a growing industry with a need for well-trained employees.

ORNAMENTAL HORTICULTURE

22-2. Floriculture is the production and use of flowers.

Ornamental horticulture is growing and using plants for decorative purposes. It includes producing many different kinds of plants, breeding and propagating plants, and arranging plants in artistic ways. Markets for ornamental crops have grown as more people have moved to cities and towns. Also, when the economy is strong, people have more money to spend on luxury items, such as ornamental crops. Although it is in the urban areas where many crops are sold, much of the production occurs in rural areas.

The ornamental horticulture field can be divided into two main areas. One of the two is floriculture, which involves the production and use of flowers. Floriculture businesses include florist shops, floral supply companies, and production greenhouses. The second area, referred to as landscape horticulture, involves the growing and use of woody and herbaceous plants for landscaping. Some businesses associated with this area are landscape construction companies, garden centers, nurseries, golf courses, and public gardens.

22-3. Landscape horticulture involves woody and herbaceous plants, physical structures, and reshaping the landscape.

FLORICULTURE

The literal definition of floriculture is the "culture of flowers." Floriculture is an international, multibillion dollar industry based on flowering plants and foliage (leafy) plants. The floriculture business involves the production of the floral crops, the distribution of the crops from the grower to the consumer, and the processing of crops before sale.

FLORAL PRODUCTION

Many floriculture crops are produced in greenhouses. A **greenhouse** is a structure enclosed by glass or plastic in which plants can be grown. Greenhouses give growers the ability to control environmental conditions affecting a crop. Greenhouses allow a high level of light to reach the plants. They can be heated or cooled depending on the needs of the plants. Watering is also controlled by the grower. Greenhouses allow plants to be grown throughout the year in a carefully controlled environment.

22-4. Greenhouse structures, such as this one at Kew Gardens in England, give growers the ability to control environmental conditions for growing plants.

CULTURE OF GREENHOUSE CROPS

Growing ornamental crops in greenhouses requires special attention to cultural practices. In nature, plants have adapted to the environment. In the greenhouse, the grower must provide plants with an environment that will promote healthy growth.

Control of greenhouse temperatures is critical. To grow the highest quality crops, it is often necessary to

22-5. Computerized control systems in the greenhouse help to maintain optimum temperatures for plant growth.

maintain temperatures within a few degrees. This involves an efficient cooling and heating system. Most new greenhouses have computerized controls that automatically regulate the cooling and heating system. During the day, cooling is a main concern. Radiant energy from the sun can quickly raise greenhouse temperatures to levels that can damage plants. Excessive heat can be a problem, even when it is cold outside. The ability to heat a greenhouse is also important. Greenhouses need additional heat on cold, cloudy days and at night. It is at those times when levels of radiant energy are low.

Most growers in greenhouses today use soilless mixes in which to grow plants. Soilless mixes have no soil. Common ingredients of soilless mixes are peat moss, perlite, vermiculite, and wood bark. Soilless mixes have some advantages. They are light in weight, making it easier to transport the final product. They are free of most soil-borne diseases and pests. Soilless mixes eliminate the labor involved in mixing and sterilizing soils. In addition, soilless mixes packaged in plastic are easily stored. On the down side, soilless mixes require more regular fertilizer applications. Also, with some crops, soilless mixes may not be heavy enough to prevent plants from falling over easily.

The most important practice in growing greenhouse crops is proper watering. Timing is critical. Water is something that must be given to plants when it is needed. Watering too often keeps a soil wet. The result may be damaged plant roots caused by the lack of good air exchange. Allowing soils to become too dry between waterings may also cause root death and lower the quality of the crop. In the greenhouse, it is usual for plants to need water more than once during a warm, sunny day. Therefore, the person in charge of watering plants must have knowledge of the crop being grown and good judgment about when to water.

New watering techniques that reduce labor have been developed. With these techniques, thousands of plants can be watered at one time. Spaghetti tubing is one method. It involves small tubes connected to a main line. Each tube is placed in an individual pot. When operating, water dribbles through the tubes, watering all the pots on a bench simultaneously. Another method involves sub-irrigation. With this method, potted plants are placed on a bench that can hold water. At regular intervals, water is pumped into the system, filling the bench. The pots sit in the water. Slowly, water is absorbed by the media in the pots. After a certain period, the water drains from the bench.

22-6. Watering is the most important job in the greenhouse.

Photoperiodism, or the response by some plants to the duration of light, is important. Growers alter the length of daylight to get plants like poinsettias and chrysanthemums to flower for a specific time. They do this by either covering the plants with a black cloth during daylight hours or turning lights on in the greenhouse during the nighttime hours. Depending on the treatment, the plants will either be kept in a vegetative stage of growth or encouraged to flower. Managing the day length is critical for getting plants to flower at a time of year they normally would not flower.

In some cases, high-intensity discharge (HID) lights may be used to increase light intensity. The addition of the artificial light during cloudy periods improves plant growth and vigor.

Floral crops

Greenhouse growers may specialize in a certain crop or grow a variety of crops for market. Of course, each type of crop requires specific growing conditions:

22-7. Sub-irrigation is being used to water these plants. Note that the bench on the right is filled with water.

22-8. Artificial light is often helpful in producing high-quality crops.

- **Cut flowers**—Some greenhouse growers focus on supplying cut flowers to wholesale florists. That is, they grow flowers in the greenhouse, cut them when they reach a certain maturity, and sell them to a wholesaler. Roses, carnations, chrysanthemums, and orchids are common cut flowers. Foliage or leaves grown for floral design work is also grown and harvested. Leatherleaf, lemonleaf, and huckleberry are common "greens" used in floral work.

- *Potted flowering plants*—A large segment of the floriculture industry is the production of flowering plants in pots. The plants may be propagated by seed or through asexual means. They are grown to the flowering stage in the greenhouse. The entire plant and the pot are then shipped to market. Some popular potted flowering plants are poinsettias, chrysanthemums, Easter lilies, and African violets.

- *Foliage plants*—Some growers produce potted plants to be sold as foliage plants. Foliage plants are grown for their leaves rather than for their flowers. They are also called house plants. Florida is the largest producer of foliage plants, which are shipped throughout the United States. Philodendrons, dieffenbachias, figs, schleffleras, and draceanas are common foliage plants produced for use in homes and offices.

22-9. Bedding plants are used to add color to the landscape.

- *Bedding plants*—Many greenhouse growers produce bedding plants in the spring for outdoor planting. Bedding plants are herbaceous annual flowers and vegetables. They lack hardiness and must be planted outdoors after the risk of a spring frost. They are usually started by seed and

CAREER PROFILE

HORTICULTURE TEACHER

Horticulture teachers combine their love of working with plants and their enjoyment of people. They like helping others to learn skills related to the horticulture industry. Horticulture teachers plan lessons, use a variety of teaching techniques, provide guidance for their students, and work with people in the horticulture industry.

Horticulture teachers must have a strong understanding of plant science and applied practices found in horticulture. First-hand experiences in the horticulture field are very helpful. For most teaching positions a bachelor degree in agricultural education is required.

Positions in teaching horticulture can be found in many locations. Most are at the high school or college level. Some can be found at botanic gardens or arboretums. Part-time teaching is common with park district programs, garden clubs, and some horticultural businesses.

grown in a greenhouse during the late winter and spring. Gardeners can then plant fairly mature plants in the landscape. Bedding plants include impatiens, petunias, marigolds, tomatoes, and many other plants.

FLORAL DISTRIBUTION

There are several channels in which floral crops can be distributed. The shortest route is from the local grower to the local consumer. The longest distribution channel involves the international market. In the international market, crops move from the producer to an exporter, to a wholesaler, to a retailer, and finally to the consumer.

Holland is recognized as the leader of the floriculture industry. Other countries that are major exporters or importers of floral crops include Columbia, Israel, Italy, Kenya, and the United States. Carnations grown in Columbia, roses grown in Israel, and exotic flowers grown in Hawaii are flown to the flower auctions in Holland. Buyers from around the world inspect and buy the flowers in the auctions. Once purchased, the flowers are shipped by air or road to wholesale businesses. The wholesaler repackages the flowers and takes orders from retail florists. Because flowers are perishable, they must be bought and shipped very quickly. An example of how amazing the market works is that a rose could be harvested in Israel one day, sold to a wholesaler in Holland the same day, and displayed in a retail floral shop in Seattle, Washington, the next day.

22-10. Greenhouse growers may sell directly to the public or to other retailers.

22-11. Holland is recognized as the leader of the floriculture industry.

22-12. Consumers inspect floriculture crops before making a purchase.

22-13. In Europe, many flowers are sold by street-side vendors.

Floriculture crops are sold to consumers through different retail outlets. The retail florist who specializes in floral arrangements is one. Cut flowers, foliage plants, and potted flowering plants are also sold in supermarkets, garden centers, gas stations, and market florists.

FLORAL PROCESSING

It is important that flowers move from the grower to the consumer as quickly as possible because flowers begin to die once they are cut from the plant. The producer, wholesaler, and retailer keep the flowers cool to extend their "vase" life. Lower temperatures slow the life

CONNECTION

AUTOMATED SYSTEMS LESSEN THE WORK

Automated systems controlled by computers lessen the workload in modern greenhouses. As a result, fewer people can grow more plants. Today, machines fill pots with growing medium, plant seeds, and automatically transplant plants. Plants can also be watered automatically. The photograph shows a greenhouse grower programming an automated watering system.

Computer-automated controls also keep the growing conditions for plants ideal. They control heaters, coolers, and energy curtains. Some greenhouses are even designed to have their roofs open to allow more light to reach the plants.

22-14. Cut flowers are placed in water with floral preservatives and cooled to extend their vase life.

22-15. People enjoy arranging and displaying flowers in their home.

processes of the flowers. Flowers are also placed in clean water with a floral preservative. The preservative provides the flower with food and reduces the growth of bacteria that can clog xylem cells. In addition, cut flowers are never stored with fruit or dead and dying plant material. Those items give off ethylene gas that speeds the death of plant cells.

Floral arrangers and designers are artists with flowers. Given a customer order, they choose flowers and the necessary floral supplies to create floral arrangements. Flowers are used to express feelings of love and sorrow, as well as to decorate homes and offices. Experienced designers offer helpful advice to customers on flower selection, care, colors, and costs. Some jobs are large, such as weddings, parties, funerals, and special events.

LANDSCAPE HORTICULTURE

The landscape horticulture industry deals with producing and using plants to make the outdoors more attractive. This branch of ornamental horticulture includes the nurseries in which plants are produced, garden centers that sell landscape products to the public, sod farms that grow grass, landscape contractors that design and install landscapes, and landscapers that maintain existing landscapes.

NURSERY PRODUCTION

A *nursery* is a place where plants are started and grown. Plants typically grown in a nursery are trees and shrubs for the landscape. In addition, fruit trees, vines, ground covers,

22-16. These trees are being grown in a nursery for later planting in a landscape.

perennials, and seedlings for reforestation are grown. It is useful to know the differences of the types of plants grown in the nursery industry:

- **Tree**—A tree can be defined as a single stem, woody, perennial plant reaching the height of at least 12 feet.

- **Shrub**—Shrubs are multistem, woody plants that do not exceed 20 feet in height.

- **Ground cover**—Ground cover may be woody or herbaceous. It forms a mat less than 1 foot high covering the ground. Grass is the most common ground cover.

- **Vine**—Vines are woody or herbaceous plants that require some type of support. They may climb other objects or creep along the ground.

- Perennials—Perennials as defined in the nursery industry are herbaceous plants with life cycles of more than two years. They are grown for their flower and foliage display.

Table 22-1. Trees are Terrific in Landscapes!

1. Trees clean the air. They remove smog, dust, and pollutants from the air we breathe. We take 23,000 breaths in a day, too!

2. Trees are a source of fruits and nuts that we eat.

3. Trees serve as sound barriers cutting noise levels where we work and live.

4. Trees have a cooling effect on hot summer days. Air temperature under a large shade tree can be 15 degrees cooler than the temperature in the sun.

5. Trees are beautiful. Many trees have showy flowers and attractive fall colors.

The wholesale nursery sells its crops to garden centers, landscapers, government agencies, arboretums, etc., for sale or planting. Most nurseries are located near their major market. This is because of the cost of transporting large and heavy nursery crops. It is also because the plants grown are those adapted to the climate in that region. A nursery in Texas would not specialize in plants for use in Minnesota due to the differences in the climates and the cost of shipping.

Nursery crops are sold in different forms:

- **Balled and Burlapped (B&B)**—The plants may be grown in the field, dug with a ball of soil around the root system, and wrapped with burlap. Plants prepared in this way are said to be balled and burlapped or B&B. The soil balls can be quite large and heavy. One advantage to balled and burlapped stock is some roots are undamaged in the transplanting process. B&B material can also be left unplanted for months.

- **Bare Root (BR)**—A second form, called bare root, involves digging the crop from the field and removing soil from the roots. Bare root plants are lighter for shipment, but are under greater stress caused by the loss of roots. Bare root material needs to be planted as soon as possible to improve plant survival.

22-17. These trees dug from the field have a soil ball wrapped with burlap.

- **Containerized**—Another option is to grow the plants in containers. These plants are said to be containerized. Use of containerized nursery stock has increased in recent years. The advantage of containerized material is the root system is left undamaged. The growing medium for containerized nursery stock often includes bark chips. The disadvantage to containerized plants is the grower has to give greater attention to the water and nutrient needs of the plants during production.

Advances in technology have cut the amount of hand labor needed in digging and moving nursery stock. Hydraulic tree spades can cut through soil and roots and lift the plant from the ground. Some tree spades are mounted on trucks. These pieces of equipment can be used to dig the plant, transport it to the site of planting, and plant the tree in the landscape. A tree spade can also be used to dig the plant and place the soil ball in burlap and a wire basket for later planting.

22-18. Trees and shrubs are being grown in containers for use in the landscape.

22-19. Hydraulic tree spades are used to dig and move trees. (Courtesy, Vermeer Manufacturing Company)

22-20. Trees, shrubs, annual flowers, and physical structures have been used in this appealing landscape.

LANDSCAPING

Landscaping is a challenging field of work centered on improving the appearance of land. Landscape companies use nursery stock in their work. They are also involved in communicating with clients, designing landscape plans, moving soil, applying mulch, ensuring proper water drainage, installing physical structures, such as walls and patios, and planting nursery stock. Landscapers have an appreciation for plants and the outdoors. They must also have an understanding of the cultural requirements for different plants.

A landscape job starts with a plan or a landscape design. Landscape designers are individuals with knowledge of plant materials, plant care, and principles of art. Climate, soil type, drainage of water, and exposure to light all influence a designer's decision in selecting plants. The time they devote to planning pays off with an attractive landscape. Landscape designs may be formal or informal. Formal designs are very structured and exact. Informal designs have a natural appearance. The landscape designs are prepared by hand or with the use of computers and software developed specifically for landscape design work.

Preparation of the soil prior to planting is very important. Traffic from heavy construction equipment damages soil structure by compacting the soil. The compaction often results in hard, poorly drained soils. With most plants, it is essential that the soil is loose and well drained. Landscapers will loosen these

22-21. An example of a formal landscape.

soils by working the soil and by adding soil amendments for the benefit of the plants. They also seek to establish a smooth grade to the land that is pleasing to look at and allows for surface drainage of water.

Landscape contractors are involved in the installation of landscape materials. Handling the plants with care and placing them in the proper locations according to the landscape plan are important. The process of moving and planting a plant grown at another location is called *transplanting*. Patios, walks,

22-22. A rotary tiller is used to loosen soil and incorporate soil amendments.

22-23. Golf course management is an area of ornamental horticulture devoted to recreation.

fences, and retaining walls or non-plant features are known as ***hardscapes***. Landscape contractors install the hardscapes as well as plant materials.

Some landscapers are also in the business of maintaining existing landscapes. They accept the task of keeping a landscape looking attractive. They may mow grass, replace damaged or dead plants, prune woody plants, irrigate, fertilize plants, and control plant pests.

The landscape horticulture industry has areas that are very specialized. Sod farms are in the business of growing and selling sod for new landscapes or athletic fields. Golf courses are designed and maintained for recreational purposes. Arborists are individuals trained in the care and maintenance of trees. Their segment of the industry is called arboriculture. Plant propagators are employed to start new plants from seeds or through asexual means. Garden centers are retail stores specializing in landscape supplies. Mail order companies sell perennials and other landscape plants using the postal services.

REVIEWING

MAIN IDEAS

Ornamental horticulture is one of the fastest growing segments of the agriculture industry. It involves growing and using plants for their appearance. The two main areas of ornamental horticulture are floriculture and landscape horticulture.

Floriculture is a multibillion dollar, international industry based on the production and sale of flowers and foliage. Many floriculture crops are produced in greenhouse structures. In a greenhouse, growers can control cultural conditions to produce quality crops. The crops produced in the floriculture industry include cut flowers, potted flowering plants, foliage plants, and bedding plants. Holland is the world center of the cut flower trade. Cut flowers are auctioned to buyers and shipped to locations around the world. Florists process and design flowers for retail sale.

Landscape horticulture involves the production and use of plants for outdoor use. Trees, shrubs, ground covers, and vines are grown in nurseries. They are sold as balled and burlapped, bare root, containerized, or mechanically dug plants. Landscapers plant nursery crops in the landscape. They also design landscapes, prepare soil for planting, improve water drainage, and install physical structures. Some landscapers are hired to maintain existing landscapes by mowing grass, pruning, watering, and fertilizing.

QUESTIONS

Answer the following questions using complete sentences and complete sentences.

1. What is ornamental horticulture?
2. What are the two major divisions of ornamental horticulture?
3. What are the important cultural practices a greenhouse grower must consider?
4. How are cut flowers distributed?

5. What types of plants are produced in a nursery?

6. What are some tasks performed by a landscaper?

EVALUATING

Match the term with the correct definition. Write the letter by the term in the blank that is provided.

a. nursery
b. foliage plant
c. tree

d. greenhouse
e. floriculture
f. balled and burlapped

g. bedding plants
h. cut flower

1. _____ plants, also called house plants, valued for their leaves

2. _____ place where landscape plants are produced

3. _____ flowers grown, harvested, and sold to florists for floral design work

4. _____ landscape plants dug with a mass of soil and wrapped

5. _____ glass or plastic-covered structure in which plants are grown

6. _____ an area of ornamental horticulture involving the production and use of flowers

7. _____ single stem, woody perennial at least 12 feet in height

8. _____ flowering annuals and vegetables grown for planting outdoors

EXPLORING

1. Maintain a list over a week's time of how many uses there are for ornamental plants. Turn this into a class project. Each student can contribute to a master list or students can develop a bulletin board display of photos showing the diversity of the ornamental horticulture industry.

2. Build your own "mini greenhouse" using a clear 2-liter bottle or clear plastic supported over a pot. Place soilless mix or potting soil in the bottle or pot. Plant up to 10 seeds of your choice and moisten the soil. Place your greenhouse next to a bright window. Observe how the environment inside your greenhouse differs from the environment in the room. Check the humidity level, temperature, need for water, etc.

3. Working with your family, school, or community, obtain a tree from a nursery or garden center. Plant the tree in a location where it will have a chance to develop.

4. Review learning materials in areas of horticulture. These include *Introduction to Horticulture*, *Introduction of Landscaping*, and *Floriculture: From Greenhouse Production to Floral Design*. These books are available from Interstate Publishers, Inc. Prepare a brief written summary of each book.

Forestry

This chapter introduces the importance of forestry and the production of trees. It has the following objectives:

1 Describe modern forestry and forest products.

2 Explain tree biology and identify the parts of a tree.

3 Explain major classes and forest types.

4 Describe how trees are planted.

5 Define silviculture and list major silviculture and protection practices.

TERMS

annual ring	log	silvics
artificial reforestation	lumber	silviculture
conifer	mixed stand	timberland
crown	plywood	trunk
diameter breast high	pulpwood	urban forestry
forest land	pure stand	veneer
hardwood	seedling	

23-1. People enjoy walking through forests such as this old growth forest in the Mount Hood National Forest in Oregon.

TREES are the largest living things on Earth! A few large trees are very old and have been growing for a long time—2,500 years for some sequoias! Other trees are young and have been planted to meet the special needs of people.

Trees are found in most places. They are in big cities, small towns, rural areas, and, definitely, in forests. Depending on where they are, trees are beneficial in so many ways. We get so many things from them!

People often think of the lumber and paper that we get from trees. We fail to remember that they shade our homes, cool the environment, provide habitat for wildlife, and help control soil erosion. The list of uses of trees is long!

MODERN FORESTRY

Forestry is the science of managing trees for the benefit of humans. It includes growing trees for wood products as well as growing trees to improve the quality of our environment. Forest land that productively grows sufficient trees for harvesting is known as **timberland**. Most wood products are from timberland. Large tree farms grow much of the wood that is now used in the United States.

23-2. Rugged trees on the Pacific Coast have withstood the ravages of bad weather.

URBAN FORESTRY

Urban forestry is the specialized area of forestry that deals with the growth and use of trees in cities. The trees are not grown for wood products, but provide important benefits to people. Trees lower the temperature along city streets and in parks. Trees shade homes and buildings in the summer and keep them cooler. They provide aesthetic qualities that help people enjoy life.

Urban forestry includes all of the activities in establishing and maintaining trees in urban areas. Planting large trees may require specialized equipment. Trimming trees away from power lines requires hydraulic buckets and power cutting equipment.

Cultural systems must be used to get the trees to grow. Urban environments are not always ideal for tree growth. Space for roots and

23-3. Urban forestry may involve using a rope system to prune and otherwise care for trees.

23-4. Tree pruning to prevent interference with power lines is a common activity in urban forestry.

23-5. Trees are carefully used in cities. This tree with lights is in front of a hotel in New York City.

limbs may be limited. Watering, pest control, and other practices must be adapted to the urban situation.

*F*OREST LAND AND TREE FARMS

Forest land is land on which trees are growing. In some cases, the trees may be producing high-quality products very rapidly. The trees are often native species that are growing with or without cultural practices. The quality and population of trees may vary. The amount of wood produced also varies. Sometimes, forest lands are dry and rocky, with limited production potential. Trees grow slowly, if at all.

To provide the vast amount of wood products needed, tree farms have been established. A tree farm is privately owned timberland planted to desired species and carefully managed for top quality and growth. Many wood products harvested today are from tree farms. Modern cultural practices are used to get good production.

23-6. Northern California forests provide important timber for lumber manufacture.

MORE THAN TREES

The forestry industry is far more than the production of trees. It includes all of the manufacturing activities that convert wood into the desired products. The major wood products are:

- Lumber—**Lumber** is timber that has been sawed into boards. Timber is hauled to sawmills as logs. A **log** is the large stem of a tree harvested by cutting the tree. Large logs can be sawed into lumber with varying dimensions, such as 2 × 4 or 1 × 6 boards. These numbers give the dimensions of the boards by thickness and width in inches. In addition, boards are measured by length, such as 10 or 14 feet. Logs are cut in lengths to get the most efficient lumber from them. Lumber is used in building construction and for many other purposes.

- Plywood and Veneer—These wood products are made by gluing thin sheets of wood together. **Plywood** is typically produced in 4 × 8 foot sheets of varying thicknesses, such as ½ inch. Three or five layers of wood are glued together. **Veneer** production is similar to that of plywood,

23-7. Lumber at a kiln ready for drying in Washington.

except the outer layer on one side is a very valuable wood. Lower cost woods are covered with a more expensive wood. This results in the beauty of the veneer without the high cost of the wood product being a solid material.

- Paper—Paper involves using pulpwood. **Pulpwood** is wood specially grown and cut for making paper and similar products. Manufacturing includes grinding and breaking pulpwood into pulp. Soft pines and similar trees are often used for pulpwood, though some green hardwoods can be used. The pulp is converted to paper as it is spread into thin sheets and the water is evaporated.

- Furniture—Fine-quality woods are often used to make furniture. The wood must be durable and have beauty that appeals to consumers. Walnut, cherry, oak, and mahogany are popular furniture woods.

- Other Products—Many products are manufactured from wood that are

23-8. Veneer made from yellow poplar is about the thickness of a quarter. This veneer will be glued over lower-grade wood to manufacture doors.

23-9. Short sticks of pulpwood on a rail car for delivery to a paper mill.

23-10. Manufactured wood products used in this construction include laminated joists and I-beams. Such products are stronger than lumber and resist warping.

not lumber, plywood, paper, or furniture. Particle board is made by separating wood into small pieces and gluing the pieces back together to form sheets. Poles, crossties, fence posts, and landscape timbers are a few other examples. Christmas trees, maple syrup, rosin, nuts, tea (from the roots of the sassafras tree), and many other products come from trees.

TREE BIOLOGY

Trees have the same life processes as other plants. They must have nutrients to grow. Roots take water and nutrients from the soil. Leaves carry out photosynthesis to make sugar—food for the tree. Trees must reproduce to perpetuate their species.

PARTS OF A TREE

A tree is a woody perennial plant with three major parts: roots, trunk, and crown. The root system may be a tap root or a fibrous system. Pine trees, for example, have tap roots. Oak trees have fibrous root systems.

The **crown** consists of limbs, leaves, flowers, and fruit or seed. It carries out photosynthesis to provide food for the rest of the tree.

The **trunk** is the main stem of a tree. It has heartwood and sapwood. Heartwood is older wood nearer the center of the trunk. Sapwood is young, growing wood. The sapwood carries sap from the roots to the leaves. Bark protects the cambium and other parts of the wood from damage. The trunk is sometimes known as the bole.

TREE GROWTH

Trees continue growing as long as they live. They usually grow in both height and

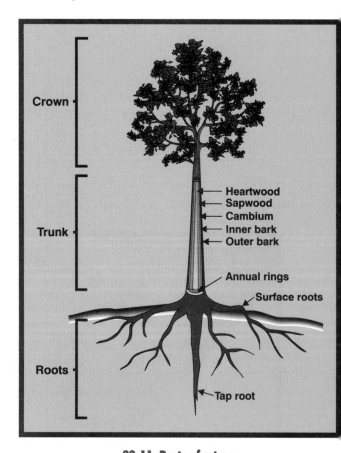

23-11. Parts of a tree.

23-12. New growth is obvious by the growing buds on this pine tree.

diameter each year. Height may be thought of as the distance from the tip of the roots to the uppermost part of the crown. Both roots and stems grow. Terminal buds in the crown send up new growth each year. With some species, this new growth is easy to see in the spring. The root tips grow additional length to support the increased crown.

Annual rings are formed each year as a tree grows. Rapid growth is desired to gain good productivity from a tree farm. Trees usually reach a height of more than 8 feet.

Trees grow in diameter by adding annual rings. An **annual ring** is formed by the growth of xylem tissue by the cambium. Each ring has both light- and dark-colored areas. The lighter

23-13. Annual rings vary in width from one year to another.

areas are made of thin-walled cells that grow in the spring. The dark areas are made of small, dense cells that grow in the summer. The difference in spring and summer growth causes the annual rings to be obvious.

Growth in diameter is important in lumber and pulp production. Larger-diameter trees produce more wood. Foresters use an increment borer to study annual rings. The width of annual rings provides information about the health and environment in which the tree was growing. Small, thin rings are a sign that the tree did not grow well. They might result from a drought or other conditions. Fire damage is also evident in annual rings.

23-14. An increment borer can be used to get a sample of annual growth from a tree. The inset photo shows a sample from a fast-growing tree.

REPRODUCTION

Trees naturally reproduce in three ways: seeds, sprouts, and suckers. Seed reproduction is most important. The process is similar to other plants with flowers. The flowers are pollinated and seeds are formed in the ovary. Pine cones, acorns, and pods are three examples of the structures in which seeds are produced.

Sprouts and suckers are asexual reproduction. The new tree is a clone of the previous tree. Some trees readily grow sprouts or suckers. Sprouts grow from the stems of trees. For example, if a small tree is cut off, it will likely send a sprout from the stump. Suckers grow

23-15. Flower on a tulip poplar.

23-16. Cones from a pine tree.

from the roots of trees. For example, cutting the roots of a black locust tree in the soil with a plow or shovel will result in new trees growing at the cuts.

FOREST CLASSES AND TYPES

Thousands of species of trees are found in the world. Fewer than 50 are economically important in the United States. Others provide habitat for wildlife and promote a quality environment.

CLASSES

Trees are classified scientifically as are other living things. All trees are members of the Plant Kingdom. Two important classes are used: Gymnospermae and Angiospermae.

23-17. A young oak tree with the acorn still attached.

The Gymnospermae class includes trees known as conifers. A **conifer** is a tree that has needles and produces seed in cones. These trees frequently grow faster and produce softer

CAREER PROFILE

FORESTER

A forester studies, establishes, and maintains trees. The work may involve planting trees, measuring timber, assessing pest damage, and designing methods of protection. Foresters often use equipment to help in their work, such as the altimeter shown here to measure the height and amount of wood in a tree.

The education of foresters varies with their level of work. Some have community college degrees in forestry technology. Many have baccalaureate degrees. Some foresters get masters and doctors degrees. Few have doctors degrees in forestry, with most doctorates being in related areas. Foresters need practical experience in all aspects of timberland work. Much of the work is outside and on forest land.

wood than the Angiospermae class. The conifers are the predominant construction lumber and pulpwood trees. Examples include many different pines, spruces, firs, and junipers.

The Angiospermae class includes the broadleaf trees. These trees are typically known as **hardwood**. Their seeds are usually produced in fruit, such as acorns, berries, and nuts. A few produce seeds in pods, such as the black locust and redbud. Some of these trees are used to produce fine-quality woods for furniture, such as walnut and cherry. Other common hardwoods include the oaks and gums.

FOREST TYPES

A stand of trees consisting primarily of one species is a **pure stand**. A pure stand is 80 percent or more of the same species. If a stand has more than 20 percent of two or more species, it is a **mixed stand**. Some natural forests are pure stands; others are mixed stands. Tree farms usually plant pure stands.

Types of forests may be pure, such as ponderosa pine, loblolly, or Douglas fir. Other types may be mixed species, such as oak-hickory forests with several species. Bottom land along creeks or rivers often grows hardwoods, such as oak and hickory. Very wet conditions may grow cypress. Hills and drier areas are more likely to have pines, spruces, and firs. Of course, the species also depends on the altitude and climate.

23-18. Pure stand of growing pines on a tree farm.

IDENTIFICATION

Tree identification is an essential skill for people in forestry. Most of the common trees are fairly easily identified. A few trees require careful study for exact identification. Trees also have common and scientific names. Common names tend to vary on a local basis. Scientific names are consistent. For example, the willow oak is known as a pin oak in some places. The scientific names help distinguish between them—the willow oak is *Quercus phellos* and the pin oak is *Quercus palustris*.

The major characteristics of trees are used in identification. Five things to look at are leaves, twigs, bark, flowers, and fruit. Some of these may be present for only a few days each year, such as flowers. Bark and twigs are always present. Leaves are always present only on evergreen trees.

Leaf type, shape, color, size, texture, arrangement, and, sometimes, odor (such as the sassafras tree) are useful in tree identification. Leaf types, shape, and arrangement were covered in Chapter 3. Examples of leaves and fruit or cones from selected trees are shown in Figure 23-21.

Measurements may also be used in identification, especially when size is important. Height is commonly used, with some trees more than 100 feet tall. Leaf, flower, and fruit size are also measurements

23-19. Mixed bottom land hardwood stand along the Soque River in Georgia.

23-20. A diameter tape (d-tape) is placed around the circumference of a tree at 4.5 feet from the ground to measure diameter breast high (DBH).

important in identification. Tree stems are measured in diameter at a standard distance above the ground—**_diameter breast high_** (DBH). DBH is 4.5 feet from the ground. For trees on hilly land, the measurement is taken on the uphill side of the tree. The measurement includes the thickness of the bark.

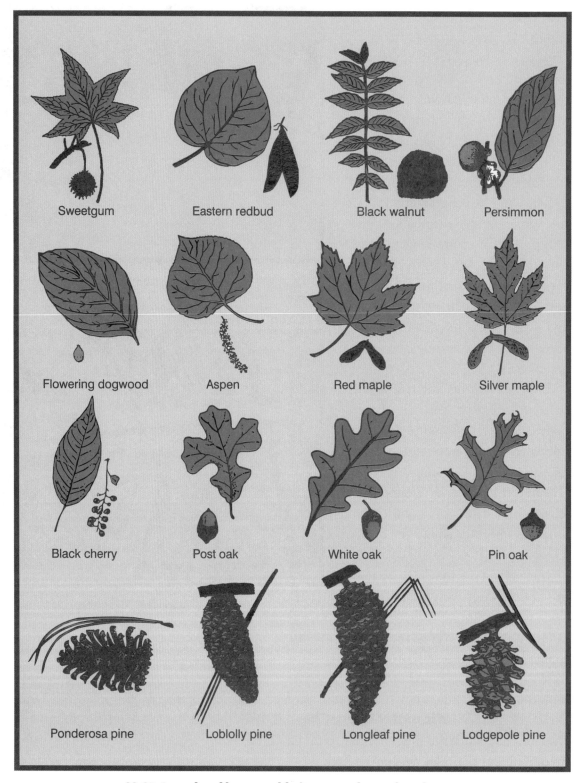

Sweetgum

Eastern redbud

Black walnut

Persimmon

Flowering dogwood

Aspen

Red maple

Silver maple

Black cherry

Post oak

White oak

Pin oak

Ponderosa pine

Loblolly pine

Longleaf pine

Lodgepole pine

23-21. Examples of leaves and fruit or cones from selected trees.

Table 23-1. Descriptions of Selected Tree Species

Common and Scientific Name	Where Grown	Description and Primary Uses
Black cherry *Prunus serotina*	eastern half of the United States	white flowers in spring; small black cherries in early summer; deciduous; excellent furniture
Black walnut *Juglans nigra*	eastern half of the United States, except southern Florida and Maine	catkins form in spring; nuts mature in fall; compound leaves are 1 to 2 feet long; makes valuable furniture
Common persimmon *Diospyros virginiana*	southeastern United States	small flowers in spring; fruit matures in fall; mature trees are medium-sized; wood specialty uses, such as golf club heads and shuttles in textile mills
Eastern redbud *Cercis canadensis*	eastern Texas to Virginia	colorful flowers in spring; reddish fruit in late summer; small-sized tree, often used as an ornamental in landscapes
Flowering dogwood *Cornus florida*	eastern United States, except southern Florida upper New England	colorful white to greenish-yellow and pink flowers; leaves and fruit are brightly colored in fall; small-sized tree; used as ornamental tree; wood is occasionally used in specialty textile mill applications
Loblolly pine *Pinus taeda*	southeastern United States	conifer, with needles 6 to 9 inches long; trees grow 90 to 110 feet tall; 2 to 4 feet at DBH; cones 2 to 6 inches long; used for lumber, pulpwood, and many other uses
Lodgepole pine *Pinus contorta*	United States, western mostly northwest of Colorado with some in California	evergreen, with two needles per cluster 1 to 3 inches long; grows small to large; trunks have pole appearance and often used for posts, lumber, and pulpwood
Longleaf pine *Pinus palustris*	southern areas of Gulf Coast states	conifer, with cones 6 to 10 inches long; needles 8 to 18 inches long; trees grow 80 to 100 feet tall; DBH is 1 to 3 feet; produces excellent poles, lumber, and naval stores (rosin)
Pin oak *Quercus palustris*	eastern-midwest through Atlantic Coast states	deciduous; deeply lobbed leaves; small acorns; numerous drooping branches; durable lumber
Ponderosa pine *Pinus ponderosa*	scattered in western United States	cones 3 to 6 inches long; 2 to 3 needles per cluster; needles 4 to 7 inches long; used for lumber, poles, toys, and many other uses
Post oak *Quercus stellata*	southeastern United States	deciduous; three lobed-leavesin shape of a cross; small to medium trees; widely used, but posts made from this tree do not last a long time
Quaking aspen *Populus tremuloides*	northern United States and at higher elevations in the west	nearly rounded leaves 1 to 3 inches in diameter; may appear to shake in a breeze; small trees; limited use, with some used as pulpwood

(Continued)

Table 23-1 (Continued)

Common and Scientific Name	Where Grown	Description and Primary Uses
Red maple *Acer rubrum*	eastern United States	deciduous, with leaves 2.5 to 4 inches long; colorful leaves in the fall; large tree with wood used in furniture and other products
Silver maple *Acer saccharinum*	eastern United United States, except Florida	leaves are bright green on top and silvery-white on the bottom; leaves turn yellow in fall; large trees are used in many ways, such as furniture
Sweetgum *Liquidambar styraciflua*	southeastern United States	deciduous; grows 130 to 150 feet tall; seed is formed in spiny balls; simple leaves with 5 to 6 lobes; many uses, including musical instruments, pulpwood, crates, and plywood
White oak *Quercus alba*	midwest and eastern United States	midwest and eastern deeply lobed leaves; fine-quality lumber; used in furniture, flooring, and cooperage

PLANTING

Trees are planted in home landscapes, on tree farms, and in other locations. Methods of planting vary with the kind of tree and purpose for which the tree is being grown.

Tree farms

On tree farms, the process of planting trees is often known as **artificial reforestation**. The trees are planted in areas where trees previously grew. It is likely that the trees were harvested or another crop has been grown on the land. In some cases, the land has been in crops and is being returned to forestry to control erosion.

Two methods of planting are commonly used: direct seeding and nursery grown seedlings. The most common method of planting is the use of seedlings. Seedlings get off to a better start growing than relying on seed scattered over the land.

A **seedling** is typically a couple of years old and 15 to 18 inches (38 to 46 cm) tall. Seedlings are grown from improved seed in nurseries. They are set in the winter or early spring on land that has been cleared and otherwise prepared for a new crop of trees. Most seedlings are moved and set in bare root form. They do not have any soil on the roots. It is important to keep roots moist and not allow them to dry out. Place the seedling approximately the same depth in the soil that it grew in the nursery. Mechanical planters can be used on land that is not too steep and rugged or wet. Otherwise, the seedlings are set by hand with a crew of planters walking the area to be planted.

Other methods of planting include using wild seedlings (trees that grow from seed on the land), using cuttings (vegetative propagation, as with willow and cottonwood), and container seedlings (seedlings grown in small pots). Using wild seedlings is not recommended because of the poor stand that may be obtained and the inability to control the species that grows.

Urban Areas

Trees are planted in urban areas for shade and beauty. People want them to grow quickly. Seedlings can be used, but most urban areas are set with larger trees. Sometimes, the trees may be 20 or more feet tall when planted. Specialized equipment is needed, as well as extra care in the planting process.

Larger trees may be in wooden planters that are removed at the time of planting or they may have a ball of soil on the roots wrapped in burlap. Some trees are moved with tree handling equipment. In digging a tree to transplant, be sure to keep as many roots intact as possible and keep soil on the roots.

Trees should be planted in a hole larger than the root system requires. In planting a medium-sized tree, make the hole at least 12 inches (30 cm) larger than the roots. The soil in the hole may have organic matter added to improve the root environment. The depth of the tree in the soil should be approximately the same depth as it was before being dug for

23-22. A pine seedling ready for planting.

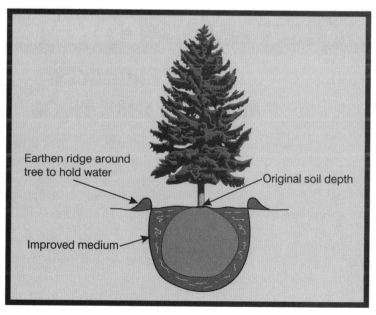

Earthen ridge around tree to hold water

Original soil depth

Improved medium

23-23. How to set a larger tree.

23-24. Two methods of bracing a tree: staking (left) and guying (right).

planting. The tree must be well watered. In some cases, trickle irrigation may be installed. Trees more than three feet (1 m) tall should be staked to prevent bending in the wind.

CONNECTION

FOREST PRODUCTS ARE MORE THAN WOOD

Trees have valuable products in addition to wood. Some have a highly desirable sap, or juice, known as rosin. Most rosin is from pine trees.

Three kinds of rosin are used—gum, wood, and sulfate. Gum rosin is made from living trees. The trees are tapped to collect the sap in a bucket for manufacturing into rosin products. Wood rosin is made by processing the stumps of dead trees. Sulfate rosin is a by-product of pulpwood manufacturing.

This photograph shows gum rosin being collected from a longleaf pine. As the sap flows in the tree, some flows out of the cut and goes into the bucket. The rate of collection of rosin is fairly slow.

SILVICULTURE AND PROTECTION

Trees live and grow in a complex environment. The science associated with the growth and development of trees is **silvics**. The important principles of silvics are the basis of silviculture.

Silviculture is the art and science of growing trees. It stresses practices that improve the productivity of the timber stand. These often involve selective cuttings. Through science, many new applications of technology have been developed. These form the basis of silviculture.

PRACTICES

The major practices in silviculture are:

- Thinning—Thinning is removing small young trees to allow other trees to grow more efficiently. Thinning is used to regulate the space between trees and produce some income from smaller trees of marketable size for posts, pulpwood, and similar uses.

- Cleaning—This silviculture practice is used in young stands to remove weak or improperly growing trees. Cleaning is also used to remove undesirable species. It increases the growth of the desired species.

- Liberation cuttings—Liberation cuttings are used to free young trees from older dominant trees. Large trees shade young trees and take up space. Removing inferior larger trees allows younger trees to grow more efficiently.

23-25. Signs are used to alert people to fire dangers. (Fires can damage and destroy growing trees.)

23-26. Feller cutting a tree.

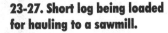
23-27. Short log being loaded for hauling to a sawmill.

23-28. A historic tree at George Washington's Mount Vernon home has a ground wire to protect it from lightening.

- Harvest cuttings—Harvest cuttings are used to remove marketable timber from a timber stand. It may include cutting older and larger seed trees and selectively cutting mature trees or those that have been damaged but have marketable value.

PROTECTION

Forests must be protected from damage that destroys trees or greatly reduces the growth of quality wood. Trees require much time to grow—20 or more years. Unprotected trees can be quickly destroyed and all of the investment lost.

The major sources of damage to trees on a tree farm are:

- Fire—Forest fires can quickly kill growing trees. Fallen leaves or needles, twigs, and other dead and dry plant materials on the ground will often readily burn. Older forests will also have materials in the tree canopies that will burn. After several years, a significant amount of material can accumulate on the ground. Controlled burns are sometimes used to rid the ground of the material before it accumulates to a point where it can make very hot fires. Constructing firebreaks (6 feet wide or 2 m) around forests helps to keep ground fires from spreading. Fires must be prevented. Prevention is the first and major step in protecting forests. No fires of any type should be allowed in forests during dry seasons. Even the exhaust systems of vehicles can start fires. Leaves or grass against a catalytic converter can easily be ignited.

23-29. An area of Yellowstone National Park several years after a fire destroyed the trees.

- Insects and disease—As with other plants, trees are subject to attack by insects and disease. A particular pest in the South and East has been the Southern pine beetle. This pest has killed millions of pine trees. Information about insect and disease problems in a local area is available from a forester.

- Human activity—Human activity in a forest can cause a great deal of damage. Roads by four-wheelers and other vehicles compact the soil and damage young trees. Sometimes, trespassers cut trees for wood or other purposes. Most forests should be posted against unwanted human activity. People should never go onto another person's land without permission. It is private property!

- Weather—Ice, wind storms, lightening, flooding, and other weather conditions can damage trees. Sometimes, large forest areas have been destroyed. To the extent possible, damaged trees should be marketed and cleaned to allow other trees to grow.

- Large animals—Cattle, deer, and other animals can damage trees. Most tree farms prohibit grazing

23-31. This pine forest has been destroyed by an ice storm.

cattle in the growing trees. The animals can trample trees and bite off leaves, twigs, and other parts.

- Air pollution—Pollution from automobiles, factories, and other sources can cause damage to forests. Leaves can drop early because of air pollution. This stunts trees and makes them susceptible to disease. Acid rain is causing damage to trees in some locations, particularly in the eastern United States.

Harvesting

Trees are harvested at different times and in different ways. How and when depends on the species and the use to be made of them. Some forests are clear-cut, which means that every tree is cut. Others are selectively harvested, which means that only trees meeting certain requirements are cut.

Before harvest, the trees in a forest are cruised. This means that measurements are made of the trees. A total volume of wood that can be harvested is determined.

23-32. An area on the side of a mountain in the Pacific Northwest has had a clear-cut harvest. (The land is very subject to erosion.)

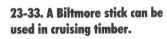

23-33. A Biltmore stick can be used in cruising timber.

23-34. A chain saw may be used in harvesting trees.

23-35. A semi-trailer loaded with several logs.

23-36. Baseball bats are made of ash and undergo a number of steps to assure a quality product. These bats are in the process of being manufactured.

Harvest is often with felling machines or by sawing. Smaller, younger trees are more easily harvested with fellers. Larger trees may require saws–typically chain saws.

Harvested trees are transported to mills, exported, or otherwise used to manufacture wood products. Some are chipped; others are sawed. Some products may require specific manufacturing activities.

REVIEWING

MAIN IDEAS

Forestry is the management of trees for the benefit of humans. Land on which sufficient trees grow for harvesting is timberland. In recent years, the role of trees in cities and residential areas has become more important. Urban forestry is the specialized area of forestry that deals with the unique needs of trees grown for benefits other than wood.

Tree farms grow improved varieties of quality trees. They produce high-quality logs, pulpwood, and other wood products. Practices are followed to assure efficient growth.

A tree carries out life processes similar to most other plants. Roots, trunk, and crown are the major parts of a tree. Growth occurs as long as a tree lives. Both crown and roots grow in height. The stem grows in diameter. Annual rings give important information about the growth of a tree.

The two major classes of trees are the conifers and broadleaf trees. Tree species may have pure or mixed stands. Most tree farms grow pure stands. Tree identification requires study of the leaves, fruit, flowers, stem bark, and, in some cases, the roots.

Trees are carefully planted and maintained. Planting methods vary with the size and use of a tree. Silviculture includes practices used to help a tree grow well. Since young trees in a forest are planted closer together than they can grow, thinning and other silviculture practices are used. Trees must be protected from fire, disease, and other damage.

QUESTIONS

Answer the following questions using complete sentences and correct spelling.

1. What is forestry? How is forest production on timberland different from urban forestry?

2. Distinguish between plywood and veneer.

3. What are the major parts of a tree?

4. How do trees grow?

5. What are the natural methods of tree reproduction?

6. Distinguish between conifer and hardwood trees.

7. What is diameter breast high?

8. How are large trees planted in urban forestry?

9. What are the major silviculture practices?

10. What are the major sources of damage in a forest?

EVALUATING

Match the term with the correct definition. Write the letter by the term in the blank that is provided.

a. silviculture
b. timberland
c. lumber
d. plywood

e. tree farm
f. conifer
g. hardwood
h. pure stand

i. mixed stand
j. pulpwood

1. _____ stand with 20 percent or more of two or more tree species

2. _____ art and science of growing trees

3. _____ Angiospermae class trees that are deciduous

4. _____ Gymnospermae class trees that are evergreen

5. _____ a stand of trees that is 80 percent or more of one species

6. _____ forest land that productively grows trees

7. _____ land planted to desired tree species and carefully managed

8. _____ wood products made by sawing logs into boards

9. _____ wood products made by gluing large sheets of wood together

10. _____ wood used for making paper and similar products

EXPLORING

1. Tour a furniture store. Read the labels on furniture to identify the woods used in making the furniture. List tree species and the kinds of furniture. Also, make a list of products made from chipped wood and other products. Prepare a report on your observations.

2. Use an increment borer to collect a sample of annual rings for study. Measure the width of the annual rings in your sample. Assess why the width of the rings varies. Prepare a report on your observations. (Be sure to properly identify the species of the tree you are studying.)

3. Arrange for a local forester to serve as a resource person in your class. Ask the forester to discuss local forestry production, including species, silviculture, and protection practices. Prepare a report on the presentation.

Appendixes

Appendix A. Common Customary and Metric Equivalents

Customary Measure	Metric Equivalent
1 ounce	28,3495 grams (g)
1 pound	453.59 g or 0.45359 kilograms (kg)
1 gallon	3.79 liters (L)
1 acre	0.40468 hectares
1 inch	2.54 centimeters (cm)
1 foot	30.38 cm or 0.348 meters (m)
1 cubic yard	0.76 cubic m

Appendix B. Common Metric and Customary Equivalents

Metric Measure	Customary Equivalent
1 kilogram	35.27 ounces or 2.2046 pounds
1 metric ton	2,204.6 pounds
1 hectare	2.47 acres
1 liter	1.057 quarts or 0.2642 gallons
1 meter	39.37 inches
1 centimeter	0.4 inch
1 cubic meter	35.135 cubic feet

Commodity	Measure	Net Weight	
		Pounds[1]	Kilograms[2]
Alfalfa seed	bushel	60	27.2
Asparagus	crate	30	13.6
Barley	bushel	48	21.8
Broccoli	crate	20–25	9.1–11.3
Broomcorn	bale	333	151
Cabbage	open mesh bag	50	22.7
Celery	crate	60	27.2
Cherries	lug	160	7.3
Clover seed	bushel	60	27.2
Coffee	bag	132.3	60
Corn			
ear (husked)	bushel	70	31.8
shelled	bushel	56	25.4
Cotton	bale (gross)	500	227
	bale (net)	480	218
Eggplant	bushel	33	15.0
Flaxseed	bushel	56	25.4
Milet	bushel	50	22.7
Milo	bushel	50–60	22.7–27.2
Mustard seed	bushel	58–60	26.3–27.2
Oats	bushel	32	14.5
Peanuts	bushel	22	10.0
Rice	bushel	45	20.4
Rosin	drum	520	236
Sorghum (grain)	bushel	56	25.4
Soybeans	bushel	60	27.2
Sunflower seed	bushel	24–32	10.9–14.5
Walnuts	sack	50	22.7
Wheat	bushel	60	27.2

[1]To convert from kilograms to pounds, multiply kilograms by 2.2046.

[2]To convert from pounds to kilograms, multiply pounds by 0.45359237.

Glossary

A

abiotic disease—disease caused by elements in the environment of a plant; not caused by a pathogen.

abiotic environment—all of the nonliving elements in a plant's environment.

abscisic acid—plant hormone that causes seed dormancy and loss of leaves by deciduous plants.

acid rain—any form of precipitation that contains acid; acid often forms in precipitation from emissions of factories and engines.

acre-foot—amount of water needed to cover an acre of land one foot deep in water.

acre-inch—amount of water needed to cover an acre of land one inch deep in water.

active ingredient—percent of poison material in a pesticide.

adsorption—bonding of soil mineral cations to soil particles.

adventitious root—root that begins growth from the stem of a plant or a leaf.

aeration—making holes or openings in turf to allow air to enter.

aerial application—using airplanes or helicopters to apply pesticides or other materials.

aeroponics—a form of hydroponics in which the roots of plants are suspended in air and sprayed with nutrient solution.

aflatoxin—a toxin formed on grain by the fungus, *Aspergillus flavus*.

aggregate culture—using aggregate or substrate materials to support plants grown with hydroponics.

aggregates—formed by soil particles clinging together to create larger particles of various sizes and shapes.

agriculture—science of growing crops and raising animals.

agroecosystem—the ecosystem created in fields, orchards, pastures, and other places where crops are grown.

agronomy—specialized area of plant science that deals with field crops.

air quality—suitability of air for use by living organisms.

alluvial soil—soil deposited by water.

angiosperms—flowering plants that produce seed protected in a fruit or pod, such as apples or beans.

animal fiber—fiber of animal origin, such as sheep, camels, and silkworms.

anion—negatively charged ion.

annual—plant that completes its life cycle in one year or growing season.

annual ring—ring formed in the cross-section of a tree trunk by the growth of dark and light xylem tissue.

annual weed—a weed that completes its life cycle in one growing season.

anther—part of a flower that produces pollen.

apical dominance—condition created by hormones in the apical meristem that prevents lateral buds from developing.

apical meristem—the primary growing point in the terminal bud.

aquaponics—the combination of fish culture or other aquatic animal organisms and plant culture in the same aquatic system.

arable land—land that can be used for row crops.

artificial reforestation—planting trees of specific species for timber production and other purposes.

asexual reproduction—plant reproduction using leaves, stems, or roots; no union of male and female gametes.

auxins—plant hormones that cause plant cells to elongate; influence apical dominance in plants.

available soil moisture—water in the soil that can be used by plants.

bactericide—a chemical used to control bacteria.

balled and burlapped (B&B)—plants dug from a nursery with a ball of soil on the roots that is wrapped with burlap.

band application—applying fertilizer or other materials in a strip on the row or field.

bare root (BR)—plants, usually small, that have no soil on the roots; careful handling is needed to keep the roots from drying out.

bark—outer layer of a stem or tree trunk; soil amendments made from bark.

bedding plants—annual plants started early in the spring for later establishing flower beds or as vegetable plants in a garden; examples of flower bedding plants are the petunia and marigold; examples of vegetable plants include tomatoes and peppers.

beet pulp—solid materials remaining after sugar has been extracted from sugar beets; used for cattle feed.

beneficial insect—an insect that is of value for the role it fills in the environment.

best land use—use of land that produces the most benefits to society.

biennial—plant that completes its life cycle in two years or growing seasons.

biennial weed—a weed with a life span of two years.

bio-diesel—fuel made by blending diesel and vegetable oil.

biological control—using the natural enemies of a pest to control it.

biotic disease—disease caused by a pathogen—living organism.

biotic environment—all of the living organisms that have an impact on the environment of a growing plant.

blackstrap molasses—syrup that remains after as much sugar as possible has been extracted; used in cattle feed.

blade—the leaf of a grass plant; flat part of a leaf.

blight—withering and death of a plant caused by a bacterial disease.

boll—fruit of the cotton plant that develops following flowering and contains seed cotton.

boot stage—stage of growth of grass plants following jointing (see tillering, jointing, and heading).

border irrigation—using small earthen ridges through pastures or fields to direct and contain flood irrigation water.

broadcast—to spread fertilizer or other materials evenly over a field.

broadleaf plants—plants with wide, flat leaves.

brown sugar—raw sugar containing a small amount of molasses.

bud—plant structure that contains undeveloped leaves, stems, and/or flowers.

bud scales—plant structures that cover and protect undeveloped parts.

bur—parts surrounding the dried and open cotton boll.

calibration—setting equipment to properly apply fertilizer, pesticide, etc., in the exact amount.

callous—group of cells with no particular function formed in tissue culture from an explant.

calyx—all of the sepals of a flower.

cambium—layer of cells where cell division and plant growth occur.

cane fruit—fruit produced on small perennial plants that grow low to the ground or no more than 4 to 6 feet high; blueberries and strawberries are examples.

canker—a disease on the bark and in the tissue beneath it, thought to be caused by fungi and viruses.

capability factors—characteristics of land that determine its best use.

capillary water—water held between soil particles against the force of gravity.

carcinogen—a pesticide or other substance suspected of causing cancer.

carrier—material in which the active ingredient of a pesticide is mixed.

cation exchange capacity—total measure of the cations that soil can hold.

cations—positively charged ions.

cell—basic unit of life containing living material bound by a membrane.

cellular respiration—using food to create energy; opposite of photosynthesis.

cellulose—structural material in the wall of plant cells; protects the cell and gives strength because of rigidity.

cell wall—the outer layer of plant cells.

cereal grain—the seed of grass-type plants grown for food and animal feed.

chemigation—using irrigation water to apply fertilizer, herbicides, and insecticides to crops.

chewing insect—insect classified on the basis of chewing mouthparts; insect bites off, chews, and swallows plant parts.

chloroplast—part of plant cell containing green pigment to trap light energy for photosynthesis.

chromosomes—arrangement of genes; made of proteins and nucleic acid and controls cell activity.

classing—process of grading cotton lint by color, staple length, and presence of trash.

clay—smallest inorganic particles in the soil.

climate—average of weather conditions over a long time.

clippings—grass blade pieces that are cut off during mowing; can be used in composting or as mulch.

clone—a genetic duplicate of a parent.

clothing—the garments, accessories, and ornaments worn on the human body to protect it and provide a certain appearance.

complete flower—a flower with four parts: sepals, petals, stamens, and pistil.

complete metamorphosis—an insect with four stages of metamorphosis: egg, larvae, pupa, and adult.

compost—soil amendments made by decomposing leaves, grass clippings, and other plant or animal materials.

compound leaf—leaf composed of petiole and two or more leaf blades called leaflets.

confectioner's sugar—finely ground granulated sugar; also known as powdered sugar.

conifer—trees in the Gymnospermae class that have needles and produce seeds in cones; fast growing and soft wood; examples are pines, spruces, and firs.

conservation—using resources to assure that some will be available in the future.

conservation tillage—using farming practices that eliminate traditional methods of soil tillage.

contact herbicide—a herbicide that kills plants by exposure to the material.

contact insecticide—a pesticide absorbed through the skin or exterior of an insect.

containerized—plants grown in containers at a nursery; root systems are not disturbed until the plants are planted, which increases the survival rate.

cooking oil—fat in which foods can be fried or used in other ways for human consumption.

cool-season vegetables—vegetable crops that can withstand a frost; examples include broccoli and onions.

cossettes—slices of sugar beets used in extracting sugar.

cotton bale—a standard quantity of lint cotton based on a net weight of 480 pounds; compacted and bound into a 500-pound gross weight bale.

cotyledon—seed leaves on an embryo.

cropland—land used for growing crops.

crop rotation—planting different crops on land in alternating years or growing seasons.

cross-pollination—when the pollen of a plant pollinates the flower of another plant.

crown—the top of a tree where leaves, limbs, twigs, and, often, flowers and fruit grow; may be a habitat for birds, insects, squirrels, and other animals.

culm—stem of a grass plant.

cultivar—a cultivated plant that has specific and distinguishable characteristics.

cultural practices—the control or manipulation of plant environment to encourage growth.

curing—process of drying hay to a moisture level that makes it safe for storage.

cut flowers—flowers grown and cut at the right maturity for the market; used to make floral arrangements; examples include roses and chrysanthemums.

cuticle—epidermis cells with a waxy coating that prevents excessive water loss.

cutting—stems, leaves, or roots of plants used for asexual reproduction.

cytokinin—plant hormone responsible for cell division and differentiation.

cytoplasm—living material inside a cell.

damping-off—disease that attacks young seedlings causing destruction of the soft plant tissue in the stem near the soil line; especially a problem in transplanting some vegetable and bedding plants.

day-neutral plants—flowering plants unaffected by length of day.

deciduous—woody perennial plant that loses its leaves in the fall.

decomposition—process by which organic matter is broken down to form soil.

defoliant—chemical applied to a plant that causes the leaves to fall off; improves fruit maturation and harvesting; used with cotton.

dent corn—the major type of corn grown; grains have a slight indention in the crown.

deoxyribonucleic acid (DNA)—long molecular chains that store genetic information in a cell.

desiccant—chemical applied to a plant that causes the entire plant to dry up; often used on cotton harvested with stripping equipment.

diameter breast high (DBH)—the standard place to measure the diameter of a standing tree; 4.5 feet above the ground on the uphill side.

dicot—a class of flowering plants; oaks, cacti, roses, and soybeans are examples.

dioecious—plant species with male and female flowers on different plants.

diploid—cell containing two sets of chromosomes.

directed application—treating only selected or target plants with herbicide; new "smart" sprayers use photo detectors to determine where to direct the spray.

direct plant source—plants are sources of food, clothing, and shelter for humans; not an animal product.

dispersal—the way weeds are spread.

disseminate—dispersion or scattering of seed from a plant.

division—cutting parts of a plant into sections for growing new plants.

double fertilization—process of fertilization in flowering plants; one sperm nucleus in a pollen grain fertilizes the egg to produce a zygote; the other sperm nucleus fuses with nuclei in the embryo sac to form endosperm.

drought—a period of insufficient rainfall.

dwarf—fruit trees less than one-third the size of a standard tree.

economic threshold—density of a pest population that will justify using control measures.

ecosystem—all the parts of the environment where a particular plant is living.

edaphic environment—soil and the area where the roots are located.

electromagnetic spectrum—measurement of light waves on the basis of wave length.

embryo—a rudimentary plant contained in a seed.

emitter—small opening that releases irrigation water.

endosperm—stored nutritive material in seed; source of energy for seed.

entrepreneurship—owning and operating a business to serve a niche market.

enzyme—a complex protein molecule that stimulates chemical reactions in plants.

epicotyl—in dicot plants, the portion of the embryo above the cotyledons.

epidermis—protective layer of cells on the outside of leaves and other organs.

erosion—loss of soil by wind, water, or other natural actions; washing or wearing away of the soil.

ethanol—kind of alcohol made from grains and oil plant seed; may be mixed with gasoline as a fuel.

ethylene—natural plant hormone that influences ripening of fruit, stem development, and other activities.

etiolation—movement of plant toward a light source when the amount of light is lower than needed by the plant.

eutrophication—condition that develops when the natural water in lakes and other bodies is changed by increased nutrients, which results in the destruction of habitat for native species.

evergreen—plants that keep their leaves year round.

experiment—a carefully planned and carried out action to demonstrate or discover a truth.

explants—small pieces of parent plant material used in tissue culture.

external feeding insect—an insect that chews or sucks from the exterior of a plant.

fertilizer—any material added to the soil to provide nutrients to increase plant growth, yield, or nutritional value of the plants.

fertilizer analysis—percentage of nutrients in fertilizer.

fiber—long strand of a substance; much longer than wide; very strong considering width.

fiber crops—plants grown for the fiber produced in their fruit, leaves, or stems.

fibrous root system—root system consisting of numerous slender roots.

field crops—plants grown in large fields for oil, fiber, grain, and similar products.

filament—stalk part of the stamen that holds the anther in a flower.

flood irrigation—flowing water onto the surface of a field and allowing it to soak in.

floriculture—production and use of plants for flowers and foliage.

flower—reproductive organ of a plant.

foliage plant—potted plants produced for attractive leaves and stems; examples include the philodendron and false aralia.

food—the solid and liquid materials that humans eat for nourishment.

food crop horticulture—growing horticultural plants for food, including olericulture and pomology.

forage—plants used for animal feed while immature or before seed maturity.

forb—flowering, broadleaf plant with a soft stem; good pasture and hay plants; examples include clovers, alfalfa, and lespedeza.

forcing—bringing a plant to a certain stage of development outside its normal season.

forest land—land on which trees are growing; the trees may be unimproved and producing low yields.

forestry—science of growing trees and producing wood products.

formulation—the way a pesticide product is prepared, such as a dust or liquid.

frost—frozen dew.

fruit—enlarged ovary containing seed; some are fleshy, such as the tomato, and others are dry, such as beans.

fruit food—the fleshy ripened ovary of a tree, shrub, or woody vine eaten raw or cooked.

fumigant—a gaseous pesticide; works through the respiratory system of pests.

fungicide—chemical compounds used to control fungi, especially those that cause plant diseases.

furrow irrigation—a form of flood irrigation that uses the rows on which crops are growing to divert and contain water.

gall—abnormal growth in a plant caused by the presence of another organism in the plant.

gene—segment of the DNA that codes for life processes and appearance.

general-use pesticide—a pesticide approved for widespread use without specialized applicator training; pesticides that pose less danger to the environment (see restricted-use pesticide).

genetic engineering—moving genetic material from one plant to another creating a plant with different genetic material.

genotype—the genetic makeup of a plant that is not readily observable; may be determined with DNA analysis.

geographic information system (GIS)—methods used in mapping fields into grids for precision farming.

germination—growth of a new plant from a seed; sprouting.

gibberellins—hormones in plants that induce stem cell elongation and cell division.

ginning—process of separating seed and lint as well as cleaning and drying cotton.

glacial soil—soil created, moved, and deposited by glaciers.

global positioning system (GPS)—methods used to connect orbiting satellites and a ground receiver to locate exact points on a field.

global warming—gradual warming of the earth's surface due to the greenhouse effect.

golgi body—part of plant cell that processes, sorts, and modifies protein.

grafting—process of getting the parts of one plant to grow on the parts of another plant; scion from one plant is placed on the stock of another plant.

grain—seeds of cereal grain plants.

grain crops—plants grown for edible seeds, not including the horticultural crops; cereal grain.

grain length—a method of classifying rice by the length of the kernel; types used are short, medium, and long.

grain marketing—processes that connect the grain producer with the consumer of grain products.

granulated sugar—common form of table sugar manufactured from crystals of raw sugar; primarily made from sugarcane and sugar beets.

grass—plant that grows leaves with parallel venation and has hollow or solid, but not solid woody, stems.

gravitational water—water pulled downward in soil by the force of gravity; may reach aquifers.

gravitropism—response of plants to gravity causing limbs to grow up and roots to grow down.

greenhouse—a structure enclosed by glass or plastic in which plants can be grown in a controlled environment and out of season.

greenhouse effect—ozone production in the lower atmosphere that holds heat next to the earth.

groat—the unground grain of oats.

ground application—treating for pests using equipment that travels on the ground to apply materials.

ground cover—woody or herbaceous plants that form a mass less than 1 foot tall covering the ground.

ground truthing—a procedure to verify the accuracy of remote sensing information.

ground water—water from aquifers; obtained by pumping or artesian pressure.

growing degree day (GDD)—measurement of temperature requirements for crop growth, especially corn.

growing season—number of days from the average date of the last freeze in the spring to the average date of the first freeze in the fall.

growth regulator herbicide—kills weeds by altering growth or metabolic processes.

guard cells—pair of cells that regulate the opening and closing of stomata.

gymnosperm—plants that have seeds not protected by fruit, such as pine cones.

haploid—reproductive cells that contain a single set of chromosomes.

hardiness zones—areas based on the average minimum temperatures that plants need; 11 zones are used in North America.

hardpan—tightly compacted layer of soil that interferes with water movement and root growth only a few inches below the surface.

hardscape—nonplant features in a landscape, such as walks, fences, and retaining walls.

hardwood—broadleaf trees in the Angiospermae class; wood is fine quality; examples include oaks, hickories, cherries, and gums.

hardy—plants with tolerance for cold weather.

harmful insect—an insect that damages plants, animals, or property.

harvest loss—losses of crops during harvesting caused by movement in the field or improperly adjusted equipment; losses caused by harvesting.

hay—cut and dried green plant material used for animal feed.

haylage—undried green plant material chopped similar to silage but with less than 50 percent moisture.

heading—when seeds form on a grass plant, including wheat, oats, and other cereal grains.

herb—non-woody plant whose leaves, seed, or other parts are used to enhance the flavor of food.

herbaceous—soft stems of some perennial plants that are killed by frost.

herbicide—chemical compound used to control weeds and other undesirable vegetative growth.

horizon—layers of soil in soil profile with different characteristics.

hormone—chemical substances that create internal messages within a plant to regulate plant growth.

horticultural crops—plants grown for food, comfort, and beauty.

host plant resistance—plants that are resistant to damage by some pests.

humus—organic matter that has reached an advanced stage of decomposition.

hunger—high need for food; food shortage; nutrient needs are not met.

hybrid—offspring from genetically different parents; produced through human manipulation.

hybridization—the process of breeding individuals from distinctly different varieties.

hybrid vigor—a condition where the offspring may have greater yield, height, disease resistance, or other traits than either parent.

hydroponics—the growing of terrestrial plants with their roots in a medium other than soil.

hygroscopic water—water that forms a thin film around soil particles and is not available to plants.

hypocotyl—the portion of the embryo below the cotyledons.

imperfect flower—a flower that lacks a stamen or pistil.

incomplete flower—a flower that lacks any one of the four parts of a complete flower.

incomplete metamorphosis—a metamorphosis with three stages of development: egg, nymph, and adult.

indirect plant source—plants used as animal feed and humans use the animals.

infiltration—process of water soaking into the soil.

injury threshold—amount of damage a potential pest causes to be classified as a pest.

inorganic fertilizer—fertilizer from nonliving sources.

insect—any of numerous small boneless animals whose bodies are divided into sections; some are beneficial, while others are pests.

insecticide—chemicals used to control insects.

integrated pest management (IPM)—using a combination of methods to control pests that result in a minimum of damage to the environment.

interiorscaping—using plants inside buildings to create an attractive environment.

internal feeding insect—an insect that chews a hole in a plant and goes inside to feed internally on plant tissue.

ions—electrically charged atoms.

irrigation—artificial application of water to encourage plant growth and productivity.

irrigation scheduling—providing the right amount of water at the right time for most efficient plant growth and production.

jointing—elongation of internodes on culms of grass plants.

kernel—part of an individual grain within the seed coat.

kernel hardness—a method of determining kind of wheat based on the hardness or softness of the endosperm.

key pest—a pest that regularly causes losses in crops and should usually be managed.

land—all of the natural and artificial characteristics of an area to be used for agricultural or other purposes.

land capability—suitability of land for agricultural uses.

land capability classification—assigning numbers to land based on eight land capability classes.

land forming—smoothing and shaping land to have a good surface free of dips and high spots.

landscape horticulture—growing and using plants to make the outdoor environment more appealing.

lateral bud—buds located along the sides of stems where the leaves are attached.

layering—method of asexual reproduction in which roots form on a stem while the stem is attached to the parent plant.

leaching—loss of nutrients that dissolve in gravitational water and move through or out of the soil.

leaf blade—large broad part of a leaf.

leaflet—two or more leaf blades.

leaf mold—soil amendment made from leaves; similar to compost.

leaf spot—disease in plant with yellowing around dead areas on leaves, stems, and fruit due to bacteria or fungi.

leaves—plant organs responsible for food production for the plant.

life cycle—time required for a plant to grow from its beginning until it dies.

lint—cotton fibers from which the seeds have been removed.

listing—planting seed in the bottoms of shallow furrows to make better use of moisture; used with several crops, including grain sorghum.

loess—soil moved and deposited by wind.

log—the large stem of a tree harvested by cutting.

long-day plants—plants that flower when the days get longer.

lowland rice—rice grown in large, flat fields.

lumber—timber that has been sawed into boards.

machinery controllers—devices used to regulate the operation of machines, especially applicators and other equipment used in precision farming.

macronutrients—the nutrients needed by plants in the largest amounts: nitrogen, phosphorus, and potassium.

mechanical weed management—using tools or equipment to control weeds, such as plows, cutters, and hoes.

medicinal plants—plants that produce products used in human medicine.

meiosis—sex cell division.

mesophyll—tissue in the middle layer of a leaf that conducts photosynthesis.

metamorphosis—the stages of development of an insect; may be complete or incomplete.

micronutrients—nutrients needed by plants in small amounts, such as boron and copper.

mineral matter—inorganic material in the soil.

miticides—chemical compounds used to kill mites.

mitochondria—converts food in a cell into energy through cellular respiration.

mitosis—cell division; how organisms grow and replace cells.

mixed stand—a stand of trees in which two or more species are more than 20 percent of the stand.

module—a specially compacted unit of seed cotton that can be easily transported.

molasses—syrup made from sugarcane.

monecious—plants that have both male and female flowers separately, such as corn.

monocot—a class of flowering plants includes lilies, grasses, corn, and palms.

mosaic diseases—grain diseases resulting from the destruction of chlorophyll and the formation of yellow spots; caused by viruses.

mulching—covering the soil with a material that smothers the growth of weeds and retains moisture; mulches include straw, grass clippings, shredded bark, and sheets of plastic.

mutation—naturally occurring genetic change in an organism.

narrowleaf plant—plants with needles or scale-shaped leaves.

natural organic fertilizer—fertilizers of plant or animal origin.

nematocide—chemical used to kill nematodes.

nonrenewable natural resources—resources that cannot be replaced when used up.

non-selective herbicide—a compound that kills all plants no matter the species.

noxious weed—a weed that creates big problems with crops; the worst kinds of weeds.

nozzle—device that dispenses a spray in the desired shape and droplet size.

nucleus—cell part containing the DNA.

nursery—place where plants are started and grown to market size; nurseries grow trees, shrubs, fruit trees, vines, ground covers, and seedlings for reforestation.

nut—a hard, bony, one-seeded fruit of a woody plant.

nutrient cycle—the recycling of essential elements used by plants.

nutrient film technique (NFT)—a hydroponics system that uses a continuous flow of solution through a series of tubes, troughs, or channels washing over the roots of plants.

oil crops—plants grown to produce vegetable oil from their seed or fruit.

olericulture—science of producing vegetable crops.

orchard—the field or area where tree fruits are grown.

organelles—specialized structures in cell cytoplasm.

organic fertilizer—fertilizer originating as plant or animal tissue, such as manure and compost.

organic matter—decayed or partially decayed remains of plants and animals found in the soil.

ornamental crops—plants grown for beauty and personal appeal; include flowers, shrubs, vines, and other species.

ornamental horticulture—growing and using plants for their beauty.

ovary—the part of a flower that contains one or more ovules where eggs are produced and seeds develop; the ovary becomes a fruit—apples are ripened ovaries.

palisade layer—layer of cells below the upper epidermis in a leaf.

parts per million (ppm)—method of measuring nutrients in hydroponics and in other applications; gives the number of parts of a substance in one million parts.

pasture—improved or unimproved plant materials on land areas where animals graze.

pathogen—living organism that causes disease, including bacteria, fungi, and viruses.

peat moss—organic soil amendment dug from peat bogs.

peds—clusters of soil particles that contribute to soil structure.

percolation—downward movement of water in the soil.

perennial—plant with a life cycle of more than two years.

perennial weed—a weed with a life span of more than two years.

perfect flower—a flower that has both a stamen and a pistil, the two parts involved in fertilization.

perlite—inorganic soil amendment made from volcanic rock.

permafrost—permanently frozen subsoil, as in the arctic zone.

permeable—characteristic of quality soil that allows water movement by infiltration and percolation.

pest—any living organism that causes loss, injury, or irritation to a plant.

pesticide—a chemical substance used to manage pests.

petal—leaf-like colorful parts of a flower.

petiole—leaf stalk; connecting structure between leaf blade and plant stem.

phenotype—the outward or physical appearance of a plant.

phloem—plant tissue that transports food made in the leaves to the remainder of the plant, including the roots and stem.

photoperiod—the amount of light in a 24-hour day.

photoperiodism—the response of a plant to light duration.

photosynthesis—process by which green plants convert solar energy into stored chemical energy.

phototropism—response of plants to light.

pistil—female part of the flower that contains the stigma, style, and ovary.

plant breeding—the systematic process of improving plants using scientific methods in achieving desired goals with plant reproduction.

plant disease—any abnormal condition of a living plant.

plant domestication—removing plants from their native wild environment and growing them under controlled conditions.

plant environment—the surroundings in which a plant grows.

plant fiber—fiber of plant origin, such as cotton, hemp, flax, and kenaf.

plant health—condition of a plant as related to disease.

plantlets—small plants that develop in the third stage of tissue culture.

plant pest—anything that causes injury or loss to a plant.

plant population—the number of plants growing on one acre (or hectare).

plant science—study of the structure, functions, growth, and protection of plants.

plumule—the first bud of the embryo consists of miniature leaves enclosing a growing point.

plywood—sheets of wood manufactured by gluing layers of wood together.

pollen—produced by the anther in the flower of a plant; contains male sex cells.

pollination—transfer of pollen from the male to the female part of a plant flower.

pollution—substances or conditions in the environment that damage the usefulness or productivity of plants.

pomologist—person who studies fruit and nuts.

pomology—science of producing fruit and nuts.

pore space—spaces in the soil between soil particles.

postemergence application—applying a herbicide or other material after a crop is up and growing.

potted flowering plants—plants grown in pots for marketing while in bloom; examples include African violets and gloxinia.

precipitation—natural way water is deposited on earth, such as rain or snow.

precision farming—an information- and technology-based crop management system.

preemergence application—applying a herbicide or other material before the crop comes up; often a part of the planting operation.

preharvest loss—losses of crops before harvesting by lodging, shattering, or other means.

preplant application—applying a herbicide or other material before a crop is planted.

prescription farming—mapping system and associated technology used in applying fertilizer and other inputs based on what is needed in particular areas of fields.

primary root—the major root of a plant; the first root developed by a seed to anchor the plant and absorb water and nutrients.

protectants—chemical materials applied to protect plants against disease.

pruning—selective removal of branches to develop a strong framework that will support a heavy load of fruit.

pulpwood—wood specially grown and cut for making paper and other pulp products.

pure stand—a stand of trees that is 80 percent or more of the same species; tree farms often grow pure stands.

quality of life—having a good environment in which to live.

quarantine—the isolation or exclusion of a pest or pest problem.

radicle—the embryo's root that develops into the primary root.

range—large open areas of pasture; vegetation is mostly native plants.

ratoon crop—sugarcane produced by sprouting of roots in successive years in tropical climates.

remote sensing—collecting information about something without being in contact with what is being studied.

renewable natural resources—resources that can be replaced when used up.

research—seeking answers to questions using systematic methods of investigation.

restricted-use pesticide—higher toxicity pesticides that pose greater risks to the environment; training required before use (see general-use pesticide).

rhizome—below-the-ground structure that sends up shoots to reproduce some plants.

rockwool—a substrate material used in some hydroponics systems; a spongy, fibrous material made from molten volcanic rock.

root cap—specialized cells on the tips of roots that protect them as they grow through the soil.

root hairs—tiny root structures that increase the area for absorbing water.

root knot nematode—a nematode that forms cysts or knots on the roots of plants.

rootstock—lower portion of a plant used in grafting.

root zone—where plant roots are found in the soil.

rot—decay of plant tissue typically caused by bacteria, but fungi may also cause rot.

rough endoplasmic reticulum—produces proteins for a plant cell.

sand—the largest-size mineral constituent of soil; used as an inorganic soil amendment.

sap—watery juice in plants.

scarification—breaking down of a seed coat to encourage germination.

science—knowledge gained by systematic study.

scion—stem or bud portion of a graft.

scouting—carefully observing crops for pests or signs of pest damage.

secondary roots—small branches formed on primary roots.

seed—fertilized, mature ovule of a plant; consists of three major parts: seed coat, embryo, and stored food.

seed coat—protective shell surrounding the embryo and endosperm of a seed.

seed cotton—cotton as picked from the field containing both lint and seed.

seedling—a small tree, often of improved genetics, 15 to 18 inches high for planting.

selection—the process of breeding plants selected for a particular characteristic.

selective herbicide—a compound that kills only certain plant species and not others.

self-pollination—when the pollen of a plant pollinates a flower on the same plant.

semi-dwarf—fruit trees about half the size of standard trees but larger than dwarf trees.

sensor—a device to remotely collect information.

sepal—green, leaf-like structures that protect a flower until it opens.

separation—a propagation method in which natural plant structures are removed from the parent plant and grown separately.

sexual reproduction—fusion of male and female sex cells to produce a new individual.

shattering—grains lost when the mature kernels fall to the ground.

shelter—protection from weather and harm; homes and the furnishings in them.

short-day plants—plants that flower when the days get shorter.

shrub—a multi-stem woody plant that does not grow more than 20 feet tall.

side-dressing—applying fertilizer or other materials along side existing crops.

sign—the presence of pathogen structures, such as rust spores on a leaf.

silage—chopped green plant materials that have fermented.

silo—upright or horizontal facility where silage is stored.

silt—middle-sized mineral particles in soil with good water and nutrient holding capacity.

silvics—the science of growing and developing trees.

silviculture—the art and science of growing trees stressing practices that improve productivity of the timber stand.

simple leaf—leaf with a single blade and petiole.

slope—rise and fall in the elevation of land.

smooth endoplasmic reticulum—produces lipids and hormones in plant cells.

smut—a fungus disease evident by dusty spores that cover grain and crops.

sod—layer of turf including stems, leaves, and roots with some soil attached.

soil—top layer of the earth's crust; renewable natural resource that supports life.

soil amendments—materials added to the soil to improve drainage, moisture holding ability, and aeration.

soil conservation—using cropping practices to assure long-term soil productivity.

soil contamination—condition resulting when pollutants get into the soil.

soil degradation—any thing or process that lowers soil productivity.

soil depth—thickness of soil layers that are important in crop production.

soil fertility—ability of soil to provide nutrients for plant growth.

soilless culture—the growing of terrestrial plants with their roots in a medium other than soil; hydroponics.

soil moisture—moisture found in soil (see available soil moisture).

soil moisture management—following practices that make good use of natural and artificial sources of water.

soil moisture tension—the force by which soil particles hold on to moisture (see available soil moisture).

soil pH—measure of acidity or alkalinity of soil.

soil profile—cross-section of soil, usually 3 to 4 feet (1 m) deep.

soil science—study of the structure, composition, fertility, and use of soil.

soil solarization—the process of using radiant heat from the sun to destroy disease or other organisms in the soil.

soil sterilant—compound that prevents the growth of plants in soil.

soil structure—the way soil aggregates arrange themselves; eight categories of structures are found.

soil test—analysis of a soil sample to determine nutrients present and those that need to be added for cropping.

soil texture—proportions of sand, silt, and clay in soil.

soluble salts—nutrient fertilizers and other minerals that dissolve in water; may damage plants by burning roots; sodium and chloride forms are particularly damaging to land with irrigation water.

spatial variability—differences found in a field or other area.

spice—parts of aromatic plants used to season food.

spongy layer—loosely arranged layer of cells between the palisade layer and mesophyll in a leaf.

spot application—treating only certain areas in a field with a pesticide.

spring wheat—wheat planted in the spring, grown in the summer, and harvested in early fall.

sprinkler irrigation—applying water through the air in droplet form.

square—cotton flower bud.

stamen—male reproductive parts of a flower made of filaments and anthers to produce pollen.

standard—a normal-sized fruit tree; contrasted with semi-dwarf and dwarf.

staple—length of individual cotton fibers.

stigma—sticky part of a flower pistil where pollen is collected.

stolon—above-ground creeping stem that sends up shoots and develops roots to reproduce a plant.

stomach insecticide—an insecticide that is effective if eaten; best with chewing insects.

stomata—pores or openings in the leaf that allow the exchange of oxygen, carbon dioxide, and water vapor.

stratification—process of a seed going through a time of cold temperature before it germinates.

stunt—a viral plant disease that causes a plant not to grow.

style—neck part of a flower pistil that connects the stigma and the ovary.

subsoil—the B horizon, or layer of soil below the top soil.

substrate—the aggregates used in a hydroponics system.

subterranean insect—species of insects that attack the roots of plants below the ground.

sucking insect—classification of insects based on mouthparts in which the insect pierces the outer layer of a plant and sucks sap from it.

sucrose—the kind of sugar in table sugar.

sugar—any food used as a sweetener.

sugar crops—crops grown as a source of sucrose (table sugar).

summer annual weed—a weed that grows during the summer.

surface runoff—water from precipitation that does not soak into the soil.

surface water—accumulated precipitation in lakes, streams, and reservoirs available for potential use in irrigation and other purposes.

surfactant—material added to herbicide mixes to have good spread of the applied product over leaves.

sustainable agriculture—using practices to assure the ability to produce crops and livestock into the future.

sustainable agriculture system—using multiple practices in combination to have sustainable agriculture.

sustainable resource use—using resources so they last a long time.

synthetic organic fertilizer—artificial carbon-based fertilizer materials.

syrup—sweet liquid usually made from the watery juices of plants.

systemics—chemical compounds applied to plants that enter the vascular system and move throughout the plant in the sap.

systemic insecticide—toxic substance absorbed and translocated by a plant that destroys an insect when it chews or sucks on the plant.

tap root system—a root system with one thick, main root that grows straight down.

technology—practical use of science; applied science.

terminal bud—large bud at the tip of a twig.

thatch—accumulated grass stems, leaves, and roots that may form a thick mat that restricts plant growth and encourages disease.

thigmotropism—response of plants to solid objects in their environment, such as tree to a rock.

tillering—growth of shoots from buds creating many branches on grass plants; gives a bunching appearance to the plant.

tilth—ease with which soil is worked; soil condition relative to a good seedbed.

timberland—forest land that productively grows sufficient trees for periodic harvesting.

tissue culture—reproducing a plant by using small pieces of tissue in an artificial medium under sterile conditions.

tissue testing—using laboratory analysis to diagnose plant disease.

top-dressing—applying fertilizer or other material broadcast over an existing crop.

topography—the form or outline of the surface of the earth.

topsoil—the A horizon or the top layer of soil.

toxicity—degree of poison in a pesticide material.

transgenic—a plant or other organism developed through genetic engineering.

translocation—movement of herbicides within a plant.

transpiration—movement of water vapor through stomata and out of a plant.

transplanting—the process of moving a plant from one location and planting it in another.

trap cropping—planting a small plot of a crop plant near a field several days before the crop itself is planted to attract insects away from the crop.

tree—a woody perennial with three major parts—roots, trunk, and crown.

tree farm—a cultured forest; where trees are planted and managed for efficient growth usually in pure stands.

tree fruits—fruits that grow on small to medium-sized trees; examples include apples, peaches, and apricots.

trickle irrigation—applying water directly to the root zones of plants using small emitters; also known as drip irrigation.

tropism—response of plants to their environment.

trunk—the main stem of a tree; cut for logs, poles, pulpwood, or other specialty products.

turf—plants used to present a pleasing appearance and protect the soil; usually low-growing, matted grasses or other plants.

turgidity—water pressure in plant cells.

upland rice—rice grown in small areas or patches on hillsides.

urban forestry—specialized applications of forestry that deal with the growth and use of trees in cities.

vacuole—part of plant cell containing water, stored foods, salts, pigments, and wastes.

vascular cambium—layer of cambium between the xylem and phloem.

vector—an organism the transmits disease-causing bacteria, fungi, and viruses; common vectors include birds, rodents, deer, insects, and humans.

vegetable—a herbaceous plant or plant part that is edible raw or cooked.

vegetable oil—fat obtained from certain plants, usually the seeds, such as soybeans and sunflowers, but some is from fruit, such as the olive.

veneer—plywood-type product with one exterior side made of a layer of high-value wood.

vermiculite—inorganic soil amendment made from a mineral known as mica; improves moisture holding capacity of soil.

viability—ability of a seed to germinate under optimum conditions.

vigor—ability of a seed to germinate under different conditions and still produce healthy seedlings.

vine—woody or herbaceous plants that require some type of support.

vine fruit—fruit that grows on wood stems that require support; the grape is the best-known example.

warm-season vegetables—vegetable plants readily killed by frost; examples include tomatoes and peppers.

water conservation—using crop production and other practices that make efficient use of available water and prevent the loss of moisture.

water culture—a hydroponics system that grows terrestrial plants without aggregates.

weather—the current condition of the atmosphere.

weathering—process by which rock is converted into soil.

weed—a plant growing where it is not wanted; a plant pest.

wilt—plant disease affecting the vascular system of plants; most wilt disease is caused by bacteria.

wilting—plant drooping due to a lack of firmness in plant tissues caused by inadequate water.

wilting point—the amount of water in the soil when a plant starts to wilt; the point at which soil moisture deficiency stresses plants.

winter annual weed—a weed that grows during the winter, maturing in the spring.

winter wheat—wheat planted in the fall, grown through the winter, and harvested in the late spring; may be used for winter grazing of cattle.

woody—stems of some perennial plants that are not killed by frost and survive from one year to the next, with trees being an example.

xylem—plant tissue that transports water and nutrients from the roots to the leaves.

zygote—cell produced by union of sperm and egg.

Bibliography

Allen, Charles T., Emory P. Boring, III, James F. Leser, and Thomas W. Fuchs. "Management of Cotton Insects." College Station: Texas A&M University, 1995, publication B-1209.

Anderson, P. M., E. A. Oelke, and S. R. Simmons. "Growth and Development Guide for Spring Barley." St. Paul: University of Minnesota, 1995, publication FO-2548-D.

Barnes, Robert F., Darrell A. Miller, and C. Jerry Nelson. *Forages: An Introduction to Grassland Agriculture*. Ames: Iowa State University Press, 1995.

Bienz, D. R. *The Why and How of Home Horticulture*. New York: W.H. Freeman and Company, 1993.

Biondo, Ronald J. and Charles B. Schroeder. *Introduction to Landscaping*, 3rd Ed. Danville, Illinois: Interstate Publishers, Inc., 2003.

Biondo, Ronald J. and Dianne A. Noland. *Floriculture*. Danville, Illinois: Interstate Publishers, Inc., 2000.

Brady, Nyle C. and Ray R. Weil. *The Nature and Properties of Soils*, 12th Ed. Upper Saddle River, New Jersey: Prentice-Hall, Inc., 1999.

Brown, Steve L., ed. "Agricultural Plant Pest Control." Athens: University of Georgia, 1993, special bulletin 8.

Burt, C. K. O'Connor, and T. Ruehr. *Fertigation*. San Luis Obispo, California: Irrigation Training and Research Center, 1998.

Colcheedas, Tom. *Hydroponics Simplified*. Victoria, Australia: Moorooduc, 1994.

Criswell, Jim T. And Jesse Campbell. "Toxicity of Pesticides." Stillwater: Oklahoma State University, n.d., no. 7457.

Dashefsky, H. Steven. *Entomology*. New York: McGraw-Hill, Inc., 1994.

Duncan, H. E. and W. M. Hagler, Jr. "Alflatoxins and Other Mycotoxins." Stillwater: Oklahoma State University, n.d., NCH-52.

Fipps, Guy. "Soil Moisture Management." College Station: Texas A&M University, 1995, publication B-1670.

Greer, Howard A. L. "Guide to Effective Weed Control." Stillwater: Oklahoma State University, n.d., publication 2750.

Grey, Gene W. *Urban Forestry*. New York: John Wiley & Sons, Inc., 1996.

Harrison, Kerry, and Anthony W. Tyson. "Factors to Consider in Selecting a Farm Irrigation System." Athens: University of Georgia, 1995, bulletin 882.

Hawkins, Stephen, Bill Massey, Ervin Williams, and Mark Hodges. "Grain Sorghum Production Calendar." Stillwater: Oklahoma State University, n.d., publication 2113.

Hill, J.E., S. R. Roberts, D. M. Brandon, S. C. Scardaci, J. F. Williams, C. M. Wick, W. M. Canevari, and B. L. Weir. *Rice Production in California*. Oakland: University of California, 1992, Publication 21498.

Rolfe, G. L., Edgington, J., and Holland, I. I.. *Forests and Forestry*, 6th Ed. Danville, Illinois: Interstate Publishers, Inc., 2003.

Janick, Jules, and James E. Simon, ed. *New Crops*. New York: John Wiley & Sons, Inc., 1993.

Johnson, Gordon and Ed Hanlon. "Classification of Irrigation Water Quality." Stillwater: Oklahoma State University, fact sheet 2401, n.d.

Kick-Raack, Joanne. *A Study Guide for Commercial Turfgrass Applicators: Ohio Pesticide Applicator Training*. Columbus: The Ohio State University, 1994, bulletin 841-8.

Klonsky, Karen, Kim Norris, and Rene Atwater. *Environmental Liability in Agriculture*. Oakland: University of California, 1994, publication 21502.

Lee, Jasper S. and Diana L. Turner. *AgriScience*, 3rd Ed. Danville, Illinois: Interstate Publishers, Inc., 2003.

Lee, Stephen J., Christy Mecey-Smith, Elizabeth Morgan, Ray Chelewski, Randi Hunewill, and Jasper S. Lee. *Biotechnology*. Danville, Illinois: Interstate Publishers, Inc., 2001.

Maynard, Donald N. and George J. Hochmuth. *Knott's Handbook for Vegetable Growers*, 4th Ed. New York: John Wiley & Sons, Inc., 1997.

Metcalf, Robert L. And William H. Luckmann, ed. *Introduction to Insect Pest Management*. New York: John Wiley & Sons, Inc., 1994.

Minor, Paul E. *Soil Science*. Columbia: University of Missouri-Columbia, March 1995.

Nechols, J.R., ed. *Biological Control in the Western United States*. Oakland: University of California, 1995.

Parker, Rick. *Introduction to Plant Science*. Albany, New York: Delmar Publishers, 2000.

Porter, Lynn, Jasper S. Lee, Diana L. Turner, and Malcolm Hillan. *Environmental Science and Technology*, 2nd Ed. Danville, Illinois: Interstate Publishers, Inc., 2003.

Resh, Howard M. *Hydroponic Food Production*. Santa Barbara, California: Woodbridge Press Publishing Company, 1993.

Schroeder, Charles B., Eddie Dean Seagle, Lorie M. Felton, John M. Ruter, William Terry Kelley, and Gerard Krewer. *Introduction to Horticulture: Science and Technology*, 3rd Ed. Danville, Illinois: Interstate Publishers, Inc., 2000.

Schroeder, Charles B. and Howard B. Sprague. *Turf Management Handbook*, 5th Ed. Danville, Illinois: Interstate Publishers, Inc., 1996.

Schwab, Glenn O., Delmar D. Fangmeier, William J. Elliot, and Richard K. Frevert. *Soil and Water Conservation Engineering*, 4th Ed. New York: John Wiley & Sons, Inc., 1995.

Sheaffer, Craig C., Russell D. Mathison, Neal P. Martin, David L. Rabas, Harlan J. Ford, and Douglas R. Swanson. *Forage Legumes*. St. Paul: University of Minnesota. 1993, bulletin 597.

Simmons, S. R., E. A. Oelke, and P. M. Anderson. "Growth and Development Guide for Spring Wheat." St. Paul: University of Minnesota. 1995, publication FO-2547-D.

Smith, C. Wayne. *Crop Production: Evolution, History, and Technology*. New York: John Wiley & Sons, Inc., 1995.

Solomon, Eldra Pearl, Linda R. Berg, Diana W. Martin, and Claude Villee. *Biology*. Orlando, Florida: Sanders College Publishing, 1993.

Swiader, John M., George W. Ware, and J. P. McCollum. *Producing Vegetable Crops*, 5th. Danville, Illinois: Interstate Publishers, Inc., 2002.

Trimmer, W. L., T. W. Ley, G. Clough, and D. Larsen. *Chemigation in the Pacific Northwest*. Corvalis: Oregon State University, 1992, Pacific Northwest Extension Publication 360.

Undersander, Dan. *Alfalfa Management Guide*. Madison: University of Wisconsin, 1994, publication NCR 547.

Weinzierl, Rick and Tess Henn. *Alternatives in Insect Management*. Urbana: University of Illinois, 1991, North Central Regional Extension Publication 401.

Wyman, Donald. *Wyman's Gardening Encyclopedia*. New York: MacMillan Publishing Company, 1977.

Young, H. M., Jr., and William A. Hayes. *No-Tillage Farming and Minimum Tillage Farming*. Brookfield, Wisconsin: No-Till Farmer, Inc., 1992.

Zuk, Judith D., ed. *A-Z Encyclopedia of Garden Plants*. New York: DK Publishing, Inc., 1997.

_____. *Small Farm Handbook*. Oakland: University of California, 1994.

_____. *Curriculum Material for Plant and Soil Science*. College Station: Texas A&M University, n.d.

_____. "Applying Pesticides Correctly." Athens: University of Georgia, 1993, special bulletin 15.

_____. *Western Fertilizer Handbook*, 8th Ed. Danville, Illinois: Interstate Publishers, Inc., 1995.

_____. *Emerging Technologies in Agriculture*. Alexandria, Virginia: National Council for Agricultural Education, 2000.

Index

DATE